# Signal Processing
## A Mathematical Approach
*Second Edition*

# MONOGRAPHS AND RESEARCH NOTES IN MATHEMATICS

## Series Editors

John A. Burns
Thomas J. Tucker
Miklos Bona
Michael Ruzhansky
Chi-Kwong Li

## Published Titles

*Iterative Optimization in Inverse Problems*, Charles L. Byrne

*Signal Processing: A Mathematical Approach, Second Edition*, Charles L. Byrne

*Modeling and Inverse Problems in the Presence of Uncertainty*, H. T. Banks, Shuhua Hu, and W. Clayton Thompson

*Sinusoids: Theory and Technological Applications*, Prem K. Kythe

*Blow-up Patterns for Higher-Order: Nonlinear Parabolic, Hyperbolic Dispersion and Schrödinger Equations*, Victor A. Galaktionov, Enzo L. Mitidieri, and Stanislav Pohozaev

*Set Theoretical Aspects of Real Analysis*, Alexander B. Kharazishvili

*Special Integrals of Gradshetyn and Ryzhik: the Proofs – Volume I*, Victor H. Moll

## Forthcoming Titles

*Stochastic Cauchy Problems in Infinite Dimensions: Generalized and Regularized Solutions*, Irina V. Melnikova and Alexei Filinkov

*Monomial Algebra, Second Edition*, Rafael Villarreal

*Groups, Designs, and Linear Algebra*, Donald L. Kreher

*Geometric Modeling and Mesh Generation from Scanned Images*, Yongjie Zhang

*Difference Equations: Theory, Applications and Advanced Topics, Third Edition*, Ronald E. Mickens

*Method of Moments in Electromagnetics, Second Edition*, Walton C. Gibson

*The Separable Galois Theory of Commutative Rings, Second Edition*, Andy R. Magid

*Dictionary of Inequalities, Second Edition*, Peter Bullen

*Actions and Invariants of Algebraic Groups, Second Edition*, Walter Ferrer Santos and Alvaro Rittatore

*Practical Guide to Geometric Regulation for Distributed Parameter Systems*, Eugenio Aulisa and David S. Gilliam

*Analytical Methods for Kolmogorov Equations, Second Edition*, Luca Lorenzi

*Handbook of the Tutte Polynomial*, Joanna Anthony Ellis-Monaghan and Iain Moffat

*Application of Fuzzy Logic to Social Choice Theory*, John N. Mordeson, Davendar Malik and Terry D. Clark

*Microlocal Analysis on $R^n$ and on NonCompact Manifolds*, Sandro Coriasco

## Forthcoming Titles (continued)

**MONOGRAPHS AND RESEARCH NOTES IN MATHEMATICS**

# Signal Processing
## A Mathematical Approach
### *Second Edition*

Charles L. Byrne

University of Massachusetts Lowell
Lowell, Massachusetts, USA

**CRC Press**
Taylor & Francis Group
Boca Raton London New York

CRC Press is an imprint of the
Taylor & Francis Group, an **informa** business

A CHAPMAN & HALL BOOK

CRC Press
Taylor & Francis Group
6000 Broken Sound Parkway NW, Suite 300
Boca Raton, FL 33487-2742

First issued in paperback 2020

ISBN-13: 978-1-4822-4184-6 (hbk)
ISBN-13: 978-0-367-65894-6 (pbk)

### Library of Congress Cataloging-in-Publication Data

Byrne, Charles L., 1947-
  Signal processing : a mathematical approach / Charles L. Byrne, Department of Mathematical Sciences, University of Massachusetts Lowell. -- Second edition.
    pages cm. -- (Monographs and research notes in mathematics)
  Includes bibliographical references and index.
  ISBN 978-1-4822-4184-6
  1. Signal processing--Mathematics. I. Title.

TK5102.9.B96 2015
621.382'20151--dc23                                                    2014028555

**Visit the Taylor & Francis Web site at**
**http://www.taylorandfrancis.com**

**and the CRC Press Web site at**
**http://www.crcpress.com**

*I dedicate this book to Eileen,*
*my wife for forty-four wonderful years.*

*My thanks to my graduate student*
*Jessica Barker, who read most of this book*
*and made many helpful suggestions.*

# Contents

## 26 Some Theory of Fourier Analysis     383

## 27 Reverberation and Echo Cancellation     391

## Bibliography     397

## Index     409

# *Preface*

In graduate school, and for the first few years as an assistant professor, my research was in pure mathematics, mainly topology and functional analysis. Around 1979 I was drawn, largely by accident, into signal processing, collaborating with friends at the Naval Research Laboratory who were working on sonar. Initially, I felt that the intersection of the mathematics that I knew and that they knew was nearly empty. After a while, I began to realize that the basic tools of signal processing are subjects with which I was already somewhat familiar, including Fourier series, matrices, and probability and statistics. Much of the jargon and notation seemed foreign to me, and I did not know much about the particular applications everyone else was working on. For a while it seemed that everyone else was speaking a foreign language. However, my knowledge of the basic mathematical tools helped me gradually to understand what was going on and, eventually, to make a contribution.

Signal processing is, in a sense, applied Fourier analysis, applied linear algebra, and some probability and statistics. I had studied Fourier series and linear algebra as an undergraduate, and had taught linear algebra several times. I had picked up some probability and statistics as a professor, although I had never had a course in that subject. Now I was beginning to see these tools in a new light; Fourier coefficients arise as measured data in array processing and tomography, eigenvectors and eigenvalues are used to locate sonar and radar targets, matrices become images and the singular-value decomposition provides data compression. For the first time, I saw Fourier series, matrices and probability and statistics used all at once, in the analysis of the sampled cross-sensor correlation matrices and the estimation of power spectra.

In my effort to learn signal processing, I consulted a wide variety of texts. Each one helped me somewhat, but I found no text that spoke directly to people in my situation. The texts I read were either too hard, too elementary, or written in what seemed to me to be a foreign language. Some texts in signal processing are written by engineers for engineering students, and necessarily rely only on those mathematical notions their students have encountered previously. In texts such as [116] basic Fourier series and transforms are employed, but there is little discussion of matrices and no mention of probability and statistics, hence no random models.

I found the book [121] by Papoulis helpful, although most of the examples deal with issues of interest primarily to electrical engineers. The books written by mathematicians tend to treat signal processing as a part of harmonic analysis or of stochastic processes. Books about Fourier analysis focus on its use in partial differential equations, or explore rigorously the mathematical aspects of the subject. I was looking for something different. It would have helped me a great deal if there had been a book addressed to people like me, people with a decent mathematical background who were trying to learn signal processing. My hope is that this book serves that purpose.

There are many opportunities for mathematically trained people to make a contribution in signal and image processing, and yet few mathematics departments offer courses in these subjects to their students, preferring to leave it to the engineering departments. One reason, I imagine, is that few mathematics professors feel qualified to teach the subject. My message here is that they probably already know a good deal of signal processing, but do not realize that they know it. This book is designed to help them come to that realization and to encourage them to include signal processing as a course for their undergraduates.

The situations of interest that serve to motivate much of what is discussed in this book can be summarized as follows: We have obtained data through some form of sensing; physical models, often simplified, describe how the data we have obtained relates to the information we seek; there usually isn't enough data and what we have is corrupted by noise, modeling errors, and other distortions. Although applications differ from one another in their details, they often make use of a common core of mathematical ideas. For example, the Fourier transform and its variants play an important role in remote sensing, and therefore in many areas of signal and image processing, as do the language and theory of matrix analysis, iterative optimization and approximation techniques, and the basics of probability and statistics. This common core provides the subject matter for this text. Applications of the core material to tomographic medical imaging, optical imaging, and acoustic signal processing are included in this book.

The term *signal processing* is used here in a somewhat restrictive sense to describe the extraction of information from measured data. I believe that to get information out we must put information in. How to use the mathematical tools to achieve this is one of the main topics of the book.

This text is designed to provide a bridge to help those with a solid mathematical background to understand and employ signal processing techniques in an applied environment. The emphasis is on a small number of fundamental problems and essential tools, as well as on applications. Certain topics that are commonly included in textbooks are touched on only briefly or in exercises or not mentioned at all. Other topics not usually considered to be part of signal processing, but which are becoming increasingly important, such as iterative optimization methods, are included. The book, then, is a rather personal view of the subject and reflects the author's interests.

The term *signal* is not meant to imply a restriction to functions of a single variable; indeed, most of what we discuss in this text applies equally to functions of one and several variables and therefore to image processing. However, there are special problems that arise in image processing, such as edge detection, and special techniques to deal with such problems; we shall not consider such techniques in this text. Topics discussed include the following: Fourier series and transforms in one and several variables; applications to acoustic and electro-magnetic propagation models, transmission and emission tomography, and image reconstruction; sampling and the limited data problem; matrix methods, singular value decomposition, and data compression; optimization techniques in signal and image reconstruction from projections; autocorrelations and power spectra; high-resolution methods; detection and optimal filtering; eigenvector-based methods for array processing and statistical filtering, time-frequency analysis, and wavelets.

The ordering of the first eighteen chapters of the book is not random; these main chapters should be read in the order of their appearance. The remaining chapters are ordered randomly and are meant to supplement the main chapters.

Reprints of my journal articles referenced here are available in pdf format at my website, http://faculty.uml.edu/cbyrne/cbyrne.html.

# Chapter 1

## Introduction

## 1.1  Chapter Summary

We begin with an overview of applications of signal processing and the variety of sensing modalities that are employed. It is typical of remote-sensing problems that what we want is not what we can measure directly, and we must obtain our information by indirect means. To illustrate that point without becoming entangled in the details of any particular application, we present a marbles-in-bowls model of remote sensing that, although simple, still manages to capture the dominate aspects of many real-world problems.

## 1.2  Aims and Topics

The term *signal processing* has broad meaning and covers a wide variety of applications. In this course we focus on those applications of signal processing that can loosely be called *remote sensing*, although the mathematics we shall study is fundamental to all areas of signal processing.

In a course in signal processing it is easy to get lost in the details and lose sight of the big picture. My main objectives here are to present the most important ideas, techniques, and methods, to describe how they relate to one another, and to illustrate their uses in several applications. For signal processing, the most important mathematical tools are Fourier series and related notions, matrices, and probability and statistics. Most students with a solid mathematical background have probably encountered each of these topics in previous courses, and therefore already know some signal processing, without realizing it.

Our discussion here will involve primarily functions of a single real variable, although most of the concepts will have multi-dimensional versions. It is not our objective to treat each topic with the utmost mathematical rigor, and we shall seek to avoid issues that are primarily of mathematical concern.

### 1.2.1  The Emphasis in This Book

This text is designed to provide the necessary mathematical background to understand and employ signal processing techniques in an applied environment. The emphasis is on a small number of fundamental problems and essential tools, as well as on applications. Certain topics that are commonly included in textbooks are touched on only briefly or in exercises or not mentioned at all. Other topics not usually considered to be part of signal processing, but which are becoming increasingly important, such as matrix theory and linear algebra, are included.

The term *signal* is not meant to imply a specific context or a restriction to functions of time, or even to functions of a single variable; indeed, most of what we discuss in this text applies equally to functions of one and several variables and therefore to image processing. However, this is in no sense an introduction to image processing. There are special problems that arise in image processing, such as edge detection, and special techniques to deal with such problems; we shall not consider such techniques in this text.

### 1.2.2 Topics Covered

Topics discussed in this text include the following: Fourier series and transforms in one and several variables; applications to acoustic and EM propagation models, transmission and emission tomography, and image reconstruction; sampling and the limited data problem; matrix methods, singular value decomposition, and data compression; optimization techniques in signal and image reconstruction from projections; autocorrelations and power spectra; high-resolution methods; detection and optimal filtering; eigenvector-based methods for array processing and statistical filtering; time-frequency analysis; and wavelets.

### 1.2.3 Limited Data

As we shall see, it is often the case that the data we measure is not sufficient to provide a single unique answer to our problem. There may be many, often quite different, answers that are consistent with what we have measured. In the absence of prior information about what the answer should look like, we do not know how to select one solution from the many possibilities. For that reason, I believe that to get information out we must put information in. How to do this is one of the main topics of the course. The example at the end of this chapter will illustrate this point.

## 1.3 Examples and Modalities

There are a wide variety of problems in which what we want to know about is not directly available to us and we need to obtain information by more indirect methods. In this section we present several examples of remote sensing. The term "modality" refers to the manner in which the desired information is obtained. Although the sensing of acoustic and electromagnetic signals is perhaps the most commonly used method, remote sensing involves a wide variety of modalities: electromagnetic waves (light, x-ray, microwave, radio); sound (sonar, ultrasound); radioactivity (positron and single-photon emission); magnetic resonance (MRI); seismic waves; and a number of others.

### 1.3.1    X-ray Crystallography

The patterns produced by the scattering of x-rays passing through various materials can be used to reveal their molecular structure.

### 1.3.2    Transmission Tomography

In transmission tomography x-rays are transmitted along line segments through the object and the drop in intensity along each line is recorded.

### 1.3.3    Emission Tomography

In emission tomography radioactive material is injected into the body of the living subject and the photons resulting from the radioactive decay are detected and recorded outside the body.

### 1.3.4    Back-Scatter Detectors

There is considerable debate at the moment about the use of so-called *full-body scanners* at airports. These are not scanners in the sense of a CAT scan; indeed, if the images were skeletons there would probably be less controversy. These are images created by the returns, or *backscatter*, of millimeter-wavelength (MMW) radio-frequency waves, or sometimes low-energy x-rays, that penetrate only the clothing and then reflect back to the machine.

The controversies are not really about safety to the passenger being imaged. The MMW imaging devices use about 10,000 times less energy than a cell phone, and the x-ray exposure is equivalent to two minutes of flying in an airplane. At present, the images are fuzzy and faces are intentionally blurred, but there is some concern that the images will get sharper, will be permanently stored, and eventually end up on the net. Given what is already available on the net, the market for these images will almost certainly be non-existent.

### 1.3.5 Cosmic-Ray Tomography

Because of their ability to penetrate granite, cosmic rays are being used to obtain transmission-tomographic three-dimensional images of the interiors of active volcanos. Where magma has replaced granite there is less attenuation of the rays, so the image can reveal the size and shape of the magma column. It is hoped that this will help to predict the size and occurrence of eruptions.

In addition to mapping the interior of volcanos, cosmic rays can also be used to detect the presence of shielding around nuclear material in a cargo container. The shielding can be sensed by the characteristic scattering by it of muons from cosmic rays; here neither we nor the objects of interest are the sources of the probing. This is about as "remote" as sensing can be.

### 1.3.6 Ocean-Acoustic Tomography

The speed of sound in the ocean varies with the temperature, among other things. By transmitting sound from known locations to known receivers and measuring the travel times we can obtain line integrals of the temperature function. Using the reconstruction methods from transmission tomography, we can estimate the temperature function. Knowledge of the temperature distribution may then be used to improve detection of sources of acoustic energy in unknown locations.

### 1.3.7 Spectral Analysis

In our detailed discussion of transmission and remote sensing we shall, for simplicity, concentrate on signals consisting of a single frequency. Nevertheless, there are many important applications of signal processing in which the signal being studied has a *broad spectrum*, indicative of the presence of many different frequencies. The purpose of the processing is often to determine which frequencies are present, or not present, and to determine their relative strengths. The hotter inner body of the sun emits radiation consisting of a continuum of frequencies. The cooler outer layer absorbs the radiation whose frequencies correspond to the elements present in that outer layer. Processing these signals reveals a spectrum with a number of missing frequencies, the so-called *Fraunhofer lines*, and provides information about the makeup of the sun's outer layers. This sort of *spectral analysis* can be used to identify the components of different materials, making it an important tool in many applications, from astronomy to forensics.

### 1.3.8   Seismic Exploration

Oil companies want to know if it is worth their while drilling in a particular place. If they go ahead and drill, they will find out, but they would like to know what is the chance of finding oil without actually drilling. Instead, they set off explosions and analyze the signals produced by the seismic waves, which will tell them something about the materials the waves encountered. Explosive charges create waves that travel through the ground and are picked up by sensors. The waves travel at different speeds through different materials. Information about the location of different materials in the ground is then extracted from the received signals.

### 1.3.9   Astronomy

Astronomers know that there are radio waves, visible-light waves, and other forms of electro-magnetic radiation coming from the sun and distant regions of space, and they would like to know precisely what is coming from which regions. They cannot go there to find out, so they set up large telescopes and antenna arrays and process the signals that they are able to measure.

### 1.3.10   Radar

Those who predict the weather use radar to help them see what is going on in the atmosphere. Radio waves are sent out and the returns are analyzed and turned into images. The location of airplanes is also determined by radar. The radar returns from different materials are different from one another and can be analyzed to determine what materials are present. Synthetic-aperture radar is used to obtain high-resolution images of regions of the earth's surface. The radar returns from different geometric shapes also differ in strength; by avoiding right angles in airplane design *stealth* technology attempts to make the plane invisible to radar.

### 1.3.11   Sonar

Features on the bottom of the ocean are imaged with sonar, in which sound waves are sent down to the bottom and the returning waves are analyzed. Sometimes near or distant objects of interest in the ocean emit their own sound, which is measured by sensors. The signals received by the sensors are processed to determine the nature and location of the objects. Even changes in the temperature at different places in the ocean can be determined by sending sound waves through the region of interest and measuring the travel times.

## 1.3.12 Gravity Maps

The pull of gravity varies with the density of the material. Features on the surface of the earth, such as craters from ancient asteroid impacts, can be imaged by mapping the variations in the pull of gravity, as measured by satellites.

Gravity, or better, changes in the pull of gravity from one location to another, was used in the discovery of the crater left behind by the asteroid strike in the Yucatan that led to the extinction of the dinosaurs. The rocks and other debris that eventually filled the crater differ in density from the surrounding material, thereby exerting a slightly different gravitational pull on other masses. This slight change in pull can be detected by sensitive instruments placed in satellites in earth orbit. When the intensity of the pull, as a function of position on the earth's surface, is displayed as a two-dimensional image, the presence of the crater is evident.

Studies of the changes in gravitational pull of the Antarctic ice between 2002 and 2005 revealed that Antarctica is losing 36 cubic miles of ice each year. By way of comparison, the city of Los Angeles uses one cubic mile of water each year. While this finding is often cited as clear evidence of global warming, it contradicts some models of climate change that indicate that global warming may lead to an increase of snowfall, and therefore more ice, in the polar regions. This does not show that global warming is not taking place, but only the inadequacies of some models [119].

## 1.3.13 Echo Cancellation

In a conference call between locations A and B, what is transmitted from A to B can get picked up by microphones in B, transmitted back to speakers in A and then retransmitted to B, producing an echo of the original transmission. Signal processing performed at the transmitter in A can reduce the strength of the second version of the transmission and decrease the echo effect.

## 1.3.14 Hearing Aids

Makers of digital hearing aids include signal processing to enhance the quality of the received sounds, as well as to improve localization, that is, the ability of the hearer to tell where the sound is coming from. When a hearing aid is used, sounds reach the ear in two ways: first, the usual route directly into the ear, and second, through the hearing aid. Because that part that passes through the hearing aid is processed, there is a slight delay. In order for the delay to go unnoticed, the processing must be very fast. When hearing aids are used in both ears, more sophisticated processing can be used.

### 1.3.15   Near-Earth Asteroids

An area of growing importance is the search for potentially damaging near-earth asteroids. These objects are initially detected by passive optical observation, as small dots of reflected sunlight; once detected, they are then imaged by active radar to determine their size, shape, rotation, path, and other important parameters. Satellite-based infrared detectors are being developed to find dark asteroids by the heat they give off. Such satellites, placed in orbit between the sun and the earth, will be able to detect asteroids hidden from earth-based telescopes by the sunlight.

### 1.3.16   Mapping the Ozone Layer

Ultraviolet light from the sun is scattered by ozone. By measuring the amount of scattered UV at various locations on the earth's surface, and with the sun in various positions, we obtain values of the Laplace transform of the function describing the density of ozone, as a function of elevation.

### 1.3.17   Ultrasound Imaging

While x-ray tomography is a powerful method for producing images of the interior of patients' bodies, the radiation involved and the expense make it unsuitable in some cases. Ultrasound imaging, making use of back-scattered sound waves, is a popular method of inexpensive preliminary screening for medical diagnostics, and for examining a developing fetus.

### 1.3.18   X-ray Vision?

The MIT computer scientist and electrical engineer Dina Katabi and her students are currently exploring new uses of wireless technologies. By combining *Wi-Fi* and *vision* into what she calls *Wi-Vi*, she has discovered a way to detect the number and approximate location of persons within a closed room and to recognize simple gestures. The scattering of reflected low-bandwidth wireless signals as they pass through the walls is processed to eliminate motionless sources of reflection from the much weaker reflections from moving objects, presumably people.

## 1.4   The Common Core

The examples just presented look quite different from one another, but the differences are often more superficial than real. As we begin to use mathematics to model these various situations we often discover a common core of mathematical tools and ideas at the heart of each of these applications. For example, the Fourier transform and its variants play an important role in many areas of signal and image processing, as do the language and theory of matrix analysis, iterative optimization and approximation techniques, and the basics of probability and statistics. This common core provides the subject matter for this book. Applications of the core material to tomographic medical imaging, optical imaging, and acoustic signal processing are among the topics to be discussed in some detail.

Although the applications of interest to us vary in their details, they have common aspects that can be summarized as follows: the data has been obtained through some form of sensing; physical models, often simplified, describe how the data we have obtained relates to the information we seek; there usually isn't enough data and what we have is corrupted by noise and other distortions.

## 1.5 Active and Passive Sensing

In some signal and image processing applications the sensing is *active*, meaning that we have initiated the process, by, say, sending an x-ray through the body of a patient, injecting a patient with a radionuclide, transmitting an acoustic signal through the ocean, as in sonar, or transmitting a radio wave, as in radar. In such cases, we are interested in measuring how the system, the patient, the quiet submarine, the ocean floor, the rain cloud, will respond to our probing. In many other applications, the sensing is *passive*, which means that the object of interest to us provides its own signal of some sort, which we then detect, analyze, image, or process in some way. Certain sonar systems operate passively, listening for sounds made by the object of interest. Optical and radio telescopes are passive, relying on the object of interest to emit or reflect light, or other electromagnetic radiation. Night-vision instruments are sensitive to lower-frequency, infrared radiation.

From the time of Aristotle and Euclid until the middle ages there was an ongoing debate concerning the active or passive nature of human sight [112]. Those like Euclid, whose interests were largely mathematical, believed that the eye emitted rays, the *extramission theory*. Aristotle and others, more interested in the physiology and anatomy of the eye than in mathematics, believed that the eye received rays from observed objects outside the body, the *intromission theory*. Finally, around 1000 AD, the Arabic mathematician and natural philosopher Alhazen demolished the extramission theory by noting the potential for bright light to hurt the eye, and combined the mathematics of the extramission theorists with a refined theory of intromission. The extramission theory has not gone away completely, however, as anyone familiar with Superman's x-ray vision knows.

---

## 1.6 Using Prior Knowledge

An important point to keep in mind when doing signal processing is that, while the data is usually limited, the information we seek may not be lost. Although processing the data in a reasonable way may suggest otherwise, other processing methods may reveal that the desired information is still available in the data. Figure 1.1 illustrates this point.

The original image on the upper right of Figure 1.1 is a discrete rectangular array of intensity values simulating the distribution of the x-ray-attenuating material in a slice of a head. The data was obtained by taking the two-dimensional discrete Fourier transform of the original image, and then discarding, that is, setting to zero, all these spatial frequency values, except for those in a smaller rectangular region around the origin. Reconstructing the image from this limited data amounts to solving a large system of linear equations. The problem is under-determined, so a minimum-norm solution would seem to be a reasonable reconstruction method. For now, "norm" means the Euclidean norm.

The minimum-norm solution is shown on the lower right. It is calculated simply by performing an inverse discrete Fourier transform on the array of modified discrete Fourier transform values. The original image has relatively large values where the skull is located, but the least-squares reconstruction does not want such high values; the norm involves the sum of squares of intensities, and high values contribute disproportionately to the norm. Consequently, the minimum-norm reconstruction chooses instead to conform to the measured data by spreading what should be the skull intensities throughout the interior of the skull. The minimum-norm reconstruction does tell us something about the original; it tells us about the existence of the skull itself, which, of course, is indeed a prominent feature of the original. However, in all likelihood, we would already know about the skull; it would be the interior that we want to know about.

Using our knowledge of the presence of a skull, which we might have obtained from the minimum-norm reconstruction itself, we construct the prior estimate shown in the upper left. Now we use the same data as before, and calculate a minimum-weighted-norm reconstruction, using as the weight vector the reciprocals of the values of the prior image. This minimum-

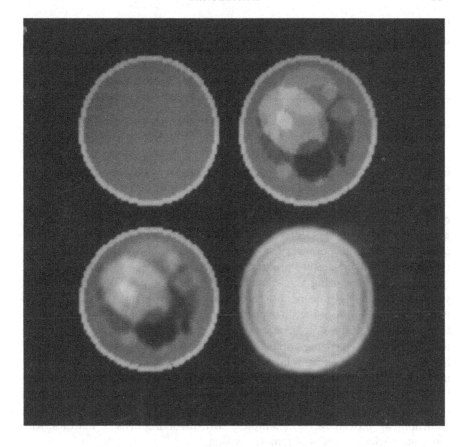

**FIGURE 1.1**: Extracting information in image reconstruction.

weighted-norm reconstruction, also called the PDFT estimator, is shown on the lower left; it is clearly almost the same as the original image. The calculation of the minimum-weighted-norm solution can be done iteratively using the ART algorithm [143].

When we weight the skull area with the inverse of the prior image, we allow the reconstruction to place higher values there without having much of an effect on the overall weighted norm. In addition, the reciprocal weighting in the interior makes spreading intensity into that region costly, so the interior remains relatively clear, allowing us to see what is really present there.

When we try to reconstruct an image from limited data, it is easy to assume that the information we seek has been lost, particularly when a reasonable reconstruction method fails to reveal what we want to know. As this example, and many others, show, the information we seek is often still in the data, but needs to be brought out in a more subtle way.

---

## 1.7   An Urn Model of Remote Sensing

Most of the signal processing that we shall discuss in this book is related to the problem of *remote sensing*, which we might also call *indirect measurement*. In such problems we do not have direct access to what we are really interested in, and must be content to measure something else that is related to, but not the same as, what interests us. For example, we want to know what is in the suitcases of airline passengers, but, for practical reasons, we cannot open every suitcase. Instead, we x-ray the suitcases. A recent paper [137] describes progress in detecting nuclear material in cargo containers by measuring the scattering, by the shielding, of cosmic rays; you can't get much more *remote* than that. Before we get into the mathematics of signal processing, it is probably a good idea to consider a model that, although quite simple, manages to capture many of the important features of remote-sensing applications. To convince the reader that this is indeed a useful model, we relate it to the problem of image reconstruction in *single-photon emission computed tomography* (SPECT). There seems to be a tradition in physics of using simple models or examples involving urns and marbles to illustrate important principles. In keeping with that tradition, we have here two examples, both involving urns of marbles, to illustrate various aspects of remote sensing.

### 1.7.1   An Urn Model

Suppose that there is a box containing a large number of small pieces of paper, and on each piece is written one of the numbers from $j = 1$ to $j = J$. I want to determine, for each $j = 1, ..., J$, the probability of selecting a piece of paper with the number $j$ written on it. Unfortunately, I am not allowed to examine the box. I am allowed, however, to set up a remote-sensing experiment to help solve my problem.

My assistant sets up $J$ urns, numbered $j = 1, ..., J$, each containing marbles of various colors. Suppose that there are $I$ colors, numbered $i = 1, ..., I$. I am allowed to examine each urn, so I know precisely the probability that a marble of color $i$ will be drawn from urn $j$. Out of my view, my assistant removes one piece of paper from the box, takes one marble from the indicated urn, announces to me the color of the marble, and then replaces both the piece of paper and the marble. This action is repeated $N$ times, at the end of which I have a long list of colors, $\mathbf{i} = \{i_1, i_2, ..., i_N\}$, where $i_n$ denotes the color of the $n$th marble drawn. This list $\mathbf{i}$ is my data, from which I must determine the contents of the box.

This is a form of remote sensing; what we have access to is related to, but not equal to, what we are interested in. What I wish I had is the list of urns used, $\mathbf{j} = \{j_1, j_2, ..., j_N\}$; instead I have $\mathbf{i}$, the list of colors. Sometimes data such as the list of colors is called "incomplete data," in contrast to the "complete data," which would be the list $\mathbf{j}$ of the actual urn numbers drawn from the box.

Using our urn model, we can begin to get a feel for the *resolution problem*. If all the marbles of one color are in a single urn, all the black marbles in urn $j = 1$, all the green in urn $j = 2$, and so on, the problem is trivial; when I hear a color, I know immediately which urn contained that marble. My list of colors is then a list of urn numbers; $\mathbf{i} = \mathbf{j}$. I have the complete data now. My estimate of the number of pieces of paper containing the urn number $j$ is then simply the proportion of draws that resulted in urn $j$ being selected.

At the other extreme, suppose two urns have identical contents. Then I cannot distinguish one urn from the other and I am unable to estimate more than the total number of pieces of paper containing either of the two urn numbers. If the two urns have nearly the same contents, we can distinguish them only by using a very large $N$. This is the resolution problem.

Generally, the more the contents of the urns differ, the easier the task of estimating the contents of the box. In remote-sensing applications, these issues affect our ability to resolve individual components contributing to the data.

## 1.7.2 Some Mathematical Notation

To introduce some mathematical notation, let us denote by $x_j$ the proportion of the pieces of paper that have the number $j$ written on them. Let $P_{ij}$ be the proportion of the marbles in urn $j$ that have the color $i$. Let $y_i$ be the proportion of times the color $i$ occurs in the list of colors. The expected proportion of times $i$ occurs in the list is $E(y_i) = \sum_{j=1}^{J} P_{ij} x_j = (Px)_i$, where $P$ is the $I$ by $J$ matrix with entries $P_{ij}$ and $x$ is the $J$ by 1 column vector with entries $x_j$. A reasonable way to estimate $x$ is to replace $E(y_i)$ with the actual $y_i$ and solve the system of linear equations $y_i = \sum_{j=1}^{J} P_{ij} x_j$, $i = 1, ..., I$. Of course, we require that the $x_j$ be nonnegative and sum to one, so special algorithms may be needed to find such solutions. In a number of applications that fit this model, such as medical tomography, the values $x_j$ are taken to be parameters, the data $y_i$ are statistics, and the $x_j$ are estimated by adopting a probabilistic model and maximizing the likelihood function. Iterative algorithms, such as the expectation maximization maximum likelihood (EMML) algorithm, are often used for such problems; see Chapter 14 for details.

### 1.7.3  An Application to SPECT Imaging

In *single-photon emission computed tomography* (SPECT) the patient is injected with a chemical to which a radioactive tracer has been attached. Once the chemical reaches its destination within the body the photons emitted by the radioactive tracer are detected by gamma cameras outside the body. The objective is to use the information from the detected photons to infer the relative concentrations of the radioactivity within the patient.

We discretize the problem and assume that the body of the patient consists of $J$ small volume elements, called *voxels*, analogous to *pixels* in digitized images. We let $x_j \geq 0$ be the unknown proportion of the radioactivity that is present in the $j$th voxel, for $j = 1, ..., J$. There are $I$ detectors, denoted $\{i = 1, 2, ..., I\}$. For each $i$ and $j$ we let $P_{ij}$ be the known probability that a photon that is emitted from voxel $j$ is detected at detector $i$; these probabilities are usually determined by examining the relative positions in space of voxel $j$ and detector $i$. We denote by $i_n$ the detector at which the $n$th emitted photon is detected. This photon was emitted at some voxel, denoted $j_n$; we wish that we had some way of learning what each $j_n$ is, but we must be content with knowing only the $i_n$. After $N$ photons have been emitted, we have as our data the list $\mathbf{i} = \{i_1, i_2, ..., i_N\}$; this is our *incomplete data*. We wish we had the *complete data*, that is, the list $\mathbf{j} = \{j_1, j_2, ..., j_N\}$, but we do not. Our goal is to estimate the frequency with which each voxel emitted a photon, which we assume, reasonably, to be proportional to the unknown proportions $x_j$, for $j = 1, ..., J$.

This problem is completely analogous to the urn problem previously discussed. Any mathematical method that solves one of these problems will solve the other one. In the urn problem, the colors were announced; here the detector numbers are announced. There, I wanted to know the urn numbers; here I want to know the voxel numbers. There, I wanted to estimate the frequency with which the $j$th urn was used; here, I want to estimate the frequency with which the $j$th voxel is the site of an emission, which is assumed to be equal to the proportion of the radionuclide within the $j$th voxel. In the urn model, two urns with nearly the same contents are hard to distinguish unless $N$ is very large; here, two neighboring voxels will be very hard to distinguish (i.e., to resolve) unless $N$ is very large. But in the SPECT case, a large $N$ means a high dosage, which will be prohibited by safety considerations. Therefore, we have a built-in resolution problem in the SPECT case.

Both problems are examples of probabilistic mixtures, in which the mixing probabilities are the $x_j$ that we seek. The *maximum likelihood* (ML) method of statistical parameter estimation can be used to solve such problems. The interested reader should consult the text [42].

---

## 1.8   Hidden Markov Models

In the urn model we just discussed, the order of the colors in the list is unimportant; we could randomly rearrange the colors on the list without affecting the nature of the problem. The probability that a green marble will be chosen next is the same, whether a blue or a red marble was just chosen the previous time. This independence from one selection to another is fine for modeling certain physical situations, such as emission tomography. However, there are other situations in which this independence does not conform to reality.

In written English, for example, knowing the current letter helps us, sometimes more, sometimes less, to predict what the next letter will be. We know that, if the current letter is a "q", then there is a high probability that the next one will be a "u". So what the current letter is affects the probabilities associated with the selection of the next one.

Spoken English is even tougher. There are many examples in which the pronunciation of a certain sound is affected, not only by the sound or sounds that preceded it, but by the sound or sounds that will follow. For example, the sound of the "e" in the word "bellow" is different from the sound of the "e" in the word "below"; the sound changes, depending on whether there is a double "l" or a single "l" following the "e". Here the entire context of the letter affects its sound.

Hidden Markov models (HMM) are increasingly important in speech processing, optical character recognition, and DNA sequence analysis. They allow us to incorporate dependence on the context into our model. In this section we illustrate HMM using a modification of the urn model.

Suppose, once again, that we have $J$ urns, indexed by $j = 1, ..., J$ and $I$ colors of marbles, indexed by $i = 1, ..., I$. Associated with each of the $J$ urns is a box, containing a large number of pieces of paper, with the number of one urn written on each piece. My assistant selects one box, say the $j_0$th box, to start the experiment. He draws a piece of paper from that box, reads the number written on it, call it $j_1$, goes to the urn with the number $j_1$ and draws out a marble. He then announces the color. He then draws a piece of paper from box number $j_1$, reads the next number, say $j_2$, proceeds to urn number $j_2$, etc. After $N$ marbles have been drawn, the only data I have is a list of colors, $\mathbf{i} = \{i_1, i_2, ..., i_N\}$.

The *transition probability* that my assistant will proceed from the urn numbered $k$ to the urn numbered $j$ is $b_{jk}$, with $\sum_{j=1}^{J} b_{jk} = 1$. The number of the current urn is the current *state*. In an ordinary *Markov chain* model, we observe directly a sequence of states governed by the transition probabilities. The Markov chain model provides a simple formalism for describing a system that moves from one state into another, as time goes on. In the hidden Markov model we are not able to observe the states directly; they are hidden from us. Instead, we have indirect observations, the colors of the marbles in our urn example.

The probability that the color numbered $i$ will be drawn from the urn numbered $j$ is $a_{ij}$, with $\sum_{i=1}^{I} a_{ij} = 1$, for all $j$. The colors announced are the *visible states*, while the unannounced urn numbers are the *hidden states*.

There are several distinct objectives one can have, when using HMM. We assume that the data is the list of colors, **i**.

- **Evaluation:** For given probabilities $a_{ij}$ and $b_{jk}$, what is the probability that the list **i** was generated according to the HMM? Here, the objective is to see if the model is a good description of the data.

- **Decoding:** Given the model, the probabilities, and the list **i**, what list $\mathbf{j} = \{j_1, j_2, ..., j_N\}$ of urns is most likely to be the list of urns actually visited? Now, we want to infer the hidden states from the visible ones.

- **Learning:** We are told that there are $J$ urns and $I$ colors, but are not told the probabilities $a_{ij}$ and $b_{jk}$. We are given several data vectors **i** generated by the HMM; these are the *training sets*. The objective is to learn the probabilities.

Once again, the ML approach can play a role in solving these problems [68]. The *Viterbi algorithm* is an important tool used for the decoding phase (see [149]).

# Chapter 2

## Fourier Series and Fourier Transforms

## 2.1 Chapter Summary

We begin with Fourier series and Fourier transforms, which are essential tools in signal processing. In this chapter we give the formulas for

Fourier series and Fourier transforms, in both trigonometric and complex-exponential form, summarize their basic properties, and give several examples of Fourier-transform pairs. We connect Fourier series to Fourier transforms using Shannon's Sampling Theorem. We solve a heat-equation problem to illustrate the use of Fourier series while introducing fundamental aspects of inverse problems. We leave to Chapter 26 the more theoretical details regarding Fourier series and Fourier transforms.

---

## 2.2   Fourier Series

Most mathematics students see Fourier series for the first time in a course on boundary-value problems. There students usually study the wave equation and the heat equation in two dimensions, using the technique of separating the space and time variables. Fourier series and Fourier transforms arise as we attempt to satisfy the initial conditions using a superposition of sine and cosine functions.

Suppose, for concreteness, that we have a function $f : [-L, L] \to \mathbb{R}$ and we want to express this function as a Fourier series. The Fourier series for $f$, relative to the interval $[-L, L]$, is

$$f(x) \approx \frac{a_0}{2} + \sum_{n=1}^{\infty} a_n \cos\left(\frac{n\pi}{L}x\right) + b_n \sin\left(\frac{n\pi}{L}x\right), \qquad (2.1)$$

where the Fourier coefficients $a_n$ and $b_n$ are

$$a_n = \frac{1}{L} \int_{-L}^{L} f(x) \cos\left(\frac{n\pi}{L}x\right) dx, \qquad (2.2)$$

and

$$b_n = \frac{1}{L} \int_{-L}^{L} f(x) \sin\left(\frac{n\pi}{L}x\right) dx. \qquad (2.3)$$

To obtain the formula for, say, $a_m$, the usual approach is to write

$$f(x) = \frac{a_0}{2} + \sum_{n=1}^{\infty} a_n \cos\left(\frac{n\pi}{L}x\right) + b_n \sin\left(\frac{n\pi}{L}x\right), \qquad (2.4)$$

for $|x| \leq L$, multiply both sides of Equation (2.4) by $\cos\left(\frac{m\pi}{L}x\right)$, and then integrate both sides, integrating term-by-term inside the sum on the right side of the equation. Orthogonality then gives the desired answer, since we have

$$\int_{-L}^{L} \cos\left(\frac{m\pi}{L}x\right) \sin\left(\frac{n\pi}{L}x\right) dx = 0,$$

$$\int_{-L}^{L} \cos\left(\frac{m\pi}{L}x\right) \cos\left(\frac{m\pi}{L}x\right) dx = L,$$

and

$$\int_{-L}^{L} \sin\left(\frac{m\pi}{L}x\right) \sin\left(\frac{m\pi}{L}x\right) dx = L,$$

for all $m$ and $n$, and, for $m \neq n$,

$$\int_{-L}^{L} \cos\left(\frac{m\pi}{L}x\right) \cos\left(\frac{n\pi}{L}x\right) dx = 0,$$

and

$$\int_{-L}^{L} \sin\left(\frac{m\pi}{L}x\right) \sin\left(\frac{n\pi}{L}x\right) dx = 0.$$

This derivation of the Fourier coefficients sweeps several important issues under the rug, so to speak.

We haven't said anything about the properties of the function $f$, so we cannot be sure that the Fourier series converges, for a given $x$, and even if it does, we cannot be sure that the sum of the series is $f(x)$. We also have not said anything about the integrability of the function $f$, and have not specified the type of integral being used in Equations (2.2) and (2.3). Finally, we have not justified integrating an infinite series term-by-term. These are not issues that are easily dealt with and it is reasonable, given our aims in this book, to leave those issues under the rug for now and to rely on the formulas above without further comment. In signal processing our primary concern is computing with measured data, in the form of finite-length vectors and matrices. Functions of continuous variables and infinite sequences guide our thinking, but enter into our calculations only as members of finite-parameter families.

There are many texts, such as [80], that the reader may consult that address the more mathematical aspects of Fourier analysis. The book [101] by Körner is a highly entertaining journey through many aspects for pure and applied Fourier analysis, while the small book [51] by Champeney summarizes, without proofs, most of the relevant theorems pertaining to Fourier series and Fourier transforms. The discussion in Chapter 26 is taken largely from [51]. For a sampling of more advanced material on signal processing and its applications, the reader may consult [3, 87].

At this early stage, it is useful to address the issue of periodicity. Clearly, the Fourier series itself can be viewed as a function of $x$ with period $2L$. Consequently, many books on the subject assume, from the start, that the function $f$ is also $2L$-periodic. We can, of course, extend the original function $f$ to the whole real line as a $2L$-period function. If $f$ is continuous on $[-L, L]$, but $f(-L) \neq f(L)$, we can preserve continuity of the periodic extension by first reflecting the function about the point $x = L$, creating a function on the interval $[-L, 3L]$ that has the same values at $-L$ and $3L$, and then extending that function as a $4L$-periodic function. However, our concern here is largely with problems that arise in remote sensing, such as radar, sonar, tomography, and the like, in which the function $f$ of interest is nonzero only on some finite interval. As we shall see, assuming a periodic extension at the start may not be a good idea.

---

## 2.3   Complex Exponential Functions

The most important functions in signal processing are the complex exponential functions. Using trigonometric identities it is easy to show that the function $h : \mathbb{R} \to \mathbb{C}$ defined by

$$h(x) = \cos x + i \sin x,$$

has the property $h(x+y) = h(x)h(y)$. Therefore, we write it in exponential form as $h(x) = c^x$, for some (necessarily complex) scalar $c$. With $x = 1$ we have

$$h(1) = \cos 1 + i \sin 1 = c.$$

Applying the Taylor series expansion

$$e^t = 1 + t + \frac{t^2}{2!} + \frac{t^3}{3!} + \dots,$$

for $t = i$ we have

$$e^i = \cos 1 + i \sin 1.$$

Consequently, we have $c = e^i$ and

$$h(x) = (e^i)^x = e^{ix}.$$

Because it is simpler to work with exponential functions than with trigonometric functions, we use the identities

$$\cos x = \frac{1}{2}(e^{ix} + e^{-ix}),$$

and

$$\sin x = \frac{1}{2i}(e^{ix} - e^{-ix})$$

to reformulate Fourier series and Fourier transforms in terms of complex exponential functions. In place of Equation (2.1) we have

$$f(x) \approx \sum_{n=-\infty}^{\infty} c_n e^{i\frac{n\pi}{L}x},$$

with

$$c_n = \frac{1}{2L} \int_{-L}^{L} f(x) e^{-i\frac{n\pi}{L}x} dx. \tag{2.5}$$

If $f$ is a continuous function, with $f(-L) = f(L)$ (so that it has a continuous $2L$-periodic extension), then $f$ is uniquely determined by its Fourier coefficients [101, Theorem 2.4], even though the Fourier series may not converge to $f(x)$ for some $x$.

## 2.4 Fourier Transforms

Suppose now that $f$ is a complex-valued function defined on the whole real line. The Fourier transform of $f$ is the function $F : \mathbb{R} \to \mathbb{C}$ given by

$$F(\gamma) = \int_{-\infty}^{\infty} f(x) e^{i\gamma x} dx. \tag{2.6}$$

Given $F$, the Fourier Inversion Formula tells us how to get back to $f(x)$:

$$f(x) = \frac{1}{2\pi} \int_{-\infty}^{\infty} F(\gamma) e^{-i\gamma x} d\gamma. \tag{2.7}$$

The function $f(x)$ is sometimes called the *inverse Fourier transform* (IFT) of $F(\gamma)$. Note that the formulas in Equations (2.6) and (2.7) are nearly identical. Because of this, the terminology in other texts may differ from ours. As was the case with Fourier series, we have again swept several issues under the rug for now. We have not specified the properties of the function $f$ that would guarantee the existence of the integrals in Equation (2.6); indeed, we have not said which definition of integration we must use. Even when we require that $f$ be sufficiently well behaved, the Fourier transform function $F$ may not be, and so the inversion formula in Equation (2.7) may require some interpretation. The functions $f(x)$ and $F(\gamma)$ are called a *Fourier-transform pair*. The definitions of the FT and IFT just given may differ slightly from the ones found elsewhere; our definitions are those of Bochner and Chandrasekharan [13] and Twomey [156]. The differences are minor and involve only the placement of the quantity $2\pi$ and of the minus sign in the exponent. One sometimes sees the Fourier transform of the function $f$ denoted $\hat{f}$, but we shall not use that notation here.

## 2.5    Basic Properties of the Fourier Transform

In this section we present the basic properties of the Fourier transform. Proofs of these assertions are left as exercises.

Let $u(x)$ be the *Heaviside function*; that is, $u(x) = 1$ if $x \geq 0$, and $u(x) = 0$ otherwise. Let $\chi_A(x)$ be the *characteristic function* of the interval $[-A, A]$; that is, $\chi_A(x) = 1$ for $x$ in $[-A, A]$ and $\chi_A(x) = 0$ otherwise. Let $\text{sgn}(x)$ be the *sign function*; that is, $\text{sgn}(x) = 1$ if $x \geq 0$, and $\text{sgn}(x) = -1$ if $x < 0$. The following are basic properties of the Fourier transform.

- **Symmetry:** The FT of the function $F(x)$ is $2\pi f(-\gamma)$. For example, the FT of the function $f(x) = \frac{\sin(\Omega x)}{\pi x}$ is $\chi_\Omega(\gamma)$, so the FT of $g(x) = \chi_\Omega(x)$ is $G(\gamma) = 2\pi \frac{\sin(\Omega \gamma)}{\pi \gamma}$.

- **Conjugation:** The FT of $\overline{f(-x)}$ is $\overline{F(\gamma)}$.

- **Scaling:** The FT of $f(ax)$ is $\frac{1}{|a|}F(\frac{\gamma}{a})$ for any nonzero constant $a$.

- **Shifting:** The FT of $f(x - a)$ is $e^{ia\gamma}F(\gamma)$.

- **Modulation:** The FT of $f(x)\cos(\gamma_0 x)$ is $\frac{1}{2}[F(\gamma + \gamma_0) + F(\gamma - \gamma_0)]$.

- **Differentiation:** The FT of the $n$th derivative, $f^{(n)}(x)$, is $(-i\gamma)^n F(\gamma)$. The IFT of $F^{(n)}(\gamma)$ is $(ix)^n f(x)$.

- **Convolution in $x$:** Let $f, F, g, G$ and $h, H$ be FT pairs, with

$$h(x) = \int f(y)g(x - y)dy, \qquad (2.8)$$

so that $h(x) = (f * g)(x)$ is the *convolution* of $f(x)$ and $g(x)$. Then $H(\gamma) = F(\gamma)G(\gamma)$. For example, if we take $g(x) = \overline{f(-x)}$, then

$$h(x) = \int f(x + y)\overline{f(y)}dy = \int f(y)\overline{f(y - x)}dy = r_f(x)$$

is the *autocorrelation function* associated with $f(x)$ and

$$H(\gamma) = |F(\gamma)|^2 = R_f(\gamma) \geq 0$$

is the *power spectrum* of $f(x)$.

• **Convolution in** $\gamma$: Let $f, F, g, G$ and $h, H$ be FT pairs, with $h(x) = f(x)g(x)$. Then $H(\gamma) = \frac{1}{2\pi}(F * G)(\gamma)$.

**Ex. 2.1** *Let $F(\gamma)$ be the FT of the function $f(x)$. Use the definitions of the FT and IFT given in Equations (2.6) and (2.7) to establish the following basic properties of the Fourier transform operation listed above. To establish the convolution formula calculate $H(\gamma)$ using Equation (2.8) and switch the order of integration.*

## 2.6 Some Fourier-Transform Pairs

The exercises in this section introduce the reader to several Fourier-transform pairs.

**Ex. 2.2** *Show that the FT of the function $f(x) = u(x)e^{-ax}$ is $F(\gamma) = \frac{1}{a-i\gamma}$, for every positive constant a, where $u(x)$ is the Heaviside function.*

**Ex. 2.3** *Show that the FT of $f(x) = \chi_A(x)$ is $F(\gamma) = 2\frac{\sin(A\gamma)}{\gamma}$. Similarly, show that the IFT of the function $F(\gamma) = \chi_\Gamma(\gamma)$ is $f(x) = \frac{\sin \Gamma x}{\pi x}$.*

**Ex. 2.4** *Show that the IFT of the function $F(\gamma) = 2i/\gamma$ is $f(x) = \text{sgn}(x)$. Hint: Write the formula for the inverse Fourier transform of $F(\gamma)$ as*

$$f(x) = \frac{1}{2\pi} \int_{-\infty}^{+\infty} \frac{2i}{\gamma} \cos \gamma x d\gamma - \frac{i}{2\pi} \int_{-\infty}^{+\infty} \frac{2i}{\gamma} \sin \gamma x d\gamma,$$

*which reduces to*

$$f(x) = \frac{1}{\pi} \int_{-\infty}^{+\infty} \frac{1}{\gamma} \sin \gamma x d\gamma,$$

*since the integrand of the first integral is odd. For $x \geq 0$ consider the Fourier transform of the function $\chi_x(t)$. For $x < 0$ perform the change of variables $u = -x$.*

Generally, the functions $f(x)$ and $F(\gamma)$ are complex-valued, so that we may speak about their real and imaginary parts. The next exercise explores the connections that hold among these real-valued functions.

**Ex. 2.5** *Let $f(x)$ be arbitrary and $F(\gamma)$ its Fourier transform. Let $F(\gamma) = R(\gamma) + iX(\gamma)$, where $R$ and $X$ are real-valued functions, and similarly, let $f(x) = f_1(x) + if_2(x)$, where $f_1$ and $f_2$ are real-valued. Find relationships between the pairs $R,X$ and $f_1,f_2$.*

**Definition 2.1** *We define the* even part *of $f(x)$ to be the function*

$$f_e(x) = \frac{f(x) + f(-x)}{2},$$

*and the* odd part *of $f(x)$ to be*

$$f_o(x) = \frac{f(x) - f(-x)}{2};$$

*define $F_e$ and $F_o$ similarly for $F$ the FT of $f$.*

**Ex. 2.6** *Show that $F(\gamma)$ is real-valued and even if and only if $f(x)$ is real-valued and even.*

**Definition 2.2** *We say that $f$ is a* causal function *if $f(x) = 0$ for all $x < 0$.*

**Definition 2.3** *The function $X$ is the* Hilbert transform *of function $R$ if*

$$X(\gamma) = \frac{1}{\pi} \int_{-\infty}^{\infty} \frac{R(\alpha)}{\gamma - \alpha} d\alpha.$$

**Ex. 2.7** *Let $F(\gamma) = R(\gamma) + iX(\gamma)$ be the decomposition of $F$ into its real and imaginary parts. Show that, if $f$ is causal, then $R$ and $X$ are related; specifically, show that $X$ is the Hilbert transform of $R$. Hint: If $f(x) = 0$ for $x < 0$ then $f(x)\operatorname{sgn}(x) = f(x)$. Apply the convolution theorem, then compare real and imaginary parts.*

**Definition 2.4** *When the Fourier transform function $F(\gamma)$ is nonzero only within a bounded interval $[-\Gamma, \Gamma]$, we say that $F$ is* support-limited, *and $f$ is $\Gamma$-band-limited.*

**Ex. 2.8** *Let $f(x), F(\gamma)$ and $g(x), G(\gamma)$ be Fourier transform pairs. Use the conjugation property of Fourier transforms and convolution to establish the Parseval–Plancherel Equation*

$$\langle f, g \rangle = \int f(x)\overline{g(x)}dx = \frac{1}{2\pi} \int F(\gamma)\overline{G(\gamma)}d\gamma. \qquad (2.9)$$

*An important particular case of the Parseval-Plancherel Equation is*

$$\|f\|^2 = \langle f, f \rangle = \int |f(x)|^2 dx = \frac{1}{2\pi} \int |F(\gamma)|^2 d\gamma. \qquad (2.10)$$

**Ex. 2.9** *The one-sided Laplace transform (LT) of $f$ is $\mathcal{F}$ given by*

$$\mathcal{F}(z) = \int_0^\infty f(x)e^{-zx}dx.$$

*Compute $\mathcal{F}(z)$ for $f(x) = u(x)$, the Heaviside function. Compare $\mathcal{F}(-i\gamma)$ with the FT of $u$.*

**Ex. 2.10** *Show that the Fourier transform of $f(x) = e^{-\alpha^2 x^2}$ is $F(\gamma) = \frac{\sqrt{\pi}}{\alpha}e^{-\left(\frac{\gamma}{2\alpha}\right)^2}$. Hints: Calculate the derivative $F'(\gamma)$ by differentiating under the integral sign in the definition of $F$ and integrating by parts. Then solve the resulting differential equation, obtaining*

$$F(\gamma) = Ke^{-\left(\frac{\gamma}{2\alpha}\right)^2},$$

*for some constant $K$ to be determined. To determine $K$, use the Parseval-Plancherel Equation (2.10) and the change of variables $t = 2\alpha^2 x$ to write*

$$\int |f(x)|^2 dx = \int e^{-2\alpha^2 x^2} dx = \frac{1}{2\alpha^2}\int e^{-\frac{t^2}{2\alpha^2}}\, dt,$$

*from which it follows that $K = \frac{\sqrt{\pi}}{\alpha}$.*

---

## 2.7   Dirac Deltas

We saw earlier that the $F(\gamma) = \chi_\Gamma(\gamma)$ has for its inverse Fourier transform the function $f(x) = \frac{\sin \Gamma x}{\pi x}$; note that $f(0) = \frac{\Gamma}{\pi}$ and $f(x) = 0$ for the first time when $\Gamma x = \pi$ or $x = \frac{\pi}{\Gamma}$. For any $\Gamma$-band-limited function $g(x)$ we have $G(\gamma) = G(\gamma)\chi_\Gamma(\gamma)$, so that, for any $x_0$, we have

$$g(x_0) = \int_{-\infty}^\infty g(x)\frac{\sin \Gamma(x - x_0)}{\pi(x - x_0)}dx.$$

We describe this by saying that the function $f(x) = \frac{\sin \Gamma x}{\pi x}$ has the *sifting property* for all $\Gamma$-band-limited functions $g(x)$.

As $\Gamma$ grows larger, $f(0)$ approaches $+\infty$, while $f(x)$ goes to zero for $x \neq 0$. The limit is therefore not a function; it is a *generalized function* called *the Dirac delta function at zero*, denoted $\delta(x)$. For this reason the function $f(x) = \frac{\sin \Gamma x}{\pi x}$ is called an *approximate delta function*. The FT of $\delta(x)$ is the function $F(\gamma) = 1$ for all $\gamma$. The Dirac delta function $\delta(x)$ enjoys the *sifting* property for all appropriate $g(x)$; that is,

$$g(x_0) = \int_{-\infty}^\infty g(x)\delta(x - x_0)dx.$$

Describing which functions $g(x)$ are appropriate is part of the theory of generalized functions and is beyond the scope of this text. It follows from the sifting and shifting properties that the FT of $\delta(x - x_0)$ is the function $e^{ix_0\gamma}$.

The formula for the inverse FT now says

$$\delta(x) = \frac{1}{2\pi} \int_{-\infty}^{\infty} e^{-ix\gamma} d\gamma. \tag{2.11}$$

If we try to make sense of this integral according to the rules of calculus we get stuck quickly. The problem is that the integral formula doesn't mean quite what it does ordinarily and the $\delta(x)$ is not really a function, but an operator on functions; it is sometimes called a *distribution*. The Dirac deltas are mathematical fictions, not in the bad sense of being lies or fakes, but in the sense of being made up for some purpose. They provide helpful descriptions of impulsive forces, probability densities in which a discrete point has nonzero probability, or, in array processing, objects far enough away to be viewed as occupying a discrete point in space.

We shall treat the relationship expressed by Equation (2.11) as a formal statement, rather than attempt to explain the use of the integral in what is surely an unconventional manner.

If we move the discussion into the $\gamma$ domain and define the Dirac delta function $\delta(\gamma)$ to be the FT of the function that has the value $\frac{1}{2\pi}$ for all $x$, then the FT of the complex exponential function $\frac{1}{2\pi}e^{-i\gamma_0 x}$ is $\delta(\gamma - \gamma_0)$, visualized as a "spike" at $\gamma_0$, that is, a generalized function that has the value $+\infty$ at $\gamma = \gamma_0$ and zero elsewhere. This is a useful result, in that it provides the motivation for considering the Fourier transform of a signal $s(t)$ containing hidden periodicities. If $s(t)$ is a sum of complex exponentials with frequencies $-\gamma_n$, then its Fourier transform will consist of Dirac delta functions $\delta(\gamma - \gamma_n)$. If we then estimate the Fourier transform of $s(t)$ from sampled data, we are looking for the peaks in the Fourier transform that approximate the infinitely high spikes of these delta functions.

**Ex. 2.11** *Use the fact that* $\operatorname{sgn}(x) = 2u(x) - 1$ *and Exercise 2.4 to show that* $f(x) = u(x)$ *has the FT* $F(\gamma) = i/\gamma + \pi\delta(\gamma)$.

**Ex. 2.12** *Let* $f, F$ *be a FT pair. Let* $g(x) = \int_{-\infty}^{x} f(y)dy$. *Show that the FT of* $g(x)$ *is* $G(\gamma) = \pi F(0)\delta(\gamma) + \frac{iF(\gamma)}{\gamma}$. *Hint: For the Heaviside function* $u(x)$ *we have*

$$\int_{-\infty}^{x} f(y)dy = \int_{-\infty}^{\infty} f(y)u(x - y)dy.$$

## 2.8 Convolution Filters

In many remote-sensing problems we want values of a function $f(x)$, but are only able to measure values of another function, $h(x)$, related to $f(x)$ in some way. For example, suppose that $x$ is time and $f(x)$ represents what a speaker says into a telephone. The phone line distorts the signal somewhat, often attenuating the higher frequencies. What the person at the other end hears is not $f(x)$, but a related signal function, $h(x)$. For another example, suppose that $f(x, y)$ is a two-dimensional picture viewed by someone with poor eyesight. What that person sees is not $f(x, y)$ but $h(x, y)$, a distorted version of the true $f(x, y)$. In both examples, our goal is to recover the original undistorted signal or image. To do this, it helps to model the distortion. Convolution is a useful tool for this purpose.

Often, the function $h(x)$ has Fourier transform

$$H(\gamma) = F(\gamma)G(\gamma),$$

so that $h(x)$ is the convolution of the desired function $f(x)$ with another function $g(x)$. The function $G(\gamma)$ describes the effects of the measuring system, the telephone line in our first example, or the weak eyes in the second example, or the refraction of light as it passes through the atmosphere, in optical imaging. If we can use our measurements of $h(x)$ to estimate $H(\gamma)$ and if we have some knowledge of the system distortion function, that is, some knowledge of $G(\gamma)$ itself, then there is a chance that we can estimate $F(\gamma)$, and thereby estimate $f(x)$.

If we apply the Fourier Inversion Formula to $H(\gamma) = F(\gamma)G(\gamma)$, we get

$$h(x) = \frac{1}{2\pi} \int F(\gamma)G(\gamma)e^{-i\gamma x} dx. \tag{2.12}$$

The function $h(x)$ that results is $h(x) = (f * g)(x)$, the convolution of the functions $f(x)$ and $g(x)$, with the latter given by

$$g(x) = \frac{1}{2\pi} \int G(\gamma)e^{-i\gamma x} dx.$$

Note that, if $f(x) = \delta(x)$, then $h(x) = g(x)$. In the image processing example, this says that, if the true picture $f$ is a single bright spot, then the blurred image $h$ is $g$ itself. For that reason, the function $g$ is called the *point-spread function* of the distorting system.

Convolution filtering refers to the process of converting any given function, say $f(x)$, into a different function, say $h(x)$, by convolving $f(x)$ with a fixed function $g(x)$. Since this process can be achieved by multiplying $F(\gamma)$ by $G(\gamma)$ and then inverse Fourier transforming, such convolution filters are studied in terms of the properties of the function $G(\gamma)$, known in this context as the *system transfer function*, or the *optical transfer function* (OTF); when $\gamma$ is a frequency, rather than a spatial frequency, $G(\gamma)$ is called the *frequency-response function* of the filter. The magnitude function $|G(\gamma)|$ is called the *modulation transfer function* (MTF). The study of convolution filters is a major part of signal processing. Such filters provide both reasonable models for the degradation that signals undergo, and useful tools for reconstruction. For an important example of the use of filtering, see Chapter 27 on Reverberation and Echo-Cancellation.

Let us rewrite Equation (2.12), replacing $F(\gamma)$ with its definition, as given by Equation (2.6). Then we have

$$h(x) = \int \left( \frac{1}{2\pi} \int f(t) e^{i\gamma t} dt \right) G(\gamma) e^{-i\gamma x} d\gamma. \qquad (2.13)$$

Interchanging the order of integration, we get

$$h(x) = \int f(t) \left( \frac{1}{2\pi} \int G(\gamma) e^{i\gamma(t-x)} d\gamma \right) dt. \qquad (2.14)$$

The inner integral is $g(x - t)$, so we have

$$h(x) = \int f(t) g(x - t) dt; \qquad (2.15)$$

this is the definition of the convolution of the functions $f$ and $g$.

If we know the nature of the blurring, then we know $G(\gamma)$, at least approximately. We can try to remove the blurring by taking measurements of $h(x)$, estimating $H(\gamma) = F(\gamma)G(\gamma)$, dividing these numbers by the value of $G(\gamma)$, and then inverse Fourier transforming. The problem is that our measurements are always noisy, and typical functions $G(\gamma)$ have many zeros and small values, making division by $G(\gamma)$ dangerous, except for those $\gamma$ where the values of $G(\gamma)$ are not too small. These latter values of $\gamma$ tend to be the smaller ones, centered around zero, so that we end up with estimates of $F(\gamma)$ itself only for the smaller values of $\gamma$. The result is a *low-pass filtering* of the object $f(x)$.

To investigate such low-pass filtering, we suppose that $G(\gamma) = 1$, for $|\gamma| \leq \Gamma$, and $G(\gamma) = 0$, otherwise. Then the filter is called the ideal $\Gamma$-low-pass filter. In the far-field propagation model, the variable $x$ is spatial, and the variable $\gamma$ is spatial frequency, related to how the function $f(x)$ changes spatially, as we move $x$. Rapid changes in $f(x)$ are associated with values of $F(\gamma)$ for large $\gamma$. For the case in which the variable $x$ is time, the variable $\gamma$ becomes frequency, and the effect of the low-pass filter on $f(x)$ is to remove its higher-frequency components.

One effect of low-pass filtering in image processing is to smooth out the more rapidly changing features of an image. This can be useful if these features are simply unwanted oscillations, but if they are important detail, such as edges, the smoothing presents a problem. Restoring such wanted detail is often viewed as removing the unwanted effects of the low-pass filtering; in other words, we try to recapture the missing high-spatial-frequency values that have been zeroed out. Such an approach to image restoration is called *frequency-domain extrapolation*. How can we hope to recover these missing spatial frequencies, when they could have been anything? To have some chance of estimating these missing values we need to have some prior information about the image being reconstructed.

## 2.9 A Discontinuous Function

Consider the function $f(x) = \frac{1}{2A}$, for $|x| \leq A$, and $f(x) = 0$, otherwise. The Fourier transform of this $f(x)$ is

$$F(\gamma) = \frac{\sin(A\gamma)}{A\gamma},$$

for all real $\gamma \neq 0$, and $F(0) = 1$. Note that $F(\gamma)$ is nonzero throughout the real line, except for isolated zeros, but that it goes to zero as we go to the infinities. This is typical behavior. Notice also that the smaller the $A$, the slower $F(\gamma)$ dies out; the first zeros of $F(\gamma)$ are at $|\gamma| = \frac{\pi}{A}$, so the main lobe widens as $A$ goes to zero. The function $f(x)$ is not continuous, so its Fourier transform cannot be absolutely integrable. In this case, the Fourier Inversion Formula must be interpreted as involving convergence in the $L^2$ norm.

## 2.10 Shannon's Sampling Theorem

As one might expect, there are connections between Fourier series and Fourier transforms, and several different ways to establish these connections. I believe the simplest way is to use Shannon's Sampling Theorem.

When the Fourier transform function $F(\gamma)$ is nonzero only within a bounded interval $[-\Gamma, \Gamma]$, we say that $F$ is *support-limited*, and $f$ is then said to be $\Gamma$-*band-limited*. Then $F$ has a Fourier series and the Fourier coefficients are

$$c_n = \frac{1}{2\Gamma} \int_{-\Gamma}^{\Gamma} F(\gamma) e^{-i\frac{n\pi}{\Gamma}\gamma} d\gamma. \tag{2.16}$$

Comparing Equations (2.7) and (2.16), we see that

$$c_n = \frac{\pi}{L} f\left(\frac{n\pi}{\Gamma}\right).$$

This tells us that whenever $F$ is determined by its Fourier coefficients, both $f$ and $F$ are determined by the values of the inverse Fourier transform function $f$ at the infinite set of points $x = \frac{n\pi}{\Gamma}$.

The Fourier coefficients $c_n$ and the inverse Fourier transform function $f$ play similar roles. When $F$ is support-limited, we attempt to represent $F$ as an infinite sum of the complex exponential functions $e^{i\frac{n\pi}{\Gamma}\gamma}$ and the $c_n$ are the complex weights associated with each of these exponential functions. More generally, when $F$ may not be support-limited, we attempt to express $F(\gamma)$ as a sum (an integral) over $x$ of all the complex exponential functions $e^{ix\gamma}$, and the complex numbers $f(x)$ are the weight associated with each exponential function.

In many signal-processing applications the variable $x$ is time and denoted $t$, while the variable $\gamma$ is frequency, and denoted $\omega$. Then Shannon's Sampling Theorem says that, whenever there is a bound on the absolute value of the frequencies involved in the function $f$, we can reconstruct $f$ completely from values (or samples) of $f$ at an infinite discrete set of values of $x$ whose spacing depends on the bound on the frequencies; the higher the bound, the smaller the spacing between samples. When our sample spacing is too large, we get *aliasing*. Aliasing is what results in the familiar "strobe-light"effect and why the wagon wheels in cowboy movies appear to revolve backwards.

If $F(\gamma)$ is supported on the interval $[-\Gamma, \Gamma]$, then $F$ and $f$ are completely determined by the values of $f(x)$ at the infinite set of points $x = \frac{n\pi}{\Gamma}$. The spacing $\Delta = \frac{\pi}{\Gamma}$ is called the *Nyquist spacing*.

**Ex. 2.13** *Let $\Gamma = \pi$, so that $\Delta = 1$, $f_m = f(m)$, and $g_m = g(m)$. Use the orthogonality of the functions $e^{im\gamma}$ on $[-\pi, \pi]$ to establish* Parseval's Equation:

$$\langle f, g \rangle = \sum_{m=-\infty}^{\infty} f_m \overline{g_m} = \int_{-\pi}^{\pi} F(\gamma)\overline{G(\gamma)}d\gamma/2\pi, \qquad (2.17)$$

*from which it follows that*

$$\langle f, f \rangle = \int_{-\infty}^{\infty} |F(\gamma)|^2 d\gamma/2\pi.$$

**Ex. 2.14** *Let $f(x)$ be defined for all real $x$ and let $F(\gamma)$ be its FT. Let*

$$g(x) = \sum_{k=-\infty}^{\infty} f(x + 2\pi k),$$

*assuming the sum exists. Show that $g$ is a $2\pi$-periodic function. Compute its Fourier series and use it to derive the* Poisson summation formula:

$$\sum_{k=-\infty}^{\infty} f(2\pi k) = \frac{1}{2\pi} \sum_{n=-\infty}^{\infty} F(n).$$

## 2.11 What Shannon Does Not Say

It is important to remember that Shannon's Sampling Theorem tells us that the *doubly infinite* sequence of values $\{f(n\Delta)\}_{n=-\infty}^{\infty}$ is sufficient to recover exactly the function $F(\gamma)$ and, thereby, the function $f(x)$. Therefore, sampling at the rate of twice the highest frequency (in Hertz) is sufficient only when we have the complete doubly infinite sequence of samples. Of course, in practice, we never have an infinite number of values of anything, so the rule of thumb expressed by Shannon's Sampling Theorem is not valid. Since we know that we will end up with only finitely many samples, each additional data value is additional information. There is no reason to stick to the sampling rate of twice the highest frequency.

## 2.12 Inverse Problems

In this section we introduce the concept of an *inverse problem*, using Fourier series to solve a heat-conduction problem. Many of the problems we study in applied mathematics are *direct problems*. For example, we imagine a ball dropped from a building of known height $h$ and we calculate the time it takes for it to hit the ground and the impact velocity. Once we make certain simplifying assumptions about gravity and air resistance, we are able to solve this problem easily. Using his inverse-square law of universal gravitation, Newton was able to show that planets move in ellipses, with the sun at one focal point. Generally, direct problems conform to the usual flow of time and seek the effects due to known causes. Problems we call inverse problems go the other way, seeking the causes of observed effects; we measure the impact velocity to determine the height $h$ of the building. Newton solved an inverse problem when he determined that Kepler's empirical laws of planetary motion follow from an inverse-square law of universal gravitation.

In each of the examples of remote sensing just presented in Chapter 1 we have measured some effects and want to know the causes. In x-ray tomography, for example, we observe that the x-rays that passed through the body of the patient come out weaker than when they went in. We know that they were weakened, or *attenuated*, because they were partially absorbed by the material they had to pass through; we want to know precisely where the attenuation took place. This is an inverse problem; we are trying to go back in time, to uncover the causes of the observed effects.

Direct problems have been studied for a long time, while the theory of inverse problems is still being developed. Generally speaking, direct problems are easier than inverse problems. Direct problems, at least those corresponding to actual physical situations, tend to be *well-posed* in the sense of Hadamard, while inverse problems are often *ill-posed*. A problem is said to be *well-posed* if there is a unique solution for each input to the problem and the solution varies continuously with the input; roughly speaking, small changes in the input lead to small changes in the solution. If we vary the height of the building slightly, the time until the ball hits the ground and its impact velocity will change only slightly. For inverse problems, there may be many solutions, or none, and slight changes in the data can cause the solutions to differ greatly. In [7] Bertero and Boccacci give a nice illustration of the difference between direct and inverse problems, using the heat equation.

Suppose that $u(x, t)$ is the temperature distribution for $x$ in the interval $[0, a]$ and $t \geq 0$. The function $u(x, t)$ satisfies the heat equation

$$\frac{\partial^2 u}{\partial x^2} = \frac{1}{D} \frac{\partial u}{\partial t},$$

where $D > 0$ is the thermal conductivity. In addition, we adopt the boundary conditions $u(x, 0) = f(x)$, and $u(0, t) = u(a, t) = 0$, for all $t$. By separating the variables, and using Fourier series, we find that, if

$$f(x) = \sum_{n=1}^{\infty} f_n \sin\left(\frac{n\pi x}{a}\right),$$

where

$$f_n = \frac{2}{a} \int_0^a f(x) \sin\left(\frac{n\pi x}{a}\right) dx,$$

then

$$u(x, t) = \sum_{n=1}^{\infty} f_n e^{-D(\frac{\pi n}{a})^2 t} \sin\left(\frac{n\pi x}{a}\right).$$

The direct problem is to find $u(x, t)$, given $f(x)$. Suppose that we know $f(x)$ with some finite precision, that is, we know those Fourier coefficients $f_n$ for which $|f_n| \geq \epsilon > 0$. Because of the decaying exponential factor, fewer

Fourier coefficients in the expansion of $u(x, t)$ will be above this threshold, and we can determine $u(x, t)$ with the same precision or better. The solution to the heat equation tends to be smoother than the input distribution.

The inverse problem is to determine the initial distribution $f(x)$ from knowledge of $u(x, t)$ at one or more times $t > 0$. As we just saw, for any fixed time $t > 0$, the Fourier coefficients of $u(x, t)$ will die off faster than the $f_n$ do, leaving fewer coefficients above the threshold of $\epsilon$. This means we can determine fewer and fewer of the $f_n$ as $t$ grows larger. For $t$ beyond some point, it will be nearly impossible to say anything about $f(x)$.

Once again, the proper interpretation of Equation (2.7) will depend on the properties of the functions involved. It may happen that one or both of these integrals will fail to be defined in the usual way and will be interpreted as the principal value of the integral [80].

## 2.13    Two-Dimensional Fourier Transforms

The Fourier transform is also defined for functions of several real variables $f(x_1, ..., x_N)$. The multidimensional FT arises in image processing, scattering, transmission tomography, and many other areas. In this section we discuss the extension of the definitions of the FT and IFT to functions of two real variables.

### 2.13.1    The Basic Formulas

For the complex-valued function $f(x, y)$ of two real variables, the Fourier transformation is

$$F(\alpha, \beta) = \int \int f(x, y) e^{i(x\alpha + y\beta)} dx dy.$$

Just as in the one-dimensional case, the Fourier transformation that produced $F(\alpha, \beta)$ can be inverted to recover the original $f(x, y)$. The Fourier Inversion Formula in this case is

$$f(x, y) = \frac{1}{4\pi^2} \int \int F(\alpha, \beta) e^{-i(\alpha x + \beta y)} d\alpha d\beta. \tag{2.18}$$

It is important to note that this procedure can be viewed as two one-dimensional Fourier inversions: First, we invert $F(\alpha, \beta)$, as a function of, say, $\beta$ only, to get the function of $\alpha$ and $y$

$$g(\alpha, y) = \frac{1}{2\pi} \int F(\alpha, \beta) e^{-i\beta y} d\beta;$$

second, we invert $g(\alpha, y)$, as a function of $\alpha$, to get

$$f(x, y) = \frac{1}{2\pi} \int g(\alpha, y) e^{-i\alpha x} d\alpha.$$

If we write the functions $f(x, y)$ and $F(\alpha, \beta)$ in polar coordinates, we obtain alternative ways to implement the two-dimensional Fourier inversion. We shall consider these other ways in Chapter 11, when we discuss the tomography problem of reconstructing a function $f(x, y)$ from line-integral data.

## 2.13.2   Radial Functions

Now we consider the two-dimensional Fourier-transform pairs in polar coordinates. We convert to polar coordinates using $(x, y) = r(\cos\theta, \sin\theta)$ and $(\alpha, \beta) = \rho(\cos\omega, \sin\omega)$. Then

$$F(\rho, \omega) = \int_0^\infty \int_{-\pi}^\pi f(r, \theta) e^{ir\rho\cos(\theta-\omega)} r \, dr \, d\theta. \tag{2.19}$$

Say that a function $f(x, y)$ of two variables is a *radial* function if $x^2 + y^2 = x_1^2 + y_1^2$ implies $f(x, y) = f(x_1, y_1)$, for all points $(x, y)$ and $(x_1, y_1)$; that is, $f(x, y) = g(\sqrt{x^2 + y^2})$ for some function $g$ of one variable.

**Ex. 2.15** *Show that if $f$ is radial then its FT $F$ is also radial. Find the FT of the radial function $f(x, y) = \frac{1}{\sqrt{x^2+y^2}}$. Hints: Insert $f(r, \theta) = g(r)$ in Equation (2.19) to obtain*

$$F(\rho, \omega) = \int_0^\infty \int_{-\pi}^\pi g(r) e^{ir\rho\cos(\theta-\omega)} r \, dr \, d\theta$$

*or*

$$F(\rho, \omega) = \int_0^\infty r g(r) \left[ \int_{-\pi}^\pi e^{ir\rho\cos(\theta-\omega)} d\theta \right] dr.$$

*Show that the inner integral is independent of $\omega$, and then use the fact that*

$$\int_{-\pi}^\pi e^{ir\rho\cos\theta} d\theta = 2\pi J_0(r\rho),$$

*with $J_0$ the 0th order Bessel function, to get*

$$F(\rho, \omega) = H(\rho) = 2\pi \int_0^\infty r g(r) J_0(r\rho) dr.$$

The function $H(\rho)$ is called the *Hankel transform* of $g(r)$. Summarizing, we say that if $f(x, y)$ is a radial function obtained using $g$ then its Fourier transform $F(\alpha, \beta)$ is also a radial function, obtained using the Hankel transform of $g$.

### 2.13.3 An Example

For example, suppose that $f(x, y) = 1$ for $\sqrt{x^2 + y^2} \leq R$, and $f(x, y) = 0$, otherwise. Then we have

$$F(\alpha, \beta) = \int_{-\pi}^{\pi} \int_0^R e^{-i(\alpha r \cos \theta + \beta r \sin \theta)} r \, dr \, d\theta.$$

In polar coordinates, with $\alpha = \rho \cos \phi$ and $\beta = \rho \sin \phi$, we have

$$F(\rho, \phi) = \int_0^R \int_{-\pi}^{\pi} e^{ir\rho \cos(\theta - \phi)} d\theta r \, dr.$$

The inner integral is well known;

$$\int_{-\pi}^{\pi} e^{ir\rho \cos(\theta - \phi)} d\theta = 2\pi J_0(r\rho),$$

where $J_0$ and $J_n$ denote the 0th order and $n$th order Bessel functions, respectively. Using the following identity

$$\int_0^z t^n J_{n-1}(t) dt = z^n J_n(z),$$

we have

$$F(\rho, \phi) = \frac{2\pi R}{\rho} J_1(\rho R).$$

Notice that, since $f(x, z)$ is a radial function, that is, dependent only on the distance from $(0, 0)$ to $(x, y)$, its Fourier transform is also radial.

The first positive zero of $J_1(t)$ is around $t = 4$, so when we measure $F$ at various locations and find $F(\rho, \phi) = 0$ for a particular $(\rho, \phi)$, we can estimate $R \approx 4/\rho$. So, even when a distant spherical object, like a star, is too far away to be imaged well, we can sometimes estimate its size by finding where the intensity of the received signal is zero [101].

In her 1953 *Nature* paper with R. G. Gosling the British scientist Rosalind Franklin presented evidence she had obtained from x-ray scattering experiments that corroborated the double-helical structure of the DNA molecule proposed a short time previously by Crick and Watson. She showed mathematically that the scattering pattern from a helical structure would be described by the Bessel functions $J_n$ and noted that the observed maximal intensities in her photographs corresponded to the zeros of these Bessel functions.

According to Lightman [111], most historians of science who have studied the work that led to the discovery of the structure of DNA agree that the contribution of Rosalind Franklin is understated in Watson's account in his book [160]. In 1962 Francis Crick and James Watson shared the Nobel Prize in Physics with Maurice Wilkins of King's College, London, who had worked with Franklin on DNA. Had she not died of cancer in 1958, it is plausible that Franklin, not Wilkins, would have shared the prize.

## 2.14    The Uncertainty Principle

We saw earlier that the Fourier transform of the function $f(x) = e^{-\alpha^2 x^2}$ is

$$F(\gamma) = \frac{\sqrt{\pi}}{\alpha} e^{-(\frac{\gamma}{2\alpha})^2}.$$

This Fourier-transform pair illustrates well the general fact that the more concentrated $f(x)$ is, the more spread out $F(\gamma)$ is. In particular, it is impossible for both $f$ and $F$ to have bounded support. We prove the following inequality:

$$\frac{\int x^2 |f(x)|^2 dx}{\int |f(x)|^2 dx} \frac{\int \gamma^2 |F(\gamma)|^2 d\gamma}{\int |F(\gamma)|^2 d\gamma} \geq \frac{1}{4}. \tag{2.20}$$

This inequality is the mathematical version of Heisenberg's Uncertainty Principle.

As we shall show in Chapter 19, the Cauchy-Schwarz Inequality holds in any vector space with an inner product. In the present situation, the Cauchy-Schwarz Inequality tells us that

$$\left| \int f(x)\overline{g(x)} dx \right|^2 \leq \int |f(x)|^2 dx \int |g(x)|^2 dx,$$

with equality if and only if $g(x) = kf(x)$, for some scalar $k$. We will need this in the proof of the inequality (2.20). We'll also need the Parseval-Plancherel Equation (2.9), as well as the fact that, for any two complex numbers $z$ and $w$, we have

$$|zw| \geq \frac{1}{2}(z\overline{w} + \overline{z}w).$$

In addition, we assume that

$$\lim_{a \to +\infty} (a(|f(a)|^2 + |f(-a)|^2) = 0,$$

so that, using integration by parts, we have

$$\int x \left( \frac{d}{dx} |f(x)|^2 \right) dx = -\int |f(x)|^2 dx.$$

The proof of Inequality (2.20) now follows:

$$\frac{1}{2\pi} \int x^2 |f(x)|^2 dx \int \gamma^2 |F(\gamma)|^2 d\gamma = \frac{1}{2\pi} \int |xf(x)|^2 dx \int |\gamma F(\gamma)|^2 d\gamma$$

$$= \frac{1}{2\pi} \int |xf(x)|^2 dx \int |f'(x)|^2 dx \geq \left( \int |xf'(x)f(x)| dx \right)^2$$

$$\geq \left( \int \frac{x}{2} [f'(x)\overline{f(x)} + f(x)\overline{f'(x)}] dx \right)^2 = \frac{1}{4} \left( \int x(\frac{d}{dx}|f(x)|^2) dx \right)^2$$

$$= \frac{1}{4} \left( \int |f(x)|^2 dx \right)^2 = \frac{1}{8\pi} \int |f(x)|^2 dx \int |F(\gamma)|^2 d\gamma.$$

This completes the proof of Inequality (2.20).

The significance of this inequality is made evident when we reformulate it in terms of the variances of probability densities. Suppose that

$$\int |f(x)|^2 dx = \frac{1}{2\pi} \int |F(\gamma)|^2 d\gamma = 1,$$

so that we may view $|f(x)|^2$ and $\frac{1}{2\pi}|F(\gamma)|^2$ as probability density functions associated with random variables $X$ and $Y$, respectively. From probability theory we know that the expected values $E(X)$ and $E(Y)$ are given by

$$m = E(X) = \int x|f(x)|^2 dx$$

and

$$M = E(Y) = \frac{1}{2\pi} \int \gamma|F(\gamma)|^2 d\gamma.$$

Let

$$g(x) = f(x+m)e^{iMx},$$

so that the Fourier transform of $g(x)$ is

$$G(\gamma) = F(\gamma + M)e^{i(M-\gamma)m}.$$

Then, $|g(x)|^2 = |f(x+m)|^2$ and $|G(\gamma)|^2 = |F(\gamma + M)|^2$; we also have

$$\int x|g(x)|^2 dx = 0$$

and

$$\int \gamma|G(\gamma)|^2 d\gamma = 0.$$

The point here is that we can assume that $m = 0$ and $M = 0$. Consequently, the variance of $X$ is

$$var(X) = \int x^2|f(x)|^2 dx$$

and the variance of $Y$ is

$$var(Y) = \frac{1}{2\pi} \int \gamma^2|F(\gamma)|^2 d\gamma.$$

The variances measure how spread out the functions $|f(x)|^2$ and $|F(\gamma)|^2$ are around their respective means. From Inequality (2.20) we know that the product of these variances is not smaller than $\frac{1}{4}$.

**Ex. 2.16** *Show, by examining the proof of Inequality (2.20), that if the inequality is an equation for some $f$ then $f'(x) = kxf(x)$, so that $f(x) = e^{-\alpha^2 x^2}$ for some $\alpha > 0$. Hint: What can be said when Cauchy's Inequality is an equality?*

---

## 2.15　Best Approximation

The basic problem here is to estimate $F(\gamma)$ from finitely many values of $f(x)$, under the assumption that $F(\gamma) = 0$ for $|\gamma| > \Gamma$, for some $\Gamma > 0$. Since we do not have all of $f(x)$, the best we can hope to do is to approximate $F(\gamma)$ in some sense. To help us understand how best approximation works, we consider the *orthogonality principle*.

### 2.15.1　The Orthogonality Principle

Imagine that you are standing and looking down at the floor. The point $B$ on the floor that is closest to the tip of your nose, which we label $F$, is the unique point on the floor such that the vector from $B$ to any other point $A$ on the floor is perpendicular to the vector from $B$ to $F$; that is, $FB \cdot AB = 0$. This is a simple illustration of the *orthogonality principle*.

When two vectors are perpendicular to one another, their dot product is zero. This idea can be extended to functions. We say that two functions $F(\gamma)$ and $G(\gamma)$ defined on the interval $[-\Gamma, \Gamma]$ are *orthogonal* if

$$\int_{-\Gamma}^{\Gamma} F(\gamma)\overline{G(\gamma)}d\gamma = 0.$$

Suppose that $G_n(\gamma)$, $n = 0, ..., N - 1$, are known functions, and

$$A(\gamma) = \sum_{n=0}^{N-1} a_n G_n(\gamma),$$

for any coefficients $a_n$. We want to minimize the approximation error

$$\int_{-\Gamma}^{\Gamma} |F(\gamma) - A(\gamma)|^2 d\gamma,$$

over all coefficients $a_n$. Suppose that the best choices are $a_n = b_n$. The orthogonality principle tells us that the best approximation

$$B(\gamma) = \sum_{n=0}^{N-1} b_n G_n(\gamma)$$

is such that the function $F(\gamma) - B(\gamma)$ is orthogonal to $A(\gamma) - B(\gamma)$ for every choice of the $a_n$.

Suppose that we fix $m$ and select $a_n = b_n$, for $n \neq m$, and $a_m = b_m + 1$. Then we have

$$\int_{-\Gamma}^{\Gamma} (F(\gamma) - B(\gamma))\overline{G_m(\gamma)}d\gamma = 0. \qquad (2.21)$$

We can use Equation (2.21) to help us find the best $b_n$.

From Equation (2.21) we have

$$\int_{-\Gamma}^{\Gamma} F(\gamma)\overline{G_m(\gamma)}d\gamma = \sum_{n=0}^{N-1} b_n \int_{-\Gamma}^{\Gamma} G_n(\gamma)\overline{G_m(\gamma)}d\gamma.$$

Since we know the $G_n(\gamma)$, we know the integrals

$$\int_{-\Gamma}^{\Gamma} G_n(\gamma)\overline{G_m(\gamma)}d\gamma.$$

If we can learn the values

$$\int_{-\Gamma}^{\Gamma} F(\gamma)\overline{G_m(\gamma)}d\gamma$$

from measurements, then we simply solve a system of linear equations to find the $b_n$.

### 2.15.2 An Example

Suppose that we have measured the values $f(x_n)$, for $n = 0, ..., N-1$, where the $x_n$ are arbitrary real numbers. Then, from these measurements, we can find the best approximation of $F(\gamma)$ of the form

$$A(\gamma) = \sum_{n=0}^{N-1} a_n G_n(\gamma),$$

if we select $G_n(\gamma) = e^{i\gamma x_n}$.

### 2.15.3  The DFT as Best Approximation

Suppose now that our data values are $f(\Delta n)$, for $n = 0, 1, ..., N - 1$, where we have chosen $\Delta = \frac{\pi}{\Gamma}$. We can view the DFT as a best approximation of the function $F(\gamma)$ over the interval $[-\Gamma, \Gamma]$, in the following sense. Consider all functions of the form

$$A(\gamma) = \sum_{n=0}^{N-1} a_n e^{in\Delta\gamma},$$

where the best coefficients $a_n = b_n$ are to be determined. Now select those $b_n$ for which the approximation error

$$\int_{-\Gamma}^{\Gamma} |F(\gamma) - A(\gamma)|^2 d\gamma$$

is minimized. Then it is easily shown that these optimal $b_n$ are precisely

$$b_n = \Delta f(n\Delta),$$

for $n = 0, 1, ..., N - 1$.

**Ex. 2.17** *Show that $b_n = \Delta f(n\Delta)$, for $n = 0, 1, ..., N - 1$, are the optimal coefficients.*

The DFT estimate is reasonably accurate when $N$ is large, but when $N$ is not large there are usually better ways to estimate $F(\gamma)$, as we shall see.

In Figure 2.1, the real-valued function $f(x)$ is the solid-line figure in both graphs. In the bottom graph, we see the true $f(x)$ and a DFT estimate. The top graph is the MDFT estimator, the result of *band-limited extrapolation*, a technique for predicting missing Fourier coefficients that we shall discuss next.

### 2.15.4  The Modified DFT (MDFT)

We suppose, as in the previous subsection, that $F(\gamma) = 0$, for $|\gamma| > \Gamma$, and that our data values are $f(n\Delta)$, for $n = 0, 1, ..., N - 1$. It is often convenient to use a sampling interval $\Delta$ that is smaller than $\frac{\pi}{\Gamma}$ in order to obtain more data values. Therefore, we assume now that $\Delta < \frac{\pi}{\Gamma}$. Once again, we seek the function of the form

$$A(\gamma) = \sum_{n=0}^{N-1} a_n e^{in\Delta\gamma},$$

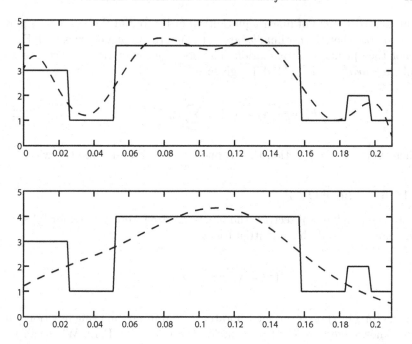

**FIGURE 2.1**: The non-iterative band-limited extrapolation method (MDFT) (top) and the DFT (bottom) for $N = 64$, 30 times over-sampled data.

defined for $|\gamma| \leq \Gamma$, for which the error measurement

$$\int_{-\Gamma}^{\Gamma} |F(\gamma) - A(\gamma)|^2 d\gamma$$

is minimized.

In the previous example, for which $\Delta = \frac{\pi}{\Gamma}$, we have

$$\int_{-\Gamma}^{\Gamma} e^{i(n-m)\Delta\gamma} d\gamma = 0,$$

for $m \neq n$. As the reader will discover in doing Exercise 2.17, this greatly simplifies the system of linear equations that we need to solve to get the optimal $b_n$. Now, because $\Delta \neq \frac{\pi}{\Gamma}$, we have

$$\int_{-\Gamma}^{\Gamma} e^{i(n-m)\Delta\gamma} d\gamma = \frac{\sin((n-m)\Delta\Gamma)}{\pi(n-m)\Delta},$$

which is not zero when $n \neq m$. This means that we have to solve a more complicated system of linear equations in order to find the $b_n$. It is important to note that the optimal $b_n$ are not equal to $\Delta f(n\Delta)$ now, so the DFT is not the optimal approximation. The best approximation in this case we call the *modified* DFT (MDFT), given by

$$F_{MDFT}(\gamma) = \chi_\Gamma(\gamma) \sum_{n=0}^{N-1} b_n e^{in\Delta\gamma}, \tag{2.22}$$

where $\chi_\Gamma(\gamma)$ is the function that is one for $|\gamma| \leq \Gamma$ and zero otherwise.

### 2.15.5   The PDFT

In the previous subsection, the functions $A(\gamma)$ were defined for $|\gamma| \leq \Gamma$. Therefore, we could have written them as

$$A(\gamma) = \chi_\Gamma(\gamma) \sum_{n=0}^{N-1} a_n e^{in\Delta\gamma}.$$

The factor $\chi_\Gamma(\gamma)$ serves to incorporate into our approximating function our prior knowledge that $F(\gamma) = 0$ outside the interval $[-\Gamma, \Gamma]$. What can we do if we have additional prior knowledge about the broad features of $F(\gamma)$ that we wish to include?

Suppose that $P(\gamma) \geq 0$ is a prior estimate of $|F(\gamma)|$. Now we approximate $F(\gamma)$ with functions of the form

$$C(\gamma) = P(\gamma) \sum_{n=0}^{N-1} c_n e^{in\Delta\gamma}.$$

As we shall see in Chapter 25, the best choices of the $c_n$ are the ones that satisfy the equations

$$f(m\Delta) = \sum_{n=0}^{N-1} c_n p((n-m)\Delta), \tag{2.23}$$

for $m = 0, 1, ..., N - 1$, where

$$p(x) = \frac{1}{2\pi} \int_{-\Gamma}^{\Gamma} P(\gamma) e^{-ix\gamma} d\gamma$$

is the inverse Fourier transform of the function $P(\gamma)$. This best approximation we call the PDFT [23, 24, 26]. The use of the PDFT was illustrated in Chapter 1, in the reconstruction of a simulated head slice.

## 2.16   Analysis of the MDFT

Let our data be $f(x_m)$, $m = 1, ..., M$, where the $x_m$ are arbitrary values of the variable $x$. If $F(\gamma)$ is zero outside $[-\Gamma, \Gamma]$, then minimizing the energy over $[-\Gamma, \Gamma]$ subject to data consistency produces an estimate of the form

$$F_{MDFT}(\gamma) = \chi_\Gamma(\gamma) \sum_{m=1}^{M} b_m \exp(ix_m\gamma),$$

with the $b_m$ satisfying the equations

$$f(x_n) = \sum_{m=1}^{M} b_m \frac{\sin(\Gamma(x_m - x_n))}{\pi(x_m - x_n)},$$

for $n = 1, ..., M$. The matrix $S_\Gamma$ with entries $\frac{\sin(\Gamma(x_m - x_n))}{\pi(x_m - x_n)}$ we call a *sinc* matrix.

### 2.16.1   Eigenvector Analysis of the MDFT

Although it seems reasonable that incorporating the additional information about the support of $F(\gamma)$ should improve the estimation, it would be more convincing if we had a more mathematical argument to make. For that we turn to an analysis of the eigenvectors of the sinc matrix. Throughout this subsection we make the simplification that $x_n = n$.

**Ex. 2.18** *The purpose of this exercise is to show that, for an Hermitian nonnegative-definite $M$ by $M$ matrix $Q$, a norm-one eigenvector $\mathbf{u}^1$ of $Q$ associated with its largest eigenvalue, $\lambda_1$, maximizes the quadratic form $\mathbf{a}^\dagger Q \mathbf{a}$ over all vectors $\mathbf{a}$ with norm one. Let $Q = ULU^\dagger$ be the eigenvector decomposition of $Q$, where the columns of $U$ are mutually orthogonal eigenvectors $\mathbf{u}^n$ with norms equal to one, so that $U^\dagger U = I$, and $L = diag\{\lambda_1, ..., \lambda_M\}$ is the diagonal matrix with the eigenvalues of $Q$ as its entries along the main diagonal. Assume that $\lambda_1 \geq \lambda_2 \geq ... \geq \lambda_M$. Then maximize*

$$\mathbf{a}^\dagger Q \mathbf{a} = \sum_{n=1}^{M} \lambda_n |\mathbf{a}^\dagger \mathbf{u}^n|^2,$$

*subject to the constraint*

$$\mathbf{a}^\dagger \mathbf{a} = \mathbf{a}^\dagger U^\dagger U \mathbf{a} = \sum_{n=1}^{M} |\mathbf{a}^\dagger \mathbf{u}^n|^2 = 1.$$

*Hint: Show $\mathbf{a}^\dagger Q \mathbf{a}$ is a convex combination of the eigenvalues of $Q$.*

**Ex. 2.19** *Show that, for the sinc matrix $Q = S_\Gamma$, the quadratic form $\mathbf{a}^\dagger Q \mathbf{a}$ in the previous exercise becomes*

$$\mathbf{a}^\dagger S_\Gamma \mathbf{a} = \frac{1}{2\pi} \int_{-\Gamma}^{\Gamma} \left| \sum_{n=1}^{M} a_n e^{in\gamma} \right|^2 d\gamma.$$

*Show that the norm of the vector $\mathbf{a}$ is the square root of the integral*

$$\frac{1}{2\pi} \int_{-\pi}^{\pi} \left| \sum_{n=1}^{M} a_n e^{in\gamma} \right|^2 d\gamma.$$

**Ex. 2.20** *For $M = 30$ compute the eigenvalues of the matrix $S_\Gamma$ for various choices of $\Gamma$, such as $\Gamma = \frac{\pi}{k}$, for $k = 2, 3, ..., 10$. For each $k$ arrange the set of eigenvalues in decreasing order and note the proportion of them that are not near zero. The set of eigenvalues of a matrix is sometimes called its eigenspectrum and the nonnegative function $\chi_\Gamma(\gamma)$ is a power spectrum; here is one time in which different notions of a spectrum are related.*

## 2.16.2 The Eigenfunctions of $S_\Gamma$

Suppose that the vector $\mathbf{u}^1 = (u_1^1, ..., u_M^1)^T$ is an eigenvector of $S_\Gamma$ corresponding to the largest eigenvalue, $\lambda_1$. Associate with $\mathbf{u}^1$ the *eigenfunction*

$$U^1(\gamma) = \sum_{n=1}^{M} u_n^1 e^{in\gamma}.$$

Then

$$\lambda_1 = \int_{-\Gamma}^{\Gamma} |U^1(\gamma)|^2 d\gamma / \int_{-\pi}^{\pi} |U^1(\gamma)|^2 d\gamma$$

and $U^1(\gamma)$ is the function of its form that is most concentrated within the interval $[-\Gamma, \Gamma]$.

Similarly, if $\mathbf{u}^M$ is an eigenvector of $S_\Gamma$ associated with the smallest eigenvalue $\lambda_M$, then the corresponding eigenfunction $U^M(\gamma)$ is the function of its form least concentrated in the interval $[-\Gamma, \Gamma]$.

**Ex. 2.21** *On the interval $|\gamma| \leq \pi$ plot the functions $|U^m(\gamma)|$ corresponding to each of the eigenvectors of the sinc matrix $S_\Gamma$. Pay particular attention to the places where each of these functions is zero.*

The eigenvectors of $S_\Gamma$ corresponding to different eigenvalues are orthogonal, that is $(\mathbf{u}^m)^\dagger \mathbf{u}^n = 0$ if $m$ is not $n$. We can write this in terms of integrals:

$$\int_{-\pi}^{\pi} U^n(\gamma)\overline{U^m(\gamma)}d\gamma = 0$$

if $m$ is not $n$. The mutual orthogonality of these eigenfunctions is related to the locations of their roots, which were studied in the previous exercise.

Any Hermitian matrix $Q$ is invertible if and only if none of its eigenvalues is zero. With $\lambda_m$ and $\mathbf{u}^m$, $m = 1, ..., M$, the eigenvalues and eigenvectors of $Q$, the inverse of $Q$ can then be written as

$$Q^{-1} = (1/\lambda_1)\mathbf{u}^1(\mathbf{u}^1)^\dagger + ... + (1/\lambda_M)\mathbf{u}^M(\mathbf{u}^M)^\dagger.$$

**Ex. 2.22** *Show that the MDFT estimator given by Equation (2.22)* $F_{MDFT}(\gamma)$ *can be written as*

$$F_{MDFT}(\gamma) = \chi_\Gamma(\gamma) \sum_{m=1}^{M} \frac{1}{\lambda_m}(\mathbf{u}^m)^\dagger \mathbf{d}\, U^m(\gamma),$$

*where* $\mathbf{d} = (f(1), f(2), ..., f(M))^T$ *is the data vector.*

**Ex. 2.23** *Show that the DFT estimate of* $F(\gamma)$, *restricted to the interval* $[-\Gamma, \Gamma]$, *is*

$$F_{DFT}(\gamma) = \chi_\Gamma(\gamma) \sum_{m=1}^{M} (\mathbf{u}^m)^\dagger \mathbf{d}\, U^m(\gamma).$$

*Hint: Use the fact that* $I = UU^\dagger$.

From these two exercises we can learn why it is that the estimate $F_{MDFT}(\gamma)$ resolves better than the DFT. The former makes more use of the eigenfunctions $U^m(\gamma)$ for higher values of $m$, since these are the ones for which $\lambda_m$ is closer to zero. Since those eigenfunctions are the ones having most of their roots within the interval $[-\Gamma, \Gamma]$, they have the most flexibility within that region and are better able to describe those features in $F(\gamma)$ that are not resolved by the DFT.

# Chapter 3

## Remote Sensing

## 3.1   Chapter Summary

A basic problem in remote sensing is to determine the nature of a distant object by measuring signals transmitted by or reflected from that object. If the object of interest is sufficiently remote, that is, is in the *far field*, the data we obtain by sampling the propagating spatio-temporal field is related, approximately, to what we want by *Fourier transformation*. In this chapter we present examples to illustrate the roles played by Fourier series and Fourier coefficients in the analysis of remote sensing and signal transmission. We use these examples to motivate several of the computational problems we shall consider in detail later in the text. We also discuss two inverse problems involving the Laplace transform.

We consider here a common problem of remote sensing of transmitted or reflected waves propagating from distant sources. Examples include optical imaging of planets and asteroids using reflected sunlight, radio-astronomy imaging of distant sources of radio waves, active and passive sonar, radar imaging using microwaves, and infrared (IR) imaging to monitor the ocean temperature. In such situations, as well as in transmission and emission tomography and magnetic-resonance imaging, what we measure are essentially the Fourier coefficients or values of the Fourier transform of the function we want to estimate. The image reconstruction problem then becomes one of estimating a function from finitely many noisy values of its Fourier transform.

## 3.2   Fourier Series and Fourier Coefficients

We suppose that $f : [-L, L] \to \mathbb{C}$, and that its Fourier series converges to $f(x)$ for all $x$ in $[-L, L]$. In the examples in this chapter, we shall see how Fourier coefficients can arise as data obtained through measurements. However, we shall be able to measure only a finite number of the Fourier coefficients. One issue that will concern us is the effect on the estimation of $f(x)$ if we use some, but not all, of its Fourier coefficients.

Suppose that we have $c_n$, as defined by Equation (2.5), for $n = 0, 1, 2, ..., N$. It is not unreasonable to try to estimate the function $f(x)$ using the *discrete Fourier transform* (DFT) estimate, which is

$$f_{DFT}(x) = \sum_{n=0}^{N} c_n e^{i \frac{n\pi}{L} x}.$$

When we know that $f(x)$ is real-valued, and so $c_{-n} = \overline{c_n}$, we naturally assume that we have the values of $c_n$ for $|n| \leq N$.

---

## 3.3 The Unknown Strength Problem

In this example, we imagine that each point $x$ in the interval $[-L, L]$ is sending out a signal that is a complex-exponential-function signal, also called a *sinusoid*, at the frequency $\omega$, each with its own strength $f(x)$; that is, the signal sent by the point $x$ is

$$f(x)e^{i\omega t}.$$

In our first example, we imagine that the strength function $f(x)$ is unknown and we want to determine it. It could be the case that the signals originate at the points $x$, as with light or radio waves from the sun, or are simply reflected from the points $x$, as is sunlight from the moon or radio waves in radar. Later in this chapter, we shall investigate a related example, in which the points $x$ transmit known signals and we want to determine what is received elsewhere.

### 3.3.1 Measurement in the Far Field

Now let us consider what is received by a point $P$ on the circumference of a circle centered at the origin and having large radius $D$. The point $P$ corresponds to the angle $\theta$ as shown in Figure 3.1; we use $\theta$ in the interval $[0, \pi]$. It takes a finite time for the signal sent from $x$ at time $t$ to reach $P$, so there is a delay.

We assume that $c$ is the speed at which the signal propagates. Because $D$ is large relative to $L$, we make the *far-field assumption*, which allows us to approximate the distance from $x$ to $P$ by $D - x\cos\theta$. Therefore, what $P$ receives at time $t$ from $x$ is approximately what was sent from $x$ at time $t - \frac{1}{c}(D - x\cos\theta)$.

**Ex. 3.1** *Show that, for any point $P$ on the circle of radius $D$ and any $x \neq 0$, the distance from $x$ to $P$ is always greater than or equal to the far-field approximation $D - x\cos\theta$, with equality if and only if $\theta = 0$ or $\theta = \pi$.*

At time $t$, the point $P$ receives from $x$ the signal

$$f(x)e^{i\omega(t-\frac{1}{c}(D-x\cos\theta))} = e^{i\omega(t-\frac{1}{c}D)}f(x)e^{i\frac{\omega\cos\theta}{c}x}.$$

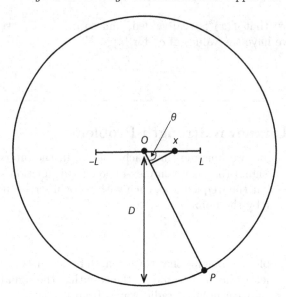

**FIGURE 3.1**: Far-field measurements.

Because the point $P$ receives signals from all $x$ in $[-L, L]$, the signal that $P$ receives at time $t$ is

$$e^{i\omega(t-\frac{1}{c}D)} \int_{-L}^{L} f(x)e^{i\frac{\omega \cos \theta}{c}x}dx.$$

Therefore, from measurements in the far field, we obtain the values

$$\int_{-L}^{L} f(x)e^{i\frac{\omega \cos \theta}{c}x}dx.$$

When $\theta$ is chosen so that

$$\frac{\omega \cos \theta}{c} = \frac{-n\pi}{L} \tag{3.1}$$

we have $c_n$.

### 3.3.2  Limited Data

Note that we will be able to solve Equation (3.1) for $\theta$ if and only if we have

$$|n| \leq \frac{L\omega}{\pi c}.$$

This tells us that we can measure only finitely many of the Fourier coefficients of $f(x)$. It is common in signal processing to speak of the *wavelength* of a sinusoidal signal; the wavelength associated with a given $\omega$ and $c$ is

$$\lambda = \frac{2\pi c}{\omega}.$$

Therefore we can measure $2N+1$ Fourier coefficients, where $N$ is the largest integer not greater than $\frac{2L}{\lambda}$, which is the length of the interval $[-L, L]$, measured in units of wavelength $\lambda$. We get more Fourier coefficients when the product $L\omega$ is larger; this means that when $L$ is small, we want $\omega$ to be large, so that $\lambda$ is small and $N$ is large. As we saw previously, using these finitely many Fourier coefficients to calculate the DFT reconstruction of $f(x)$ can lead to a poor estimate of $f(x)$, particularly when $N$ is small.

Consider the situation in which the points $x$ are reflecting signals that are sent to probe the structure of an object described by the function $f$, as in radar. This relationship between the number $L\omega$ and the number of Fourier coefficients we can measure amounts to a connection between the frequency of the probing signal and the resolution attainable; finer detail is available only if the frequency is high enough.

The wavelengths used in primitive early radar at the start of World War II were several meters long. Since resolution is proportional to aperture, that is, the length of the array measured in units of wavelength, antennas for such radar needed to be quite large. As Körner notes in [102], the general feeling at the time was that the side with the shortest wavelength would win the war. The cavity magnetron, invented during the war by British scientists, made possible microwave radar having a wavelength of 10 cm, which could then be mounted easily on planes.

### 3.3.3 Can We Get More Data?

As we just saw, we can make measurements at any points $P$ in the far field; perhaps we do not need to limit ourselves to just those angles that lead to the $c_n$. It may come as somewhat of a surprise, but from the theory of complex analytic functions we can prove that there is enough data available to us here to reconstruct $f(x)$ perfectly, at least in principle. The drawback, in practice, is that the measurements would have to be free of noise and impossibly accurate. All is not lost, however.

### 3.3.4 Measuring the Fourier Transform

If $\theta$ is chosen so that

$$\frac{\omega \cos \theta}{c} = \frac{-n\pi}{L},$$

then our measurement gives us the Fourier coefficients $c_n$. But we can select any angle $\theta$ and use any $P$ we want. In other words, we can obtain the values

$$\int_{-L}^{L} f(x)e^{i\frac{\omega \cos \theta}{c}x}dx,$$

for any angle $\theta$. With the change of variable

$$\gamma = \frac{\omega \cos \theta}{c},$$

we can obtain the value of the Fourier transform,

$$F(\gamma) = \int_{-L}^{L} f(x)e^{i\gamma x}dx,$$

for any $\gamma$ in the interval $[-\frac{\omega}{c}, \frac{\omega}{c}]$.

We are free to measure at any $P$ and therefore to obtain values of $F(\gamma)$ for any value of $\gamma$ in the interval $[-\frac{\omega}{c}, \frac{\omega}{c}]$. We need to be careful how we process the resulting data, however.

### 3.3.5  Over-Sampling

Suppose, for the sake of illustration, that we measure the far-field signals at points $P$ corresponding to angles $\theta$ that satisfy

$$\frac{\omega \cos \theta}{c} = \frac{-n\pi}{2L},$$

instead of

$$\frac{\omega \cos \theta}{c} = \frac{-n\pi}{L}.$$

Now we have twice as many data points and from these new measurements we can obtain

$$d_n = \int_{-L}^{L} f(x)e^{-i\frac{n\pi}{2L}x}dx,$$

for $|n| \leq 2N$. We say now that our data is *twice over-sampled*. Note that we call it *over-sampled* because the rate at which we are sampling is higher, even though the distance between samples is shorter. The values $d_n$ are not simply more of the Fourier coeffcients of $f$. The question now is: What are we to do with these extra data values?

The values $d_n$ are, in fact, Fourier coefficients, but not of $f$; they are Fourier coefficients of the function $g : [-2L, 2L] \to \mathbb{C}$, where $g(x) = f(x)$ for $|x| \leq L$, and $g(x) = 0$, otherwise. If we simply use the $d_n$ as Fourier coefficients of the function $g(x)$ and compute the resulting DFT estimate of $g(x)$,

$$g_{DFT}(x) = \sum_{n=-2N}^{2N} d_n e^{i\frac{n\pi}{2L}x},$$

this function estimates $f(x)$ for $|x| \leq L$, but it also estimates $g(x) = 0$ for the other values of $x$ in $[-2L, 2L]$. When we graph $g_{DFT}(x)$ for $|x| \leq L$ we find that we have no improvement over what we got with the previous estimate $f_{DFT}$. The problem is that we have wasted the extra data by estimating $g(x) = 0$ where we already knew that it was zero. To make good use of the extra data we need to incorporate this prior information about the function $g$. The MDFT and PDFT algorithms provide estimates of $f(x)$ that incorporate prior information.

### 3.3.6 The Modified DFT

The modified DFT (MDFT) estimate was first presented in [22]. For our example of twice over-sampled data, the MDFT is defined for $|x| \leq L$ and has the algebraic form

$$f_{MDFT}(x) = \sum_{n=-2N}^{2N} a_n e^{i\frac{n\pi}{2L}x}, \qquad (3.2)$$

for $|x| \leq L$. The coefficients $a_n$ are not the $d_n$. The $a_n$ are determined by requiring that the function $f_{MDFT}$ be consistent with the measured data, the $d_n$. In other words, we must have

$$d_n = \int_{-L}^{L} f_{MDFT}(x) e^{-i\frac{n\pi}{2L}x} dx. \qquad (3.3)$$

When we insert $f_{MDFT}(x)$ as given in Equation (3.2) into Equation (3.3) we get a system of $2N+1$ linear equations in $2N+1$ unknowns, the $a_n$. We then solve this system for the $a_n$ and use them in Equation (3.2). Figure 2.1 shows the improvement we can achieve using the MDFT. The data used to construct the graphs in that figure was thirty times over-sampled. We note here that, had we extended $f$ initially as a $2L$-periodic function, it would be difficult to imagine the function $g(x)$ and we would have a hard time figuring out what to do with the $d_n$.

In this example we measured twice as much data as previously. We can, of course, measure even more data, and it need not correspond to the Fourier coefficients of any function. The potential drawback is that, as we use more data, the system of linear equations that we must solve to obtain the MDFT estimate becomes increasingly sensitive to noise and round-off error in the data. It is possible to lessen this effect by *regularization*, but not to eliminate it entirely. Regularization can be introduced here simply by multiplying by, say, 1.01, the entries of the main diagonal of the matrix of the linear system. This makes the matrix less *ill-conditioned*.

In our example, we used the prior knowledge that $f(x) = 0$ for $|x| > L$. Now, we shall describe in detail the use of other forms of prior knowledge about $f(x)$ to obtain reconstructions that are better than the DFT.

### 3.3.7 Other Forms of Prior Knowledge

As we just showed, knowing that we have over-sampled in our measurements can help us improve the resolution in our estimate of $f(x)$. We may have other forms of prior knowledge about $f(x)$ that we can use. If we know something about large-scale features of $f(x)$, but not about finer details, we can use the PDFT estimate, which is a generalization of the MDFT. In Chapter 1 the PDFT was compared to the DFT in a two-dimensional example of simulated head slices.

The MDFT estimator can be written as

$$f_{MDFT}(x) = \chi_L(x) \sum_{n=-2N}^{2N} a_n e^{i\frac{n\pi}{2L}x}.$$

We include the prior information that $f(x)$ is supported on the interval $[-L, L]$ through the factor $\chi_L(x)$. If we select a function $p(x) \geq 0$ that describes our prior estimate of the shape of $|f(x)|$, we can then estimate $f(x)$ using the PDFT estimator, which, in this case of twice over-sampled data, takes the form

$$f_{PDFT}(x) = p(x) \sum_{n=-2N}^{2N} b_n e^{i\frac{n\pi}{2L}x}.$$

As with the MDFT estimator, we determine the coefficients $b_n$ by requiring that $f_{PDFT}(x)$ be consistent with the measured data.

There are other things we may know about $f(x)$. We may know that $f(x)$ is nonnegative, or we may know that $f(x)$ is approximately zero for most $x$, but contains very sharp peaks at a few places. In more formal language, we may be willing to assume that $f(x)$ contains a few Dirac delta functions in a flat background. There are nonlinear methods, such as the maximum entropy method, the indirect PDFT (IPDFT), and eigenvector methods, that can be used to advantage in such cases; these methods are often called *high-resolution methods*.

## 3.4 Generalizing the MDFT and PDFT

In our discussion so far the data we have obtained are values of the Fourier transform of the support-limited function $f(x)$. The MDFT and PDFT can be extended to handle those cases in which the data we have are more general linear-functional values pertaining to $f(x)$.

Suppose that our data values are finitely many linear-functional values,

$$d_n = \int_{-L}^{L} f(x)\overline{g_n(x)}dx,$$

for $n = 1, ..., N$, where the $g_n(x)$ are known functions. The extended MDFT estimate of $f(x)$ is

$$f_{MDFT}(x) = \chi_L(x) \sum_{m=1}^{N} a_m g_m(x),$$

where the coefficients $a_m$ are chosen so that $f_{MDFT}$ is consistent with the measured data; that is,

$$d_n = \int_{-L}^{L} f_{MDFT}(x)\overline{g_n(x)}dx,$$

for each $n$. To find the $a_m$ we need to solve a system of $N$ equations in $N$ unknowns.

The PDFT can be extended in a similar way. The extended PDFT estimate of $f(x)$ is

$$f_{PDFT}(x) = p(x) \sum_{m=1}^{N} b_m g_m(x),$$

where, as previously, the coefficients $b_m$ are chosen by forcing the estimate of $f(x)$ to be consistent with the measured data. Again, we need to solve a system of $N$ equations in $N$ unknowns to find the coefficients.

For large values of $N$, setting up and solving the required systems of linear equations can involve considerable effort. If we discretize the functions $f(x)$ and $g_n(x)$, we can obtain good approximations of the extended MDFT and PDFT using the iterative ART algorithm [142, 143].

## 3.5   One-Dimensional Arrays

In this section we consider the reversed situation in which the sources of the signals are the points on the circumference of the large circle and we are measuring the received signals at points of the $x$-axis. The objective is to determine the relative strengths of the signals coming to us from various angles.

People with sight in only one eye have a difficult time perceiving depth in their visual field, unless they move their heads. Having two functioning ears helps us determine the direction from which sound is coming; blind people, who are more than usually dependent on their hearing, often move their heads to get a better sense of where the source of sound is. Snakes who smell with their tongues often have forked tongues, the better to detect the direction of the sources of different smells. In certain remote-sensing situations the sensors respond equally to arrivals from all directions. One then obtains the needed directionality by using multiple sensors, laid out in some spatial configuration called the sensor *array*. The simplest configuration is to have the sensors placed in a straight line, as in a sonar towed array.

Now we imagine that the points $P = P(\theta)$ in the far field are the sources of the signals and we are able to measure the transmissions received at points $x$ on the $x$-axis; we no longer assume that these points are confined to the interval $[-L, L]$. The $P$ corresponding to the angle $\theta$ sends $f(\theta)e^{i\omega t}$, where the absolute value of $f(\theta)$ is the strength of the signal coming from $P$. We allow $f(\theta)$ to be complex, so that it has both magnitude and phase, which means that we do not assume that the signals from the different angles are in phase with one another; that is, we do not assume that they all begin at the same time.

In narrow-band passive sonar, for example, we may have hydrophone sensors placed at various points $x$ and our goal is to determine how much acoustic energy at a specified frequency is coming from different directions. There may be only a few directions contributing significant energy at the frequency of interest, in which case $f(\theta)$ is nearly zero for all but a few values of $\theta$.

### 3.5.1   Measuring Fourier Coefficients

At time $t$ the point $x$ on the $x$-axis receives from $P = P(\theta)$ what $P$ sent at time $t - (D - x\cos\theta)/c$; so, at time $t$, $x$ receives from $P$

$$e^{i\omega(t-D/c)} f(\theta) e^{i\frac{\omega x}{c}\cos\theta}.$$

Since $x$ receives signals from all the angles, what $x$ receives at time $t$ is

$$e^{i\omega(t-D/c)} \int_0^\pi f(\theta)e^{i\frac{\omega x}{c}\cos\theta}d\theta.$$

We limit the angle $\theta$ to the interval $[0, \pi]$ because, in this sensing model, we cannot distinguish receptions from $\theta$ and from $2\pi - \theta$.

To simplify notation, we shall introduce the variable $u = \cos\theta$. We then have

$$\frac{du}{d\theta} = -\sin(\theta) = -\sqrt{1-u^2},$$

so that

$$d\theta = -\frac{1}{\sqrt{1-u^2}}du.$$

Now let $g(u)$ be the function

$$g(u) = \frac{f(\arccos(u))}{\sqrt{1-u^2}},$$

defined for $u$ in the interval $(-1, 1)$. Since

$$\int_0^\pi f(\theta)e^{i\frac{\omega x}{c}\cos\theta}d\theta = \int_{-1}^1 g(u)e^{i\frac{\omega x}{c}u}du,$$

we find that, from our measurement at $x$, we obtain $G(\gamma)$, the value of the Fourier transform of $g(u)$ at $\gamma$, for

$$\gamma = \frac{\omega x}{c}.$$

Since $g(u)$ is limited to the interval $(-1, 1)$, its Fourier coefficients are

$$a_n = \frac{1}{2}\int_{-1}^1 g(u)e^{-in\pi u}du.$$

Therefore, if we select $x$ so that

$$\gamma = \frac{\omega x}{c} = -n\pi,$$

we have $a_n$. Consequently, we want to measure at the points $x$ such that

$$x = -n\frac{\pi c}{\omega} = -n\frac{\lambda}{2} = -n\Delta, \tag{3.4}$$

where $\lambda = \frac{2\pi c}{\omega}$ is the wavelength and $\Delta = \frac{\lambda}{2}$ is the *Nyquist spacing*.

A one-dimensional array consists of measuring devices placed along a straight line (the $x$-axis here). Obviously, there must be some smallest bounded interval, say $[A, B]$, that contains all these measuring devices. The *aperture* of the array is $\frac{B-A}{\lambda}$, the length of the interval $[A, B]$, in units of wavelength. As we just saw, the aperture is directly related to the number of Fourier coefficients of the function $g(u)$ that we are measuring, and therefore, to the accuracy of the DFT reconstruction of $g(u)$. This is usually described by saying that aperture determines resolution. As we saw, a one-dimensional array involves an inherent ambiguity, in that we cannot distinguish a signal from the angle $\theta$ from one from the angle $2\pi - \theta$. In practice a two-dimensional configuration of sensors is sometimes used to eliminate this ambiguity.

In numerous applications, such as astronomy, it is more realistic to assume that the sources of the signals are on the surface of a large sphere, rather than on the circumference of a large circle. In such cases, a one-dimensional array of sensors does not provide sufficient information and two- or three-dimensional sensor configurations are used.

The number of Fourier coefficients of $g(u)$ that we can measure, and therefore the resolution of the resulting reconstruction of $f(\theta)$, is limited by the aperture. One way to improve resolution is to make the array of sensors longer, which is more easily said than done. However, *synthetic-aperture radar* (SAR) effectively does this. The idea of SAR is to employ the array of sensors on a moving airplane. As the plane moves, it effectively creates a longer array of sensors, a *virtual array* if you will. The one drawback is that the sensors in this virtual array are not all present at the same time, as in a normal array. Consequently, the data must be modified to approximate what would have been received at other times.

The far-field approximation tells us that, at time $t$, every point $x$ receives from $P(\frac{\pi}{2})$ the same signal

$$e^{i\omega(t - D/c)} f\left(\frac{\pi}{2}\right).$$

Since there is nothing special about the angle $\frac{\pi}{2}$, we can say that the signal arriving from any angle $\theta$, which originally spread out as concentric circles of constant value, has flattened out to the extent that, by the time it reaches our line of sensors, it is essentially constant on straight lines. This suggests the *plane-wave approximation* for signals propagating in three-dimensional space. As we shall see in Chapter 24, these plane-wave approximations are solutions to the three-dimensional wave equation. Much of array processing is based on such models of far-field propagation.

As in the examples discussed previously, we do have more measurements we can take, if we use values of $x$ other than those described by Equation (3.4). The issue will be what to do with these *over-sampled* measurements.

## 3.5.2 Over-Sampling

One situation in which over-sampling arises naturally occurs in sonar array processing. Suppose that an array of sensors has been built to operate at a *design frequency* of $\omega_0$, which means that we have placed sensors a distance of $\Delta_0$ apart in $[A, B]$, where $\lambda_0$ is the wavelength corresponding to the frequency $\omega_0$ and $\Delta_0 = \frac{\lambda_0}{2}$ is the Nyquist spacing for frequency $\omega_0$. For simplicity, we assume that the sensors are placed at points $x$ that satisfy the equation

$$x = -n\frac{\pi c}{\omega_0} = -n\frac{\lambda_0}{2} = -n\Delta_0,$$

for $|n| \leq N$. Now suppose that we want to operate the sensing at another frequency, say $\omega$. The sensors cannot be moved, so we must make do with sensors at the points $x$ determined by the design frequency.

Consider, first, the case in which the second frequency $\omega$ is less than the design frequency $\omega_0$. Then its wavelength $\lambda$ is larger than $\lambda_0$, and the Nyquist spacing $\Delta = \frac{\lambda}{2}$ for $\omega$ is larger than $\Delta_0$. So we have over-sampled.

The measurements taken at the sensors provide us with the integrals

$$\int_{-1}^{1} g(u)e^{i\frac{n\pi}{K}u}du,$$

where $K = \frac{\omega_0}{\omega} > 1$. These are Fourier coefficients of the function $g(u)$, viewed as defined on the interval $[-K, K]$, which is larger than $[-1, 1]$, and taking the value zero outside $[-1, 1]$. If we then use the DFT estimate of $g(u)$, it will estimate $g(u)$ for the values of $u$ within $[-1, 1]$, which is what we want, as well as for the values of $u$ outside $[-1, 1]$, where we already know $g(u)$ to be zero. Once again, we can use the MDFT, the modified DFT, to include the prior knowledge that $g(u) = 0$ for $u$ outside $[-1, 1]$ to improve our reconstruction of $g(u)$ and $f(\theta)$. In sonar, for the over-sampled case, the interval $[-1, 1]$ is called *the visible region* (although *audible region* seems more appropriate for sonar), since it contains all the values of $u$ that can correspond to actual angles of plane-wave arrivals of acoustic energy. In practice, of course, the measured data may well contain components that are not plane-wave arrivals, such as localized noises near individual sensors, or near-field sounds, so our estimate of the function $g(u)$ should be regularized to allow for these non-plane-wave components.

## 3.5.3 Under-Sampling

Now suppose that the frequency $\omega$ that we want to consider is greater than the design frequency $\omega_0$. This means that the spacing between the sensors is too large; we have *under-sampled*. Once again, however, we cannot move the sensors and must make do with what we have.

Now the measurements at the sensors provide us with the integrals

$$\int_{-1}^{1} g(u) e^{i\frac{n\pi}{K}u} du,$$

where $K = \frac{\omega_0}{\omega} < 1$. These are Fourier coefficients of the function $g(u)$, viewed as defined on the interval $[-K, K]$, which is smaller than $[-1, 1]$, and taking the value zero outside $[-K, K]$. Since $g(u)$ is not necessarily zero outside $[-K, K]$, treating it as if it were zero there results in a type of error known as *aliasing*, in which energy corresponding to angles whose $u$ lies outside $[-K, K]$ is mistakenly assigned to values of $u$ that lie within $[-K, K]$. Aliasing is a common phenomenon; the strobe-light effect is aliasing, as is the apparent backward motion of the wheels of stagecoaches in cowboy movies. In the case of the strobe light, we are permitted to view the scene at times too far apart for us to sense continuous, smooth motion. In the case of the wagon wheels, the frames of the film capture instants of time too far apart for us to see the true rotation of the wheels.

---

## 3.6   Resolution Limitations

As we have seen, in the unknown-strength problem the number of Fourier coefficients we can measure is limited by the ratio $\frac{L}{\lambda}$. Additional measurements in the far field can provide additional information about the function $f(x)$, but extracting that information becomes an increasingly ill-conditioned problem, one more sensitive to noise the more data we gather.

In the line-array problem just considered, there is, in principle, no limit to the number of Fourier coefficients we can obtain by measuring at the points $n\Delta$ for integer values of $n$; the limitation here is of a more practical nature.

In sonar, the speed of sound in the ocean is about 1500 meters per second, so the wavelength associated with 50 Hz is $\lambda = 30$ meters. The Nyquist spacing is then 15 meters. A towed array is a line array of sensors towed behind a ship. The length of the array, and therefore the number of Nyquist-spaced sensors for passive sensing at 50 Hz, is, in principle, unlimited. In practice, however, cost is always a factor. In addition, when the array becomes too long, it is difficult to maintain it in a straight-line position.

Radar imaging uses microwaves with a wavelength of about one inch, which is not a problem; synthetic-aperture radar can also be used to simulate a longer array. In radio astronomy, however, the wavelengths can be more than a kilometer, which is why radio-astronomy arrays have to be enormous. For radio-wave imaging at very low frequencies, a sort of synthetic-aperture approach has been taken, with individual antennas located in different parts of the globe.

---

## 3.7 Using Matched Filtering

We saw previously that the signal that $x$ receives from $P(\frac{\pi}{2})$ at time $t$ is the same for all $x$. If we could turn the $x$-axis counter-clockwise through an angle of $\phi$, then the signals received from $P(\frac{\pi}{2} + \phi)$ at time $t$ would be the same for all $x$. Of course, we usually cannot turn the array physically in this way; however, we can *steer* the array mathematically. This mathematical steering makes use of *matched filtering*. In certain applications it is reasonable to assume that only relatively few values of the function $f(\theta)$ are significantly nonzero. Matched filtering is a commonly used method for dealing with such cases.

### 3.7.1 A Single Source

To take an extreme case, suppose that $f(\theta_0) > 0$ and $f(\theta) = 0$, for all $\theta \neq \theta_0$. The signal received at time $t$ at $x$ is then

$$s(x,t) = e^{i\omega(t-D/c)} f(\theta_0) e^{i\frac{\omega x}{c} \cos \theta_0}.$$

Our objective is to determine $\theta_0$.

Suppose that we multiply $s(x,t)$ by $e^{-i\frac{\omega x}{c} \cos \theta}$, for arbitrary values of $\theta$. When one of the arbitrary values is $\theta = \theta_0$, the product is no longer dependent on the value of $x$; that is, the resulting product is the same for all $x$. In practice, we can place sensors at some finite number of points $x$, and then sum the resulting products over the $x$. When the arbitrary $\theta$ is not $\theta_0$, we are adding up complex exponentials with distinct phase angles, so destructive interference takes place and the magnitude of the sum is not large. In contrast, when $\theta = \theta_0$, all the products are the same and the sum is relatively large. This is *matched filtering*, which is commonly used to determine the true value of $\theta_0$.

### 3.7.2 Multiple Sources

Having only one signal source is the extreme case; having two or more signal sources, perhaps not far apart in angle, is an important situation, as well. Then resolution becomes a problem. When we calculate the matched filter in the single-source case, the largest magnitude will occur when $\theta =$

$\theta_0$, but the magnitudes at other nearby values of $\theta$ will not be zero. How quickly the values fall off as we move away from $\theta_0$ will depend on the aperture of the array; the larger the aperture, the faster the fall-off. When we have two signal sources near to one another, say $\theta_1$ and $\theta_2$, the matched-filter output can have its largest magnitude at a value of $\theta$ between the two angles $\theta_1$ and $\theta_2$, causing a loss of resolution. Again, having a larger aperture will improve the resolution.

## 3.8    An Example: The Solar-Emission Problem

In [15] Bracewell discusses the *solar-emission* problem. In 1942, it was observed that radio-wave emissions in the one-meter wavelength range were arriving from the sun. Were they coming from the entire disk of the sun or were the sources more localized, in sunspots, for example? The problem then was to view each location on the sun's surface as a potential source of these radio waves and to determine the intensity of emission corresponding to each location.

For electromagnetic waves the propagation speed is the speed of light in a vacuum, which we shall take here to be $c = 3 \times 10^8$ meters per second. The wavelength $\lambda$ for gamma rays is around one Angstrom, that is, $10^{-10}$ meters, which is about the diameter of an atom; for x-rays it is about one millimicron, or $10^{-9}$ meters. The visible spectrum has wavelengths that are a little less than one micron, that is, $10^{-6}$ meters, while infrared radiation (IR), predominantly associated with heat, has a wavelength somewhat longer. Infrared radiation with a wavelength around 6 or 7 microns can be used to detect water vapor; we use near IR, with a wavelength near that of visible light, to change the channels on our TV sets. Shortwave radio has a wavelength around one millimeter. Microwaves have wavelengths between one centimeter and one meter; those used in radar imaging have a wavelength about one inch and can penetrate clouds and thin layers of leaves. Broadcast radio has a $\lambda$ running from about 10 meters to 1000 meters. The so-called long radio waves can have wavelengths several thousand meters long, necessitating clever methods of large-antenna design for radio astronomy.

The sun has an angular diameter of 30 min. of arc, or one-half of a degree, when viewed from earth, but the needed resolution was more like 3 min. of arc. Such resolution requires a larger aperture, a radio telescope 1000 wavelengths across, which means a diameter of 1km at a wavelength of 1 meter; in 1942 the largest military radar antennas were less than 5 meters across. A solution was found, using the method of reconstructing an object from line-integral data, a technique that surfaced again in tomography.

---

## 3.9 Estimating the Size of Distant Objects

Suppose, in the previous example of the unknown strength problem, we assume that $f(x) = B$, for all $x$ in the interval $[-L, L]$, where $B > 0$ is the unknown *brightness* constant, and we don't know $L$. More realistic, two-dimensional versions of this problem arise in astronomy, when we want to estimate the diameter of a distant star.

In this case, the measurement of the signal at the point $P$ gives us

$$\int_{-L}^{L} f(x) \cos\left(\frac{\omega \cos\theta}{c} x\right) dx$$

$$= B \int_{-L}^{L} \cos\left(\frac{\omega \cos\theta}{c} x\right) dx = \frac{2Bc}{\omega \cos\theta} \sin\left(\frac{L\omega \cos\theta}{c}\right),$$

when $\cos\theta \neq 0$, whose absolute value is then the strength of the signal at $P$. Notice that we have zero signal strength at $P$ when the angle $\theta$ associated with $P$ satisfies the equation

$$\sin\left(\frac{L\omega \cos\theta}{c}\right) = 0,$$

without

$$\cos\theta = 0.$$

But we know that the first positive zero of the sine function is at $\pi$, so the signal strength at $P$ is zero when $\theta$ is such that

$$\frac{L\omega \cos\theta}{c} = \pi.$$

If

$$\frac{L\omega}{c} \geq \pi,$$

then we can solve for $L$ and get

$$L = \frac{\pi c}{\omega \cos\theta}.$$

When $L\omega$ is too small, there will be no angle $\theta$ for which the received signal strength at $P$ is zero. If the signals being sent are actually *broadband*, meaning that the signals are made up of components at many different frequencies, not just one $\omega$, which is usually the case, then we might be able to filter our measured data, keep only the component at a sufficiently high frequency, and then proceed as before.

But even when we have only a single frequency $\omega$ and $L\omega$ is too small, there is something we can do. The received strength at $\theta = \frac{\pi}{2}$ is

$$F_c(0) = B \int_{-L}^{L} dx = 2BL.$$

If we knew $B$, this measurement alone would give us $L$, but we do not assume that we know $B$. At any other angle, the received strength is

$$F_c(\gamma) = \frac{2Bc}{\omega \cos \theta} \sin \left( \frac{L\omega \cos \theta}{c} \right).$$

Therefore,

$$F_c(\gamma)/F_c(0) = \frac{\sin(H(\theta))}{H(\theta)},$$

where

$$H(\theta) = \frac{L\omega \cos \theta}{c}.$$

From the measured value $F_c(\gamma)/F_c(0)$ we can solve for $H(\theta)$ and then for $L$. In actual optical astronomy, atmospheric distortions make these measurements noisy and the estimates have to be performed more carefully. This issue is discussed in more detail in Chapter 2, in Section 2.13 on Two-Dimensional Fourier Transforms.

There is a simple relationship involving the intrinsic luminosity of a star, its distance from earth, and its apparent brightness; knowing any two of these, we can calculate the third. Once we know these values, we can figure out how large the visible universe is. Unfortunately, only the apparent brightness is easily determined. As Alan Lightman relates in [111], it was Henrietta Leavitt's ground-breaking discovery, in 1912, of the "period-luminosity" law of variable Cepheid stars that eventually revealed just how enormous the universe really is. Cepheid stars are found in many parts of the sky. Their apparent brightness varies periodically. As Leavitt, working at the Harvard College Observatory, discovered, the greater the intrinsic luminosity of the star, the longer the period of variable brightness. The final step of calibration was achieved in 1913 by the Danish astronomer Ejnar Hertzsprung, when he was able to establish the actual distance to a relatively nearby Cepheid star, essentially by parallax methods.

There is a wonderful article by Eddington [69], in which he discusses the use of signal processing methods to discover the properties of the star Algol. This star, formally Algol (Beta Persei) in the constellation Perseus, turns out to be three stars, two revolving around the third, with both of the first two taking turns eclipsing the other. The stars rotate around their own axes, as our star, the sun, does, and the speed of rotation can be estimated by calculating the Doppler shift in frequency, as one side of the star comes toward us and the other side moves away. It is possible to measure one side at a time only because of the eclipse caused by the other revolving star.

---

## 3.10   The Transmission Problem

Now we change the situation and suppose that we are designing a broad-casting system, using transmitters at each $x$ in the interval $[-L, L]$.

### 3.10.1   Directionality

At each $x$ we will transmit $f(x)e^{i\omega t}$, where both $f(x)$ and $\omega$ are chosen by us. We now want to calculate what will be received at each point $P$ in the far field. We may wish to design the system so that the strengths of the signals received at the various $P$ are not all the same. For example, if we are broadcasting from Los Angeles, we may well want a strong signal in the north and south directions, but weak signals east and west, where there are fewer people to receive the signal. Clearly, our model of a single-frequency signal is too simple, but it does allow us to illustrate several important points about directionality in array processing.

### 3.10.2   The Case of Uniform Strength

For concreteness, we investigate the case in which $f(x) = 1$ for $|x| \leq L$. In this case, the measurement of the signal at the point $P$ gives us

$$
\begin{aligned}
F(P) &= \int_{-L}^{L} f(x) \cos\left(\frac{\omega \cos\theta}{c} x\right) dx \\
&= \int_{-L}^{L} \cos\left(\frac{\omega \cos\theta}{c} x\right) dx \\
&= \frac{2c}{\omega \cos\theta} \sin\left(\frac{L\omega \cos\theta}{c}\right),
\end{aligned}
$$

when $\cos\theta \neq 0$. The absolute value of $F(P)$ is then the strength of the signal at $P$. In Figures 3.2 through 3.7 we see the plots of the function $\frac{1}{2L}F(P)$, for various values of the aperture

$$
A = \frac{L\omega}{\pi c} = \frac{2L}{\lambda}.
$$

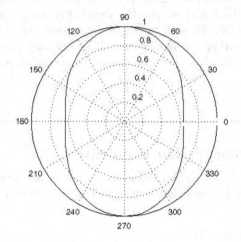

**FIGURE 3.2**: Relative strength at $P$ for $A = 0.5$.

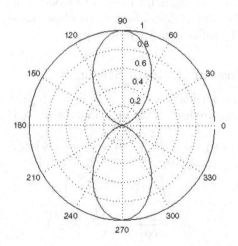

**FIGURE 3.3**: Relative strength at $P$ for $A = 1.0$.

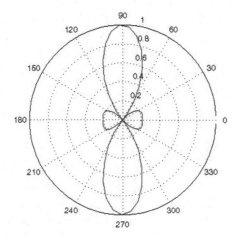

**FIGURE 3.4**: Relative strength at $P$ for $A = 1.5$.

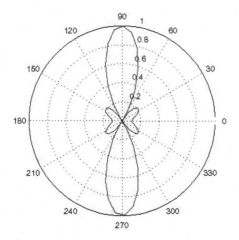

**FIGURE 3.5**: Relative strength at $P$ for $A = 1.8$.

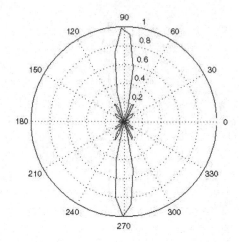

**FIGURE 3.6**: Relative strength at $P$ for $A = 3.2$.

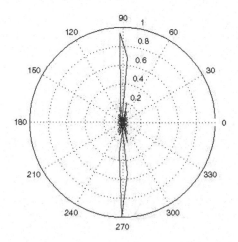

**FIGURE 3.7**: Relative strength at $P$ for $A = 6.5$.

### 3.10.2.1 Beam-Pattern Nulls

Is it possible for the strength of the signal received at some $P$ to be zero? As we saw in the previous section, to have zero signal strength, that is, to have $F(P) = 0$, we need

$$\sin\left(\frac{L\omega\cos\theta}{c}\right) = 0,$$

without

$$\cos\theta = 0.$$

Therefore, we need

$$\frac{L\omega\cos\theta}{c} = n\pi,$$

for some positive integers $n \geq 1$. Notice that this can happen only if

$$n \leq \frac{L\omega\pi}{c} = \frac{2L}{\lambda}.$$

Therefore, if $2L < \lambda$, there can be no $P$ with signal strength zero. The larger $2L$ is, with respect to the wavelength $\lambda$, the more angles at which the signal strength is zero.

### 3.10.2.2 Local Maxima

Is it possible for the strength of the signal received at some $P$ to be a local maximum, relative to nearby points in the far field? We write

$$F(P) = \frac{2c}{\omega\cos\theta}\sin\left(\frac{L\omega\cos\theta}{c}\right) = 2L\text{sinc}\left(H(\theta)\right),$$

where

$$H(\theta) = \frac{L\omega\cos\theta}{c}$$

and

$$\text{sinc}\left(H(\theta)\right) = \frac{\sin H(\theta)}{H(\theta)},$$

for $H(\theta) \neq 0$, and equals one for $H(\theta) = 0$. The value of $A$ used previously is then $A = H(0)$.

Local maxima or minima of $F(P)$ occur when the derivative of $\text{sinc}\left(H(\theta)\right)$ equals zero, which means that

$$H(\theta)\cos H(\theta) - \sin H(\theta) = 0,$$

or

$$\tan H(\theta) = H(\theta).$$

If we can solve this equation for $H(\theta)$ and then for $\theta$, we will have found angles corresponding to local maxima of the received signal strength. The largest value of $F(P)$ occurs when $\theta = \frac{\pi}{2}$, and the peak in the plot of $F(P)$ centered at $\theta = \frac{\pi}{2}$ is called the *main lobe*. The smaller peaks on either side are called the *grating lobes*. We can see grating lobes in some of the polar plots.

---

## 3.11    The Laplace Transform and the Ozone Layer

We have seen how values of the Fourier transform can arise as measured data. The following examples, the first taken from Twomey's book [156], show that values of the Laplace transform can arise in this way as well.

### 3.11.1    The Laplace Transform

The Laplace transform of the function $f(x)$, defined for $0 \leq x < +\infty$, is the function

$$\mathcal{F}(s) = \int_0^{+\infty} f(x)e^{-sx}dx.$$

### 3.11.2    Scattering of Ultraviolet Radiation

The sun emits ultraviolet (UV) radiation that enters the earth's atmosphere at an angle $\theta_0$ that depends on the sun's position, and with intensity $I(0)$. Let the $x$-axis be vertical, with $x = 0$ at the top of the atmosphere and $x$ increasing as we move down to the earth's surface, at $x = X$. The intensity at $x$ is given by

$$I(x) = I(0)e^{-kx/\cos\theta_0}.$$

Within the ozone layer, the amount of UV radiation scattered in the direction $\theta$ is given by

$$S(\theta, \theta_0)I(0)e^{-kx/\cos\theta_0}\Delta p,$$

where $S(\theta, \theta_0)$ is a known parameter, and $\Delta p$ is the change in the pressure of the ozone within the infinitesimal layer $[x, x+\Delta x]$, and so is proportional to the concentration of ozone within that layer.

### 3.11.3    Measuring the Scattered Intensity

The radiation scattered at the angle $\theta$ then travels to the ground, a distance of $X - x$, weakened along the way, and reaches the ground with intensity

$$S(\theta, \theta_0)I(0)e^{-kx/\cos\theta_0}e^{-k(X-x)/\cos\theta}\Delta p.$$

The total scattered intensity at angle $\theta$ is then a superposition of the intensities due to scattering at each of the thin layers, and is then

$$S(\theta, \theta_0)I(0)e^{-kX/\cos\theta_0}\int_0^X e^{-x\beta}dp,$$

where

$$\beta = k\left(\frac{1}{\cos\theta_0} - \frac{1}{\cos\theta}\right).$$

This superposition of intensity can then be written as

$$S(\theta, \theta_0)I(0)e^{-kX/\cos\theta_0}\int_0^X e^{-x\beta}p'(x)dx.$$

### 3.11.4 The Laplace Transform Data

Using integration by parts, we get

$$\int_0^X e^{-x\beta}p'(x)dx = p(X)e^{-\beta X} - p(0) + \beta\int_0^X e^{-\beta x}p(x)dx.$$

Since $p(0) = 0$ and $p(X)$ can be measured, our data is then the Laplace transform value

$$\int_0^{+\infty} e^{-\beta x}p(x)dx;$$

note that we can replace the upper limit $X$ with $+\infty$ if we extend $p(x)$ as zero beyond $x = X$.

The variable $\beta$ depends on the two angles $\theta$ and $\theta_0$. We can alter $\theta$ as we measure and $\theta_0$ changes as the sun moves relative to the earth. In this way we get values of the Laplace transform of $p(x)$ for various values of $\beta$. The problem then is to recover $p(x)$ from these values. Because the Laplace transform involves a smoothing of the function $p(x)$, recovering $p(x)$ from its Laplace transform is more ill-conditioned than is the Fourier transform inversion problem.

---

## 3.12 The Laplace Transform and Energy Spectral Estimation

In x-ray transmission tomography, x-ray beams are sent through the object and the drop in intensity is measured. These measurements are then used to estimate the distribution of attenuating material within the object. A typical x-ray beam contains components with different energy levels. Because components at different energy levels will be attenuated differently, it is important to know the relative contribution of each energy level to the entering beam. The energy spectrum is the function $f(E)$ that describes the intensity of the components at each energy level $E > 0$.

### 3.12.1  The Attenuation Coefficient Function

Each specific material, say aluminum, for example, is associated with attenuation coefficients, which is a function of energy, which we shall denote by $\mu(E)$. A beam with the single energy $E$ passing through a thickness $x$ of the material will be weakened by the factor $e^{-\mu(E)x}$. By passing the beam through various thicknesses $x$ of aluminum and registering the intensity drops, one obtains values of the absorption function

$$R(x) = \int_0^\infty f(E)e^{-\mu(E)x}dE. \tag{3.5}$$

Using a change of variable, we can write $R(x)$ as a Laplace transform.

### 3.12.2  The Absorption Function as a Laplace Transform

For each material, the attenuation function $\mu(E)$ is a strictly decreasing function of $E$, so $\mu(E)$ has an inverse, which we denote by $g$; that is, $g(t) = E$, for $t = \mu(E)$. Equation (3.5) can then be rewritten as

$$R(x) = \int_0^\infty f(g(t))e^{-tx}g'(t)dt.$$

We see then that $R(x)$ is the Laplace transform of the function $r(t) = f(g(t))g'(t)$. Our measurements of the intensity drops provide values of $R(x)$, for various values of $x$, from which we must estimate the functions $r(t)$, and, ultimately, $f(E)$.

# Chapter 4

## Finite-Parameter Models

## 4.1 Chapter Summary

All of the techniques discussed in this book deal, in one way or another, with one fundamental problem: Estimate the values of a function $f(x)$ from finitely many (usually noisy) measurements related to $f(x)$; here $x$ can be a multi-dimensional vector, so that $f$ can be a function of more than one variable. To keep the notation relatively simple here, we shall assume, throughout this chapter, that $x$ is a real variable, but all of what we shall say applies to multi-variate functions as well. In this chapter we begin our discussion of the use of finite-parameter models, a topic to which we shall return several times throughout this book.

---

## 4.2    Finite Fourier Series

In this section we present one of the most useful finite-parameter model, the finite Fourier series. The notation may seem unusual, but it is chosen for convenience later, when we discuss the Fast Fourier Transform (FFT).

Let $f : [0, N] \to \mathbb{C}$ have Fourier series

$$f(x) \approx \frac{1}{N} \sum_{k=-\infty}^{\infty} F_k e^{-i\frac{2\pi k}{N}x},$$

where

$$F_k = \int_0^N f(x) e^{i\frac{2\pi k}{N}x} dx. \tag{4.1}$$

Note that

$$F_k = F\left(\frac{2\pi k}{N}\right),$$

where $F(\gamma)$ is the Fourier transform of $f(x)$. In order to calculate any $F_k$ we need all of $f(x)$.

Suppose that we model $f(x)$ on $[0, N]$ using a finite Fourier series

$$f(x) \approx \frac{1}{N} \sum_{k=0}^{N-1} F_k e^{-i\frac{2\pi k}{N}x}.$$

We can still calculate the $F_k$ using Equation (4.1), but now there are other ways.

Suppose we obtain $N$ values of $f(x)$, say $f(x_n)$, for $n = 0, 1, ..., N-1$. Such situations arise, for example, in time-series analysis, where $x$ represents time and we are able to measure the function $f(x)$ at some finitely many different times. The function $f(x)$ could represent acoustic pressure coming from speech, current values of a particular stock on the Stock Exchange, the temperature at time $x$ in a particular place, and so on. We may want to model $f(x)$ to estimate values of $f(x)$ we were unable to measure, perhaps for prediction, or to break $f(x)$ up into finitely many sinusoidal components. This latter problem is important in digital sound recording and speech recognition.

Once we have the data $f(x_n)$, for $n = 0, 1, ..., N - 1$, we can then get the $F_k$ by solving a system of $N$ linear equations in $N$ unknowns:

$$f(x_n) = \frac{1}{N} \sum_{k=0}^{N-1} F_k e^{-i\frac{2\pi k}{N}x_n}. \tag{4.2}$$

Solving this system typically requires roughly $N^3$ complex multiplications, which, for many applications in which $N$ is in the thousands, is prohibitively expensive and time-consuming. However, if we have the freedom to select the $x_n$ and choose $x_n = n$, then solving the system becomes much simpler, because of *discrete orthogonality*.

With $x_n = n$, the solution of the system of linear equations

$$f(n) = \frac{1}{N} \sum_{k=0}^{N-1} F_k e^{-i\frac{2\pi kn}{N}} \tag{4.3}$$

is

$$F_k = \sum_{n=0}^{N-1} f(n) e^{i\frac{2\pi kn}{N}}, \tag{4.4}$$

for $k = 0, 1, ..., N - 1$. The proof of this assertion is contained in the following exercises.

**Ex. 4.1** *Use the formula for the sum of a finite geometric progression to show that*

$$\sum_{n=0}^{N-1} e^{int} = e^{i\frac{(N-1)t}{2}} \frac{\sin \frac{Nt}{2}}{\sin \frac{t}{2}}. \tag{4.5}$$

**Ex. 4.2** *Prove the assertion in Equation (4.4) by multiplying both sides of Equation (4.3) by $e^{i\frac{2\pi jn}{N}}$, and summing over $n$. Interchange the order of summation and use Equation (4.5).*

The formula in Equation (4.5) is perhaps the most important in signal processing and we shall encounter it several times later in this book. It describes coherent summation, the phenomenon of constructive and destructive interference, and is the basic formula in sonar and radar. It also arises in matched filtering, optimal detection theory, and the DFT estimation of the Fourier transform.

## 4.3    The DFT and the Finite Fourier Series

In the unknown strength problem we saw that measurements in the far field could give us finitely many values of the Fourier coefficients of a support-limited function. Suppose now that $f : [0, N] \to \mathbb{C}$ is such an unknown function, and we have obtained the Fourier-transform values $F(\frac{2\pi k}{N})$, for $k = 0, 1, ..., N - 1$. It is reasonable to use the DFT to estimate $f(x)$:

$$f(x) \approx f_{DFT}(x) = \frac{1}{N} \sum_{k=0}^{N-1} F_k e^{-i\frac{2\pi k}{N}x}.$$

The DFT looks just like the finite Fourier series we discussed previously. We can calculate the $N$ values of $f_{DFT}(x)$ at the points $x = n$ using the formula in Equation (4.3):

$$f_{DFT}(n) = \frac{1}{N} \sum_{k=0}^{N-1} F_k e^{-i\frac{2\pi kn}{N}}. \tag{4.6}$$

Note, however, that the context has changed. Previously, we assumed that we had actual values of $f(x)$ at the points $x = n$ and we used the finite Fourier series to model $f(x)$. Now we are assuming that it is finitely many values of the actual Fourier transform, $F(\gamma)$, that we have obtained, and we want to use those values to estimate $f(x)$. What we are getting when we use Equation (4.6) are not actual values of $f(x)$ itself, but of the DFT estimator of $f(x)$.

As we noted previously, solving for the $F_k$ using the system described by Equation (4.2) would require roughly $N^3$ complex multiplications. When we select $x_n = n$ we can solve the system in Equation (4.3) in $N^2$ complex multiplications. But for very large $N$, even $N^2$ is too large. Fortunately, there is the Fast Fourier Transform (FFT), which we shall consider in detail in Chapter 8. The FFT reduces the computational cost to roughly $N \log_2 N$ complex multiplications.

## 4.4    The Vector DFT

The discussion in the previous sections motivates the definition of the *vector* DFT (vDFT). Given any column vector **f** in $\mathbb{C}^N$ with entries $f_0, f_1, ..., f_{N-1}$, we define the vector DFT (vDFT) of **f** to be the complex vector $\mathbf{F}$ in $\mathbb{C}^N$ having the entries

$$F_k = \sum_{n=0}^{N-1} f_n e^{i\frac{2\pi nk}{N}}, \tag{4.7}$$

for $k = 0, 1, ..., N - 1$. From our previous discussion, we know that we then have

$$f_n = \frac{1}{N} \sum_{k=0}^{N-1} F_k e^{-i\frac{2\pi nk}{N}},$$

for $n = 0, 1, ..., N - 1$.

Most texts on signal processing call the vector $\mathbf{F}$ the DFT of the vector $\mathbf{f}$. I think this is bad terminology, as I shall explain. Suppose we have data $f(n)$, for $n = 0, 1, ..., N - 1$, and the Fourier transform function $F(\gamma)$ is unknown, but known to be supported on the interval $[0, 2\pi]$. We want to estimate $F(\gamma)$ using the data. One way is to use the DFT estimate,

$$F_{DFT}(\gamma) = \sum_{n=0}^{N-1} f(n) e^{in\gamma}.$$

The next step would be to plot our estimate. To do this we select some finitely many values of $\gamma$, say $\gamma_k$, for $k = 0, 1, ..., K - 1$, and evaluate $F_{DFT}(\gamma_k)$. If we choose $K = N$ and $\gamma_k = \frac{2\pi k}{N}$, we get Equation (4.7), with $F_k = F(\frac{2\pi k}{N})$. If we use the FFT, we can calculate all the $F_k$ quickly. However, the FFT prefers to have $N$ equal to some power of two. If, for example, we have $N = 250$, we can trick the FFT by defining $f(250) = f(251) = ... = f(255) = 0$, and changing $N = 250$ to $N = 256$. The DFT estimate is still the same function of the continuous variable $\gamma$, but now the FFT will evaluate the DFT at 256 equi-spaced points with the interval $[0, 2\pi)$. In fact, if we should want to generate a plot of the DFT that had, say, 1024 grid points, we could simply augment our original data set with sufficiently many zero values, and then perform the FFT; this is called *zero-padding*. In each case, we calculate a vector $\mathbf{F}$, but the sizes change as we augment the data with more zero values. To call each of these vectors $\mathbf{F}$ *the* DFT seems to me to be wrong. Each one is a vDFT of a certain set of data, original or augmented, while *the* DFT remains the same function of the continuous variable $\gamma$. It is important to remember that the values $F_k$ we calculate are not values of the actual $F(\gamma)$, but of the DFT estimator of $F(\gamma)$. This point is sometimes missed in the literature on the subject.

## 4.5   The Vector DFT in Two Dimensions

We consider now a complex-valued function $f(x, y)$ of two real variables, with Fourier transformation

$$F(\alpha, \beta) = \int \int f(x, y) e^{i(x\alpha + y\beta)} dx dy.$$

Suppose that $F(\alpha, \beta) = 0$, except for $\alpha$ and $\beta$ in the interval $[0, 2\pi]$; this means that the function $F(\alpha, \beta)$ represents a two-dimensional object with bounded support, such as a picture. Then $F(\alpha, \beta)$ has a Fourier series expansion

$$F(\alpha, \beta) = \sum_{m=-\infty}^{\infty} \sum_{n=-\infty}^{\infty} f(m, n) e^{im\alpha} e^{in\beta} \qquad (4.8)$$

for $0 \leq \alpha \leq 2\pi$ and $0 \leq \beta \leq 2\pi$.

In image processing, $F(\alpha, \beta)$ is our two-dimensional analogue image, where $\alpha$ and $\beta$ are continuous variables. The first step in digital image processing is to digitize the image, which means forming a two-dimensional array of numbers $F_{j,k}$, for $j, k = 0, 1, ..., N - 1$. For concreteness, we let the $F_{j,k}$ be the values $F(\frac{2\pi}{N} j, \frac{2\pi}{N} k)$.

From Equation (4.8) we can write

$$F_{j,k} = F\left(\frac{2\pi}{N} j, \frac{2\pi}{N} k\right) = \sum_{m=-\infty}^{\infty} \sum_{n=-\infty}^{\infty} f(m, n) e^{i\frac{2\pi}{N} jm} e^{i\frac{2\pi}{N} kn},$$

for $j, k = 0, 1, ..., N - 1$.

We can also find coefficients $f_{m,n}$, for $m, n = 0, 1, ..., N - 1$, such that

$$F_{j,k} = F\left(\frac{2\pi}{N} j, \frac{2\pi}{N} k\right) = \sum_{m=0}^{N-1} \sum_{n=0}^{N-1} f_{m,n} e^{i\frac{2\pi}{N} jm} e^{i\frac{2\pi}{N} kn},$$

for $j, k = 0, 1, ..., N - 1$. These $f_{m,n}$ are only approximations of the values $f(m, n)$, as we shall see.

Just as in the one-dimensional case, we can make use of orthogonality to find the coefficients $f_{m,n}$. We have

$$f_{m,n} = \frac{1}{N^2} \sum_{j=0}^{N-1} \sum_{k=0}^{N-1} F\left(\frac{2\pi}{N} jm, \frac{2\pi}{N} kn\right) e^{-i\frac{2\pi}{N} jm} e^{-i\frac{2\pi}{N} kn}, \qquad (4.9)$$

for $m, n = 0, 1, ..., N - 1$. Now we show how the $f_{m,n}$ can be thought of as approximations of the $f(m, n)$.

We know from the Fourier Inversion Formula in two dimensions, Equation (2.18), that

$$f(m,n) = \frac{1}{4\pi^2} \int_0^{2\pi} \int_0^{2\pi} F(\alpha, \beta) e^{-i(\alpha m + \beta n)} d\alpha d\beta. \tag{4.10}$$

When we replace the right side of Equation (4.10) with a Riemann sum, we get

$$f(m,n) \approx \frac{1}{N^2} \sum_{j=0}^{N-1} \sum_{k=0}^{N-1} F\left(\frac{2\pi}{N} jm, \frac{2\pi}{N} kn\right) e^{-i\frac{2\pi}{N} jm} e^{-i\frac{2\pi}{N} kn};$$

the right side is precisely $f_{m,n}$, according to Equation (4.9).

Notice that we can compute the $f_{m,n}$ from the $F_{j,k}$ using one-dimensional vDFTs. For each fixed $j$ we compute the one-dimensional vDFT

$$G_{j,n} = \frac{1}{N} \sum_{k=0}^{N-1} F_{j,k} e^{-i\frac{2\pi}{N} kn},$$

for $n = 0, 1, ..., N-1$. Then for each fixed $n$ we compute the one-dimensional vDFT

$$f_{m,n} = \sum_{j=0}^{N-1} G_{j,n} e^{-i\frac{2\pi}{N} jm},$$

for $m = 0, 1, ..., N - 1$. From this, we see that estimating $f(x,y)$ by calculating the two-dimensional vDFT of the values from $F(\alpha, \beta)$ requires us to obtain $2N$ one-dimensional vector DFTs.

Calculating the $f_{m,n}$ from the pixel values $F_{j,k}$ is the main operation in digital image processing. The $f_{m,n}$ approximate the spatial frequencies in the image and modifications to the image, such as smoothing or edge enhancement, can be made by modifying the values $f_{m,n}$. Improving the resolution of the image can be done by extrapolating the $f_{m,n}$, that is, by approximating values of $f(x,y)$ other than $x = m$ and $y = n$. Once we have modified the $f_{m,n}$, we return to the new values of $F_{j,k}$, so calculating $F_{j,k}$ from the $f_{m,n}$ is also an important step in image processing.

In some areas of medical imaging, such as transmission tomography and magnetic-resonance imaging, the scanners provide the $f_{m,n}$. Then the desired digitized image of the patient is the array $F_{j,k}$. In such cases, the $f_{m,n}$ are considered to be approximate values of $f(m,n)$. For more on the role of the two-dimensional Fourier transform in medical imaging, see Chapter 11 on transmission tomography.

Even if we managed to have the true values, that is, even if $f_{m,n} = f(m,n)$, the values $F_{j,k}$ are not the true values $F(\frac{2\pi}{N} m, \frac{2\pi}{N} n)$. The number

$F_{j,k}$ is a value of the DFT approximation of $F(\alpha, \beta)$. This DFT approximation is the function given by

$$F_{DFT}(\alpha, \beta) = \sum_{m=0}^{N-1} \sum_{n=0}^{N-1} f_{m,n} e^{i\alpha m} e^{i\beta n}.$$

The number $F_{j,k}$ is the value of this approximation at the point $\alpha = \frac{2\pi}{N} j$ and $\beta = \frac{2\pi}{N} k$. In other words,

$$F_{j,k} = F_{DFT} \left( \frac{2\pi}{N} j, \frac{2\pi}{N} k \right),$$

for $j, k = 0, 1, ..., N - 1$. How good this discrete image is as an approximation of the true $F(\alpha, \beta)$ depends primarily on two things: first, how accurate an approximation of the numbers $f(m, n)$ the numbers $f_{m,n}$ are; and second, how good an approximation of the function $F(\alpha, \beta)$ the function $F_{DFT}(\alpha, \beta)$ is.

We can easily see now how important the Fast Fourier Transform algorithm is. Without the Fast Fourier Transform to accelerate the calculations, obtaining a two-dimensional vDFT would be prohibitively expensive.

---

## 4.6   The Issue of Units

When we write $\cos \pi = -1$, it is with the understanding that $\pi$ is a measure of angle, in radians; the function cos will always have an independent variable in units of radians. Therefore, when we write $\cos(x\omega)$, we understand the product $x\omega$ to be in units of radians. If $x$ is measured in seconds, then $\omega$ is in units of radians per second; if $x$ is in meters, then $\omega$ is in units of radians per meter. When $x$ is in seconds, we sometimes use the variable $\frac{\omega}{2\pi}$; since $2\pi$ is then in units of radians per cycle, the variable $\frac{\omega}{2\pi}$ is in units of cycles per second, or Hertz. When we sample $f(x)$ at values of $x$ spaced $\Delta$ apart, the $\Delta$ is in units of $x$-units per sample, and the reciprocal, $\frac{1}{\Delta}$, which is called the *sampling frequency*, is in units of samples per $x$-units. If $x$ is in seconds, then $\Delta$ is in units of seconds per sample, and $\frac{1}{\Delta}$ is in units of samples per second.

## 4.7   Approximation, Models, or Truth?

We mentioned previously that, when we model $f(x)$ using a finite Fourier series, we may want to analyze $f(x)$ to determine its sinusoidal components. But does $f(x)$ actually contain these sinusoidal components in any real sense? An example from Fourier-series expansion will clarify this issue.

Consider the function $f(x) = \sin x$, for $0 \leq x \leq \pi$. The function $g(x)$, defined by $g(x) = f(x)$, for $0 \leq x \leq \pi$, and $g(x) = f(-x)$, for $-\pi \leq x \leq 0$, can be extended to a continuous even function with period $2\pi$. The Fourier series for $g(x)$ is

$$g(x) = \frac{2}{\pi} - \frac{2}{\pi} \sum_{n=2}^{\infty} \frac{1 + \cos n\pi}{n^2 - 1} \cos nx.$$

When we restrict our attention to $x$ in the interval $[0, \pi]$, we have the function $\sin x$ expressed as an infinite sum of cosine functions. It is true, in a sense, that the sine function on $[0, \pi]$ is made up of infinitely many cosines, and any partial sum of this infinite cosine series can be viewed as an approximation of the function $\sin x$ on $[0, \pi]$. However, is it really the kind of truth about the function $f(x)$ that we are seeking?

## 4.8   Modeling the Data

In time-series analysis, we have some unknown function of time, $f(t)$, and we measure its values $f(t_n)$ at the $N$ sampling points $t = t_n$, $n = 1, ..., N$. There are several different possible objectives that we may have at this point.

### 4.8.1   Extrapolation

We may want to estimate values of $f(t)$ at points $t$ at which we do not have measurements; these other points may represent time in the future, for example, and we are trying to predict future values of $f(t)$. In such cases, it is common to adopt a model for $f(t)$, which is typically some function of $t$ with finitely many as yet undetermined parameters, such as a polynomial or a sum of trig functions. We must select our model with care, particularly if the data is assumed to be noisy, as most data is. Even though we may have a large number of measurements, it may be a mistake to model $f(t)$ with as many parameters as we have data.

We do not really believe that $f(t)$ is a polynomial or a finite Fourier series. We may not even believe that the model is a good approximation of $f(t)$ for all values of $t$. We do believe, however, that adopting such a model will enable us to carry out our prediction task in a reasonably accurate way. The task may be something like predicting the temperature at noon tomorrow, on the basis of noon-time temperatures for the previous five days.

### 4.8.2   Filtering the Data

Suppose that the values $f(t_n)$ are sampled data from an old recording of a singer. We may want to clean up this digitized data, in order to be able to recapture the original sound. Now we may only desire to modify each of the values $f(t_n)$ in some way, to improve the quality. To perform this restoring task, we may model the data as samples of a finite Fourier series, or, more generally, as a finite sum of sinusoids in which the frequencies $\gamma_k$ are chosen by us. We then solve for the parameters.

To clean up the sound, we may modify the values of some of the parameters. For example, we may believe that certain of the frequencies come primarily from a noise component in the recording. To remove, or at least diminish, this component, we can reduce the associated coefficients. We may feel that the original recording technology failed to capture some of the higher notes sung by the soprano. Then we can increase the values of those coefficients associated with those frequencies that need to be restored. Obviously, restoring old recordings of opera singers is more involved than this, but you get the idea.

The point here is that we need not believe that the entire recording can be accurately described, or even approximated, by a finite sum of complex exponential functions. But using a finite sum of sinusoids does give another way to describe the measured data, and as such, another way to modify this data, namely by modifying the coefficients of the sinusoids. We do not need to believe that the entire opera can be accurately approximated by such a sum in order for this restoring procedure to be helpful.

Note that if our goal is to recapture a high note sung by the soprano, we do not really need to use samples of the function $f(t)$ that correspond to times when only the tenor was on stage singing. It would make more sense to process only those measurements taken right around the time the high note was sung by the soprano. This is *short-time* Fourier analysis, an issue that we deal with when we discuss time-frequency analysis and wavelets.

## 4.9 More on Coherent Summation

We begin this section with an exercise.

**Ex. 4.3** *On a blank sheet of paper, draw a horizontal and vertical axis. Starting at the origin, draw a vector with length one unit (a unit can be, say, one inch), in an arbitrary direction. Now, from the tip of the first vector, draw another vector of length one, again in an arbitrary direction. Repeat this process several times, using M vectors in all. Now measure the distance from the origin to the tip of the last vector drawn. Compare this length with the number M, which would be the distance from the origin to the tip of the last vector, if all the vectors had had the same direction.*

This exercise reveals the important difference between *coherent* and *incoherent summation*, or, if you will, between constructive and destructive interference. Each of the unit vectors drawn can be thought of as a complex number $e^{i\theta_m}$, where $\theta_m$ is its arbitrary angle. The distance from the origin to the tip of the last vector drawn is then

$$\left| e^{i\theta_1} + e^{i\theta_2} + ... + e^{i\theta_M} \right|.$$

If all the angles $\theta_m$ are equal, then this distance is $M$; in all other cases the distance is quite a bit less than $M$. The distinction between coherent and incoherent summation plays a central role in signal processing, as well as in quantum physics, as we discuss briefly in the next section.

## 4.10 Uses in Quantum Electrodynamics

In his experiments with light, Newton discovered the phenomenon of *partial reflection*. The proportion of the light incident on a glass surface that is reflected varies with the thickness of the glass, but the proportion oscillates between zero and about sixteen percent as the glass thickens. He tried to explain this puzzling behavior, but realized that he had not obtained a satisfactory explanation. In his beautiful small book "QED: The Strange Theory of Light and Matter" [71], the physicist Richard Feynman illustrates how the quantum theory applied to light, *quantum electrodynamics* or *QED*, can be used to unravel many phenomena involving the interaction of light with matter, including the partial reflection observed by Newton, the *least time* principle, the array of colors we see on the surface of an oily mud puddle, and so on. He is addressing an audience of non-physicists, including even some non-scientists, and avoids mathematics as much as possible. The one mathematical notion that he uses repeatedly is the addition of two-dimensional vectors pointing in a variety of directions, that is, coherent and incoherent summation. The vector sum is the *probability amplitude* of the event being discussed, and the square of its length is the probability of the event.

## 4.11 Using Coherence and Incoherence

Suppose we are given as data the $M$ complex numbers $d_m = e^{im\gamma}$, for $m = 1, ..., M$, and we are asked to find the real number $\gamma$. We can exploit coherent summation to get our answer.

First of all, from the data we have been given, we cannot distinguish $\gamma$ from $\gamma + 2\pi$, since, for all integers $m$

$$e^{im(\gamma+2\pi)} = e^{im\gamma}e^{2m\pi i} = e^{im\gamma}(1) = e^{im\gamma}.$$

Therefore, we assume, from the beginning, that the $\gamma$ we want to find lies in the interval $[-\pi, \pi)$. Note that we could have selected any interval of length $2\pi$, not necessarily $[-\pi, \pi)$; if we have no prior knowledge of where $\gamma$ is located, the intervals $[-\pi, \pi)$ or $[0, 2\pi)$ are the most obvious choices.

### 4.11.1 The Discrete Fourier Transform

Now we take any value $\omega$ in the interval $[-\pi, \pi)$, multiply each of the numbers $d_m$ by $e^{-im\omega}$, and sum over $m$ to get

$$DFT_\mathbf{d}(\omega) = \sum_{m=1}^{M} d_m e^{-im\omega}. \tag{4.11}$$

The sum we denote by $DFT_\mathbf{d}$ will be called the *discrete Fourier transform* (DFT) of the data (column) vector $\mathbf{d} = (d_1, ..., d_M)^T$. We define the column vector $\mathbf{e}_\omega$ to be

$$\mathbf{e}_\omega = (e^{i\omega}, e^{2i\omega}, ..., e^{iM\omega})^T,$$

which allows us to write $DFT_\mathbf{d} = \mathbf{e}_\omega^\dagger \mathbf{d}$, where the dagger denotes conjugate transformation of a matrix or vector.

Rewriting the exponential terms in the sum in Equation (4.11), we obtain

$$DFT_\mathbf{d}(\omega) = \sum_{m=1}^{M} d_m e^{-im\omega} = \sum_{m=1}^{M} e^{im(\gamma-\omega)}.$$

Performing this calculation for each $\omega$ in the interval $[-\pi, \pi)$, we obtain the function $DFT_\mathbf{d}(\omega)$. For each $\omega$, the complex number $DFT_\mathbf{d}(\omega)$ is the sum of $M$ complex numbers, each having length one, and angle $\theta_m = m(\gamma - \omega)$. So long as $\omega$ is not equal to $\gamma$, these $\theta_m$ are all different, and $DFT_\mathbf{d}(\omega)$ is an incoherent sum; consequently, $|DFT_\mathbf{d}(\omega)|$ will be smaller than $M$. However, when $\omega = \gamma$, each $\theta_m$ equals zero, and $DFT_\mathbf{d}(\omega) = |DFT_\mathbf{d}(\omega)| = M$; the reason for putting the minus sign in the exponent $e^{-im\omega}$ is so that we get the term $\gamma - \omega$, which is zero when $\gamma = \omega$. We find the true $\gamma$ by computing the value $|DFT_\mathbf{d}(\omega)|$ for finitely many values of $\omega$, plot the result and look for the highest value. Of course, it may well happen that the true value $\omega = \gamma$ is not exactly one of the points we choose to plot; it may happen that the true $\gamma$ is half way between two of the plot's grid points, for example. Nevertheless, if we know in advance that there is only one true $\gamma$, this approach will give us a good idea of its value.

In many applications, the number $M$ will be quite large, as will be the number of grid points we wish to use for the plot. This means that the number $DFT_\mathbf{d}(\omega)$ is a sum of a large number of terms, and that we must calculate this sum for many values of $\omega$. Fortunately, we can use the FFT for this.

**Ex. 4.4** *The* Dirichlet kernel *of size $M$ is defined as*

$$D_M(x) = \sum_{m=-M}^{M} e^{imx}.$$

*Use Equation (4.5) to obtain the closed-form expression*

$$D_M(x) = \frac{\sin((M + \frac{1}{2})x)}{\sin(\frac{x}{2})};$$

*note that $D_M(x)$ is real-valued.*

**Ex. 4.5** *Obtain the closed-form expressions*

$$\sum_{m=N}^{M} \cos mx = \cos\left(\frac{M+N}{2}x\right)\frac{\sin(\frac{M-N+1}{2}x)}{\sin\frac{x}{2}} \tag{4.12}$$

*and*

$$\sum_{m=N}^{M} \sin mx = \sin\left(\frac{M+N}{2}x\right)\frac{\sin(\frac{M-N+1}{2}x)}{\sin\frac{x}{2}}. \tag{4.13}$$

*Hint: Recall that $\cos mx$ and $\sin mx$ are the real and imaginary parts of $e^{imx}$.*

**Ex. 4.6** *Obtain the formulas in the previous exercise using the trigonometric identity*

$$\sin\left(\left(n + \frac{1}{2}\right)x\right) - \sin\left(\left(n - \frac{1}{2}\right)x\right) = 2\sin\left(\frac{x}{2}\right)\cos(nx).$$

**Ex. 4.7** *Graph the function $D_M(x)$ for various values of $M$.*

We note in passing that the function $D_M(x)$ equals $2M + 1$ for $x = 0$ and equals zero for the first time at $x = \frac{2\pi}{2M+1}$. This means that the *main lobe* of $D_M(x)$, the inverted parabola-like portion of the graph centered at $x = 0$, crosses the $x$-axis at $x = 2\pi/(2M + 1)$ and $x = -2\pi/(2M + 1)$, so its height is $2M + 1$ and its width is $4\pi/(2M + 1)$. As $M$ grows larger the main lobe of $D_M(x)$ gets higher and thinner.

In the exercise that follows we examine the resolving ability of the DFT. Suppose we have $M$ equi-spaced samples of a function $f(x)$ having the form

$$f(x) = e^{ix\gamma_1} + e^{ix\gamma_2},$$

where $\gamma_1$ and $\gamma_2$ are in the interval $(-\pi, \pi)$. If $M$ is sufficiently large, the DFT should show two peaks, at roughly the values $\omega = \gamma_1$ and $\omega = \gamma_2$. As the distance $|\gamma_2 - \gamma_1|$ grows smaller, it will require a larger value of $M$ for the DFT to show two peaks.

**Ex. 4.8** *For this exercise, we take $\gamma_1 = -\alpha$ and $\gamma_2 = \alpha$, for some $\alpha$ in the interval $(0, \pi)$. Select a value of $M$ that is greater than two and calculate the values $f(m)$ for $m = 1, ..., M$. Plot the graph of the function $|DFT_\mathbf{d}(\omega)|$ on $(-\pi, \pi)$. Repeat the exercise for various values of $M$ and values of $\alpha$ closer to zero. Notice how $DFT_\mathbf{d}(0)$ behaves as $\alpha$ goes to zero. For each fixed value of $M$ there will be a critical value of $\alpha$ such that, for any smaller values of $\alpha$, $DFT_\mathbf{d}(0)$ will be larger than $DFT_\mathbf{d}(\alpha)$. This is loss of resolution.*

## 4.12 Complications

In the real world, of course, things are not so simple. In most applications, the data comes from measurements, and so contains errors, also called *noise*. The noise terms that appear in each $d_m$ are usually viewed as random variables, and they may or may not be independent. If the noise terms are not independent, we say that we have *correlated noise*. If we know something about the statistics of the noises, we may wish to process the data using statistical estimation methods, such as the *best linear unbiased estimator* (BLUE).

## 4.12.1 Multiple Signal Components

It sometimes happens that there are two or more distinct values of $\omega$ that we seek. For example, suppose the data is

$$d_m = e^{im\alpha} + e^{im\beta},$$

for $m = 1, ..., M$, where $\alpha$ and $\beta$ are two distinct numbers in the interval $[0, 2\pi)$, and we need to find both $\alpha$ and $\beta$. Now the function $DFT_{\mathbf{d}}(\omega)$ will be

$$DFT_{\mathbf{d}}(\omega) = \sum_{m=1}^{M} (e^{im\alpha} + e^{im\beta}) e^{-im\omega} = \sum_{m=1}^{M} e^{im\alpha} e^{-im\omega} + \sum_{m=1}^{M} e^{im\beta} e^{-im\omega},$$

so that

$$DFT_{\mathbf{d}}(\omega) = \sum_{m=1}^{M} e^{im(\alpha-\omega)} + \sum_{m=1}^{M} e^{im(\beta-\omega)}.$$

So the function $DFT_{\mathbf{d}}(\omega)$ is the sum of the $DFT_{\mathbf{d}}(\omega)$ that we would have obtained separately if we had had only $\alpha$ and only $\beta$.

## 4.12.2 Resolution

If the numbers $\alpha$ and $\beta$ are well separated in the interval $[0, 2\pi)$ or $M$ is very large, the plot of $|DFT\mathbf{d}(\omega)|$ will show two high values, one near $\omega = \alpha$ and one near $\omega = \beta$. However, if the $M$ is smaller or the $\alpha$ and $\beta$ are too close together, the plot of $|DFT\mathbf{d}(\omega)|$ may show only one broader high bump, centered between $\alpha$ and $\beta$; this is loss of resolution. How close is too close will depend on the value of $M$.

## 4.12.3 Unequal Amplitudes and Complex Amplitudes

It is also often the case that the two signal components, the one from $\alpha$ and the one from $\beta$, are not equally strong. We could have

$$d_m = Ae^{im\alpha} + Be^{im\beta},$$

where $A > B > 0$. In fact, both $A$ and $B$ could be complex numbers, that is, $A = |A|e^{i\theta_1}$ and $B = |B|e^{i\theta_2}$, so that

$$d_m = |A|e^{im\alpha+\theta_1} + |B|e^{im\beta+\theta_2}.$$

In stochastic signal processing, the $A$ and $B$ are viewed as random variables; $A$ and $B$ may or may not be mutually independent.

### 4.12.4  Phase Errors

It sometimes happens that the hardware that provides the measured data is imperfect and instead of giving us the values $d_m = e^{im\alpha}$, we get $d_m = e^{im\alpha + \phi_m}$. Now each *phase error* $\phi_m$ depends on $m$, which makes matters worse than when we had $\theta_1$ and $\theta_2$ previously, neither depending on the index $m$.

---

## 4.13  Undetermined Exponential Models

In our previous discussion, we assumed that the frequencies were known and only the coefficients needed to be determined. The problem was then a linear one. It is sometimes the case that we also want to estimate the frequencies from the data. This is computationally more difficult and is a nonlinear problem. Prony's method is one approach to this problem.

The date of publication of [130] is often taken by editors to be a typographical error and is replaced by 1995; or, since it is not written in English, perhaps 1895. But the 1795 date is the correct one. The mathematical problem Prony solved arises also in signal processing, and his method for solving it is still used today. Prony's method is also the inspiration for the eigenvector methods described in Chapter 14.

### 4.13.1  Prony's Problem

Prony considers a function of the form

$$f(x) = \sum_{n=1}^{N} a_n e^{\gamma_n x}, \tag{4.14}$$

where we allow the $a_n$ and the $\gamma_n$ to be complex. If we take the $\gamma_n = i\omega_n$ to be imaginary, $f(x)$ becomes the sum of complex exponentials, which we discuss later; if we take $\gamma_n$ to be real, then $f(x)$ is the sum of real exponentials, either increasing or decreasing. The problem is to determine the number $N$, the $\gamma_n$, and the $a_n$ from samples of $f(x)$.

### 4.13.2  Prony's Method

Suppose that we have data $f_m = f(m\Delta)$, for some $\Delta > 0$ and for $m = 1, ..., M$, where we assume that $M = 2N$. We seek a vector $\mathbf{c}$ with entries $c_j$, $j = 0, ..., N$ such that

$$c_0 f_{k+1} + c_1 f_{k+2} + c_2 f_{k+3} + ... + c_N f_{k+N+1} = 0, \tag{4.15}$$

for $k = 0, 1, ..., M - N - 1$. So, we want a complex vector $\mathbf{c}$ in $\mathbb{C}^{N+1}$ orthogonal to $M - N = N$ other vectors. In matrix-vector notation we are solving the linear system

$$
\begin{bmatrix}
f_1 & f_2 & \cdots & f_{N+1} \\
f_2 & f_3 & \cdots & f_{N+2} \\
\cdot & & & \\
\cdot & & & \\
\cdot & & & \\
f_N & f_{N+1} & \cdots & f_M
\end{bmatrix}
\begin{bmatrix}
c_0 \\
c_1 \\
\cdot \\
\cdot \\
\cdot \\
c_N
\end{bmatrix}
=
\begin{bmatrix}
0 \\
0 \\
\cdot \\
\cdot \\
\cdot \\
0
\end{bmatrix},
$$

which we write as $F\mathbf{c} = \mathbf{0}$. Since $F^\dagger F \mathbf{c} = \mathbf{0}$ also, we see that $\mathbf{c}$ is an eigenvector associated with the eigenvalue zero of the hermitian nonnegative definite matrix $F^\dagger F$; here $F^\dagger$ denotes the conjugate transpose of the matrix $F$.

Fix a value of $k$ and replace each of the $f_{k+j}$ in Equation (4.15) with the value given by Equation (4.14) to get

$$
\begin{aligned}
0 &= \sum_{n=1}^{N} a_n \left( \sum_{j=0}^{N} c_j e^{\gamma_n (k+j+1)\Delta} \right) \\
&= \sum_{n=1}^{N} \left( a_n e^{\gamma_n (k+1)\Delta} \left( \sum_{j=0}^{N} c_j (e^{\gamma_n \Delta})^j \right) \right).
\end{aligned}
$$

Since this is true for each of the $N$ fixed values of $k$, we conclude that the inner sum is zero for each $n$; that is,

$$
\sum_{j=0}^{N} c_j (e^{\gamma_n \Delta})^j = 0,
$$

for each $n$. Therefore, the polynomial

$$
C(z) = \sum_{j=0}^{N} c_j z^j
$$

has for its roots the $N$ values $z = e^{\gamma_n \Delta}$. Once we find the roots of this polynomial we have the values of $e^{\gamma_n \Delta}$. If the $\gamma_n$ are real, they are uniquely determined from the values $e^{\gamma_n \Delta}$, whereas, for non-real $\gamma_n$, this is not the case, as we saw when we studied the complex exponential functions.

Then, we obtain the $a_n$ by solving a linear system of equations. In practice we would not know $N$ so would overestimate $N$ somewhat in selecting $M$. As a result, some of the $a_n$ would be zero.

If we believe that the number $N$ is considerably smaller than $M$, we do not assume that $2N = M$. Instead, we select $L$ somewhat larger than we believe $N$ is and then solve the linear system

$$
\begin{bmatrix}
f_1 & f_2 & \cdots & f_{L+1} \\
f_2 & f_3 & \cdots & f_{L+2} \\
\cdot & & & \\
\cdot & & & \\
\cdot & & & \\
\cdot & & & \\
f_{M-L} & f_{M-L+1} & \cdots & f_M
\end{bmatrix}
\begin{bmatrix}
c_0 \\
c_1 \\
\cdot \\
\cdot \\
\cdot \\
c_L
\end{bmatrix}
=
\begin{bmatrix}
0 \\
0 \\
\cdot \\
\cdot \\
0 \\
0
\end{bmatrix}.
$$

This system has $M - L$ equations and $L + 1$ unknowns, so is quite overdetermined. We would then use the least-squares approach to obtain the vector $\mathbf{c}$. Again writing the system as $F\mathbf{c} = \mathbf{0}$, we note that the matrix $F^\dagger F$ is $L + 1$ by $L + 1$ and has $\lambda = 0$ for its lowest eigenvalue; therefore, it is not invertible. When there is noise in the measurements, this matrix may become invertible, but will still have at least one very small eigenvalue.

Finding the vector $\mathbf{c}$ in either case can be tricky because we are looking for a nonzero solution of a homogeneous system of linear equations. For a discussion of the numerical issues involved in these calculations, the interested reader should consult the book by Therrien [153].

# Chapter 5

# Transmission and Remote Sensing

## 5.1  Chapter Summary

An important example of the use of the DFT is the design of directional transmitting or receiving arrays of antennas. In this chapter we revisit transmission and remote sensing, this time with emphasis on the roles played by complex exponential functions and the DFT.

## 5.2  Directional Transmission

Parabolic mirrors behind car headlamps reflect the light from the bulb, concentrating it directly ahead. Whispering at one focal point of an elliptical room can be heard clearly at the other focal point. When I call to someone across the street, I cup my hands in the form of a megaphone to concentrate the sound in that direction. In all these cases the transmitted signal has acquired *directionality*. In the case of the elliptical room, not only does the soft whispering reflect off the walls toward the opposite focal point, but the travel times are independent of where on the wall the reflections occur; otherwise, the differences in time would make the received sound unintelligible. Parabolic satellite dishes perform much the same function, concentrating incoming signals coherently. In this chapter we discuss the use of amplitude and phase modulation of transmitted signals to concentrate the signal power in certain directions. Following the lead of Richard Feynman in [72], we use radio broadcasting as a concrete example of the use of directional transmission.

Radio broadcasts are meant to be received and the amount of energy that reaches the receiver depends on the amount of energy put into the transmission as well as on the distance from the transmitter to the receiver. If the transmitter broadcasts a spherical wave front, with equal power in all directions, the energy in the signal is the same over the spherical wavefronts, so that the energy per unit area is proportional to the reciprocal of the surface area of the front. This means that, for omni-directional broadcasting, the energy per unit area, that is, the energy supplied to any receiver, falls off as the distance squared. The amplitude of the received signal is then proportional to the reciprocal of the distance.

Returning to the example we studied previously, suppose that you own a radio station in Los Angeles. Most of the population resides along the north-south coast, with fewer to the east, in the desert, and fewer still to the west, in the Pacific Ocean. You might well want to transmit the radio signal in a way that concentrates most of the power north and south. But how can you do this? The answer is to broadcast directionally. By shaping the wavefront to have most of its surface area north and south you will have the broadcast heard by more people without increasing the total energy in the transmission. To achieve this shaping you can use an array of multiple antennas.

## 5.3    Multiple-Antenna Arrays

### 5.3.1    The Array of Equi-Spaced Antennas

We place $2N + 1$ transmitting antennas a distance $\Delta > 0$ apart along an east-west axis, as shown in Figure 5.1. For convenience, let the locations of the antennas be $n\Delta$, $n = -N, ..., N$. To begin with, let us suppose that we have a fixed frequency $\omega$ and each of the transmitting antennas sends out the same signal $f_n(t) = \frac{1}{\sqrt{2N+1}} \cos(\omega t)$. With this normalization the total energy is independent of $N$. Let $(x, y)$ be an arbitrary location on the ground, and let $\mathbf{s}$ be the vector from the origin to the point $(x, y)$. Let $\theta$ be the angle measured clockwise from the positive horizontal axis to the vector $\mathbf{s}$. Let $D$ be the distance from $(x, y)$ to the origin. Then, if

$(x, y)$ is sufficiently distant from the antennas, the distance from $n\Delta$ on the horizontal axis to $(x, y)$ is approximately $D - n\Delta \cos(\theta)$. The signals arriving at $(x, y)$ from the various antennas will have traveled for different times and so will be out of phase with one another to a degree that depends on the location of $(x, y)$.

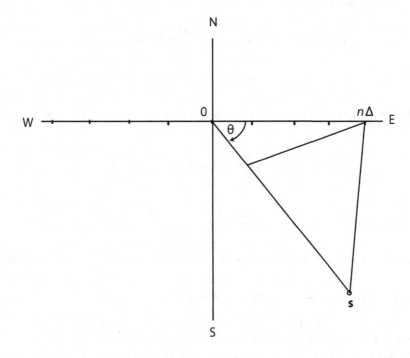

**FIGURE 5.1**: Antenna array and far-field receiver.

## 5.3.2 The Far-Field Strength Pattern

Since we are concerned only with wavefront shape, we omit for now the distance-dependence in the amplitude of the received signal. The signal received at $(x, y)$ is proportional to

$$f(\mathbf{s}, t) = \frac{1}{\sqrt{2N + 1}} \sum_{n=-N}^{N} \cos(\omega(t - t_n)),$$

where

$$t_n = \frac{1}{c}(D - n\Delta \cos(\theta))$$

and $c$ is the speed of propagation of the signal. Writing

$$\cos(\omega(t - t_n)) = \cos\left(\omega\left(t - \frac{D}{c}\right) + n\gamma\cos(\theta)\right)$$

for $\gamma = \frac{\omega\Delta}{c}$, we have

$$\cos(\omega(t - t_n)) = \cos\left(\omega\left(t - \frac{D}{c}\right)\right)\cos(n\gamma\cos(\theta))$$

$$- \sin\left(\omega\left(t - \frac{D}{c}\right)\right)\sin(n\gamma\cos(\theta)).$$

Using Equations (4.12) and (4.13), we find that the signal received at $(x, y)$ is

$$f(\mathbf{s}, t) = \frac{1}{\sqrt{2N + 1}} H(\theta)\cos\left(\omega\left(t - \frac{D}{c}\right)\right) \qquad (5.1)$$

for

$$H(\theta) = \frac{\sin((N + \frac{1}{2})\gamma\cos(\theta))}{\sin(\frac{1}{2}\gamma\cos(\theta))};$$

when the denominator equals zero the signal equals $\sqrt{2N + 1}\cos(\omega(t - \frac{D}{c}))$.

### 5.3.3 Can the Strength Be Zero?

We see from Equation (5.1) that the maximum power is in the north-south direction. What about the east-west direction? In order to have negligible signal power wasted in the east-west direction, we want the numerator, but not the denominator, in Equation (5.1) to be zero when $\theta = 0$. This means that $\Delta = m\lambda/(2N + 1)$, where $\lambda = 2\pi c/\omega$ is the wavelength and $m$ is some positive integer less than $2N + 1$. Recall that the wavelength for broadcast radio is tens to hundreds of meters.

**Ex. 5.1** *Graph the function $H(\theta)$ in polar coordinates for various choices of $N$ and $\Delta$.*

Figures 5.2, 5.3, and 5.4 show that transmission pattern $H(\theta)$ for various choices of $m$ and $N$. In Figure 5.2 $N = 5$ for each plot and the $m$ changes, illustrating the effect of changing the spacing of the array elements. The plots in Figure 5.3 differ from those in Figure 5.2 only in that $N = 21$ now. In Figure 5.4 we allow the $m$ to be less than one, showing the loss of the nulls in the east and west directions.

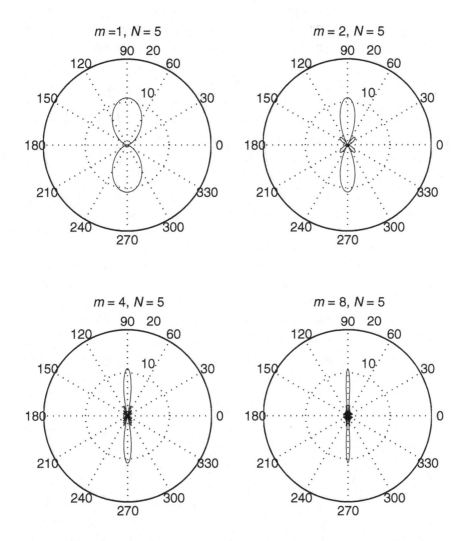

**FIGURE 5.2**: Transmission Pattern $H(\theta)$: $m = 1, 2, 4, 8$ and $N = 5$.

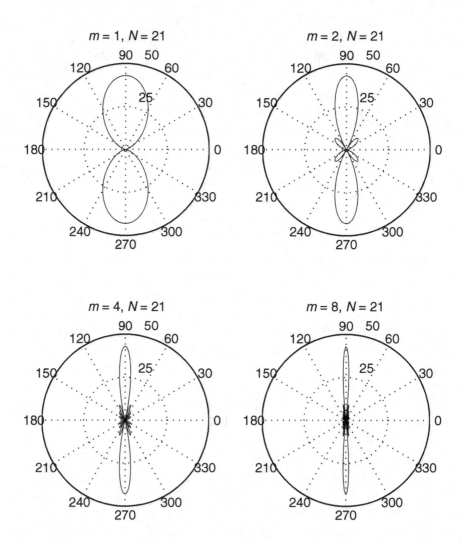

**FIGURE 5.3**: Transmission Pattern $H(\theta)$: $m = 1, 2, 4, 8$ and $N = 21$.

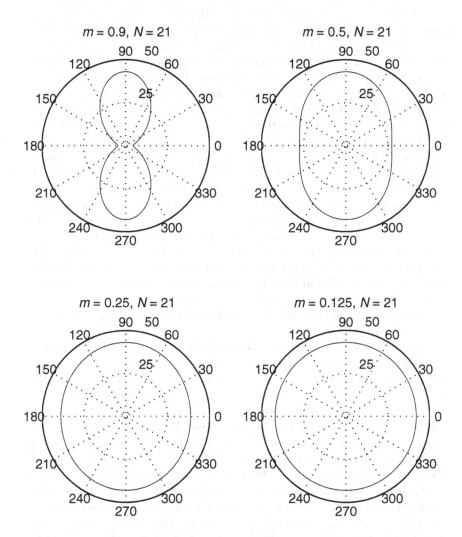

**FIGURE 5.4**: Transmission Pattern $H(\theta)$: $m = 0.9, 0.5, 0.25, 0.125$ and $N = 21$.

### 5.3.4 Diffraction Gratings

I have just placed on the table next to me a CD, with the shinier side up. Beyond it is a lamp. The CD acts as a mirror, and I see in the CD the reflection of the lamp. Every point of the lamp seems to be copied in a particular point on the surface of the CD, as if the ambient light that illuminates a particular point of the lamp travels only to a single point on the CD and then is reflected on into my eye. Each point of the lamp has its own special point on the CD. We know from basic optics that that point is such that the angle of incidence equals the angle of reflection, and the path (apparently) taken by the light beam is the shortest path the light can take to get from the lamp to the CD and then on to my eye. But how does the light know where to go?

In fact, what happens is that light beams take many paths from each particular point on the lamp to the CD and on to my eye. The reason I see only the one path is that all the other paths require different travel times, and so light beams on different paths arrive at my eye out of phase with one another. Only those paths very close to the one I see have travel times sufficiently similar to avoid this destructive interference. Speaking a bit more mathematically, if we define the function that associates with each path the time to travel along that path, then, at the shortest path, the *first derivative* of this function, in the sense of the calculus of variations, is zero. Therefore deviations from the shortest path correspond only to second-order changes in travel time, not first-order ones, which reduces the destructive interference.

But, as I look at the CD on the table, I see more than the reflection of the lamp. I see streaks of color also. There is a window off to the side and the sun is shining into the room through this window. When I place my hand between the CD and the window, some of the colored streaks disappear, and other colored streaks seem to appear. I am not seeing a direct reflection of the sun; it is off to the side. What is happening is that the grooves on the surface of the CD are each reflecting sunlight and acting as little transmitters. Each color in the spectrum corresponds to a particular frequency $\omega$ of light and at just the proper angle the spacing between the grooves on the CD leads to coherent transmission of the reflected light in the direction of my eye. The combination of frequency and spacing between the grooves determines what color I see and at what angle. When I reach over and tilt the CD off the table, the colors of the streaks change, because I have changed the spacing of the little transmitters, relative to my eye. An arrangement like this is called a *diffraction grating* and has many uses in physics. For a wonderful, and largely math-free, introduction to these ideas, see the book by Feynman [71].

## 5.4 Phase and Amplitude Modulation

In the previous section the signal broadcast from each of the antennas was the same. Now we look at what directionality can be obtained by using different amplitudes and phases at each of the antennas. Let the signal broadcast from the antenna at $n\Delta$ be

$$f_n(t) = |A_n| \cos(\omega t - \phi_n) = |A_n| \cos(\omega(t - \tau_n)),$$

for some amplitude $|A_n| > 0$ and phase $\phi_n = \omega\tau_n$. Now the signal received at $\mathbf{s}$ is proportional to

$$f(\mathbf{s},t) = \sum_{n=-N}^{N} |A_n| \cos(\omega(t - t_n - \tau_n)).$$

If we wish, we can repeat the calculations done earlier to see what the effect of the amplitude and phase changes is. Using complex notation simplifies things somewhat.

Let us consider a complex signal; suppose that the signal transmitted from the antenna at $n\Delta$ is $g_n(t) = |A_n| e^{i\omega(t - \tau_n)}$. Then, the signal received at location $\mathbf{s}$ is proportional to

$$g(\mathbf{s},t) = \sum_{n=-N}^{N} |A_n| e^{i\omega(t - t_n - \tau_n)}.$$

Then we have
$$g(\mathbf{s},t) = B(\theta) e^{i\omega(t - \frac{D}{c})}$$
for $A_n = |A_n| e^{-i\phi_n}$, $x = \frac{\omega\Delta}{c} \cos(\theta)$, and

$$B(\theta) = \sum_{n=-N}^{N} A_n e^{inx}.$$

Note that the complex amplitude function $B(\theta)$ depends on our choices of $N$ and $\Delta$ and takes the form of a finite Fourier series or DFT. We can design $B(\theta)$ to approximate the desired directionality by choosing the appropriate complex coefficients $A_n$ and selecting the amplitudes $|A_n|$ and phases $\phi_n$ accordingly. We can generalize further by allowing the antennas to be spaced irregularly along the east-west axis, or even distributed irregularly over a two-dimensional area on the ground.

## 5.5   Steering the Array

In our previous discussion, we selected $A_n = 1$ and $\phi_n = 0$ for all $n$ and saw that the maximum transmitted power was along the north-to-south axis. Suppose that we want to design a transmitting array that maximally concentrates signal power in another direction. Theoretically, we could physically rotate or steer the array until it ran along a different axis, and then proceed as before, with $A_n = 1$ and $\phi_n = 0$. This is not practical, in most cases. There is an alternative, fortunately. We can "steer" the array mathematically.

If $A_n = 1$, and

$$\phi_n = -\frac{n\Delta\omega}{c}\cos\alpha,$$

for some angle $\alpha$, then, for $x = \frac{\omega\Delta}{c}\cos(\theta)$, we have

$$B(\theta) = \sum_{n=-N}^{N} e^{inx} e^{i\phi_n} = \sum_{n=-N}^{N} e^{in\frac{\omega\Delta}{c}(\cos\theta - \cos\alpha)}.$$

The maximum absolute value of $B(\theta)$ occurs when $\cos\theta = \cos\alpha$, or when $\theta = \alpha$ or $\theta = -\alpha$. Now the greatest power is concentrated in these directions. The point here is that we have altered the directionality of the transmission, not by physically moving the array of antennas, but by changing the phases of the transmitted signals. This approach is sometimes called *phase steering*. The same basic idea applies when we are receiving signals, rather than sending them. In radar and sonar, the array of sensors is steered mathematically, by modifying the phases of the measured data, to focus the sensitivity of the detecting array in a particular direction.

## 5.6   Maximal Concentration in a Sector

In this section we take $\Delta = \frac{\pi c}{\omega}$, so that $\frac{\omega\Delta}{c} = \pi$. Suppose that we want to concentrate the transmitted power in the directions $\theta$ corresponding to $x = \frac{\omega\Delta}{c}\cos(\theta)$ in the sub-interval $[a, b]$ of the interval $[-\frac{\omega\Delta}{c}, \frac{\omega\Delta}{c}]$. Let $\mathbf{u} = (A_{-N}, ..., A_N)^T$ be the vector of coefficients for the function

$$B(x) = \sum_{n=-N}^{N} A_n e^{-inx}.$$

We want $|B(x)|$ to be concentrated in the interval $a \le x \le b$.

**Ex. 5.2** *Show that*

$$\frac{1}{2\pi} \int_{-\frac{\omega\Delta}{c}}^{\frac{\omega\Delta}{c}} |B(x)|^2 dx = \mathbf{u}^\dagger \mathbf{u},$$

*and*

$$\frac{1}{2\pi} \int_a^b |B(x)|^2 dx = \mathbf{u}^\dagger Q \mathbf{u},$$

*where Q is the matrix with entries*

$$Q_{mn} = \frac{1}{2\pi} \int_a^b \exp(i(m-n)x)\, dx.$$

Maximizing the concentration of power within the interval $[a, b]$ is then equivalent to finding the vector $\mathbf{u}$ that maximizes the ratio $\mathbf{u}^\dagger Q \mathbf{u} / \mathbf{u}^\dagger \mathbf{u}$. The matrix $Q$ is positive-definite, all its eigenvalues are positive, and the optimal $\mathbf{u}$ is the eigenvector of $Q$ associated with the largest eigenvalue. This largest eigenvalue is the desired ratio and is always less than one. As $N$ increases this ratio approaches one, for any fixed sub-interval $[a, b]$.

---

## 5.7 Scattering in Crystallography

When x-rays are passed through certain materials they are scattered, which means retransmitted in various directions. As W. L. Bragg discovered, by analyzing the distinctive pattern of the scattering the molecular structure of the material can be determined. This technique was used by Rosalind Franklin, a physicist at King's College, London, to analyze DNA and her work contributed greatly to the discovery, by Francis Crick and James Watson, of the double-helix structure of that molecule.

In 1964 the British scientist Dorothy Hodgkin won the Nobel Prize for her extension of this technique to reveal the structure of compounds more complex than any previously analyzed. Her most important work was on the structure of cholesterol, vitamin D, penicillin, vitamin $B_{12}$, and insulin, where she was able to uncover, by physical methods, chemical features not encountered before, and thereby to extend the bounds of chemistry itself. One of Dorothy Hodgkin's students at Oxford was Margaret Roberts, later Margaret Thatcher, Prime Minister of Great Britain throughout the 1980's.

In [101] Körner reveals how surprised he was when he heard that large amounts of computer time are spent by crystallographers computing Fourier transforms numerically. He goes on to describe this application.

The structure to be analyzed consists of some finite number of particles that will scatter in all directions any electromagnetic radiation that hits them. A beam of monochromatic light with unit strength and frequency $\omega$ is sent into the structure and the resulting scattered beams are measured at some number of observation points.

We say that the scattering particles are located in space at the points $\mathbf{r}_m$, $m = 1, ..., M$, and that the incoming light arrives as a planewave with wavevector $\mathbf{k}_0$. Then the planewave field generated by the incoming light is

$$g(\mathbf{s}, t) = e^{i\omega t} e^{i\mathbf{k}_0 \cdot \mathbf{s}}.$$

What is received at each $\mathbf{r}_m$ is then

$$g(\mathbf{r}_m, t) = e^{i\omega t} e^{i\mathbf{k}_0 \cdot \mathbf{r}_m}.$$

We observe the scattered signals at $\mathbf{s}$, where the retransmitted signal coming from $\mathbf{r}_m$ is

$$f(\mathbf{s}, t) = e^{i\omega t} e^{i\mathbf{k}_0 \cdot \mathbf{r}_m} e^{i\|\mathbf{s} - \mathbf{r}_m\|}.$$

When $\mathbf{s}$ is sufficiently remote from the scattering particles, the retransmitted signal from $\mathbf{r}_m$ arrives at $\mathbf{s}$ as a planewave with wavevector

$$\mathbf{k}_m = \frac{\omega}{c}(\mathbf{s} - \mathbf{r}_m)/\|\mathbf{s} - \mathbf{r}_m\|.$$

Therefore, at $\mathbf{s}$ we receive

$$u(\mathbf{s}, t) = e^{i\omega t} \sum_{m=1}^{M} e^{i\mathbf{k}_m \cdot \mathbf{s}}.$$

The objective is to determine the $\mathbf{k}_m$, which will then tell us the locations $\mathbf{r}_m$ of the scattering particles. To do this, we imagine an infinity of possible locations $\mathbf{r}$ for the particles and define $a(\mathbf{r}) = 1$ if $\mathbf{r} = \mathbf{r}_m$ for some $m$, and $a(\mathbf{r}) = 0$ otherwise. More precisely, we define $a(\mathbf{r})$ as a sum of unit-strength Dirac delta functions supported at the $\mathbf{r}_m$, a topic we shall deal with later. At each $\mathbf{r}$ we obtain (in theory) a value of the function $A(\mathbf{k})$, the Fourier transform of the function $a(\mathbf{r})$.

In practice, the crystallographers cannot measure the complex numbers $A(\mathbf{k})$, but only the magnitudes $|A(\mathbf{k})|$; the phase angle of $A(\mathbf{k})$ is lost. This presents the crystallographers with the *phase problem*, in which we must estimate a function from values of the magnitude of its Fourier transform. For a detailed discussion of the phase problem see Chapter 10.

In 1985, Hauptman and Karle won the Nobel Prize in Chemistry for developing a new method for finding $a(\mathbf{s})$ from measurements. Their technique is highly mathematical. It is comforting to know that, although there is no Nobel Prize in Mathematics, it is still possible to win the prize for doing mathematics.

# Chapter 6

## The Fourier Transform and Convolution Filtering

## 6.1   Chapter Summary

A major application of the Fourier transform is in the study of systems. We may think of a system as a device that accepts functions as input and produces functions as output. For example, the *differentiation system* accepts a differentiable function $f(x)$ as input and produces its derivative function $f'(x)$ as output. If the input is the function $f(x) = 5f_1(x) + 3f_2(x)$, then the output is $5f_1'(x) + 3f_2'(x)$; the differentiation system is *linear*. We shall describe systems algebraically by $h = Tf$, where $f$ is any input function, $h$ is the resulting output function from the system, and $T$ is the operator that represents the operation performed by the system on any input. For the differentiation system we would write the differentiation operator as $Tf = f'$.

## 6.2    Linear Filters

The system operator $T$ is *linear* if

$$T(af_1 + bf_2) = aT(f_1) + bT(f_2),$$

for any scalars $a$ and $b$ and functions $f_1$ and $f_2$. We shall be interested only in linear systems.

## 6.3    Shift-Invariant Filters

We denote by $S_a$ the system that shifts an input function by $a$; that is, if $f(x)$ is the input to system $S_a$, then $f(x - a)$ is the output. A system operator $T$ is said to be *shift-invariant* if

$$T(S_a(f)) = S_a(T(f)),$$

which means that, if input $f(x)$ leads to output $h(x)$, then input $f(x - a)$ leads to output $h(x - a)$; shifting the input just shifts the output. When the variable $x$ is time, we speak of *time-invariant* systems. When $T$ is a shift-invariant linear system operator we say that $T$ is a SILO.

## 6.4    Some Properties of a SILO

We show first that $(Tf)' = Tf'$. Suppose that $h(x) = (Tf)(x)$. For any $\Delta x$ we can write

$$f(x + \Delta x) = (S_{-\Delta x}f)(x)$$

and

$$(TS_{-\Delta x}f)(x) = (S_{-\Delta x}Tf)(x) = (S_{-\Delta x}h)(x) = h(x + \Delta x).$$

When the input to the system is

$$\frac{1}{\Delta x}\Big(f(x + \Delta x) - f(x)\Big),$$

the output is

$$\frac{1}{\Delta x}\Big(h(x + \Delta x) - h(x)\Big).$$

Now we take limits, as $\Delta x \to 0$, so that, assuming continuity, we can conclude that $Tf' = h'$. We apply this now to the case in which $f(x) = e^{-ix\omega}$ for some real constant $\omega$.

Since $f'(x) = -i\omega f(x)$ and $f(x) = \frac{i}{\omega} f'(x)$ in this case, we have

$$h(x) = (Tf)(x) = \frac{i}{\omega}(Tf')(x) = \frac{i}{\omega} h'(x),$$

so that

$$h'(x) = -i\omega h(x).$$

Solving this differential equation, we obtain

$$h(x) = ce^{-ix\omega},$$

for some constant $c$. Note that since the $c$ may vary when we vary the selected $\omega$, we must write $c = c(\omega)$. The main point here is that, when $T$ is a SILO and the input function is a complex exponential with frequency $\omega$, then the output is again a complex exponential with the same frequency $\omega$, multiplied by a complex number $c(\omega)$. This multiplication by $c(\omega)$ only modifies the amplitude and phase of the exponential function; it does not alter its frequency. So SILOs do not change the input frequencies, but only modify their strengths and phases.

**Ex. 6.1** *Let $T$ be a SILO. Show that $T$ is a convolution operator by showing that, for each input function $f$, the output function $h = Tf$ is the convolution of $f$ with $g$, where $g(x)$ is the inverse FT of the function $c(\omega)$ obtained above. Hint: Write the input function $f(x)$ as*

$$f(x) = \frac{1}{2\pi} \int_{-\infty}^{\infty} F(\omega)e^{-ix\omega} d\omega,$$

*and assume that*

$$(Tf)(x) = \frac{1}{2\pi} \int_{-\infty}^{\infty} F(\omega)(Te^{-ix\omega}) d\omega.$$

Now that we know that a SILO is a convolution filter, the obvious question to ask is What is $g(x)$? This is the *system identification* problem. One way to solve this problem is to consider what the output is when the input is the Heaviside function $u(x)$. In that case, we have

$$h(x) = \int_{-\infty}^{\infty} u(y)g(x-y)dy = \int_{0}^{\infty} g(x-y)dy = \int_{-\infty}^{x} g(t)dt.$$

Therefore, $h'(x) = g(x)$.

## 6.5    The Dirac Delta

The *Dirac delta*, denoted $\delta(x)$, is not truly a function. Its job is best described by its *sifting property*: for any fixed value of $x$,

$$f(x) = \int f(y)\delta(x-y)dy.$$

In order for the Dirac delta to perform the sifting operator on any $f(x)$ it would have to be zero, except at $x = 0$, where it would have to be infinitely large. It is possible to give a rigorous treatment of the Dirac delta, using *generalized functions*, but that is beyond the scope of this course. The Dirac delta is useful in our discussion of filters, which is why it is used.

## 6.6    The Impulse-Response Function

We can solve the system identification problem by seeing what the output is when the input is the Dirac delta; as we shall see, the output is $g(x)$; that is, $T\delta = g$. Since the SILO $T$ is a convolution operator, we know that

$$h(x) = \int_{-\infty}^{\infty} \delta(y)g(x-y)dy = g(x).$$

For this reason, the function $g(x)$ is called the *impulse-response function* of the system.

## 6.7    Using the Impulse-Response Function

Suppose now that we take as our input the function $f(x)$, but write it as

$$f(x) = \int f(y)\delta(x-y)dy.$$

Then, since $T$ is linear, and the integral is more or less a big sum, we have

$$T(f)(x) = \int f(y)T(\delta(x-y))dy = \int f(y)g(x-y)dy.$$

The function on the right side of this equation is the *convolution* of the functions $f$ and $g$, written $f * g$. This shows, as we have seen, that $T$ does its job by convolving any input function $f$ with its impulse-response function $g$, to get the output function $h = Tf = f * g$. It is useful to remember that order does not matter in convolution:

$$\int f(y)g(x-y)dy = \int g(y)f(x-y)dy.$$

## 6.8 The Filter Transfer Function

Now let us take as input the complex exponential $f(x) = e^{-ix\omega}$, where $\omega$ is fixed. Then the output is

$$h(x) = T(f)(x) = \int e^{-iy\omega}g(x-y)dy = \int g(y)e^{-i(x-y)\omega}dy = e^{-ix\omega}G(\omega),$$

where $G(\omega)$ is the Fourier transform of the impulse-response function $g(x)$; note that $G(\omega) = c(\omega)$ from Exercise 6.1. This tells us that when the input to $T$ is a complex exponential function with "frequency" $\omega$, the output is the same complex exponential function, the "frequency" is unchanged, but multiplied by a complex number $G(\omega)$. This multiplication by $G(\omega)$ can change both the amplitude and phase of the complex exponential, but the "frequency" $\omega$ does not change. In filtering, this function $G(\omega)$ is called the *transfer function* of the filter, or sometimes the *frequency-response function*.

## 6.9 The Multiplication Theorem for Convolution

Now let's take as input a function $f(x)$, but now write it, using Equation (2.7), as

$$f(x) = \frac{1}{2\pi} \int F(\omega)e^{-ix\omega}d\omega.$$

Then, taking the operator inside the integral, we find that the output is

$$h(x) = T(f)(x) = \frac{1}{2\pi} \int F(\omega)T(e^{-ix\omega})d\omega = \frac{1}{2\pi} \int e^{-ix\omega}F(\omega)G(\omega)d\omega.$$

But, from Equation (2.7), we know that

$$h(x) = \frac{1}{2\pi} \int e^{-ix\omega}H(\omega)d\omega.$$

This tells us that the Fourier transform $H(\omega)$ of the function $h = f * g$ is simply the product of $F(\omega)$ and $G(\omega)$; this is the most important property of convolution.

---

## 6.10 Summing Up

It is helpful to take stock of what we have just discovered:

1. if $h = T(f)$ then $h' = T(f')$;

2. $T(e^{-i\omega x}) = G(\omega)e^{-i\omega x}$;

3. writing

$$f(x) = \frac{1}{2\pi} \int F(\omega)e^{-i\omega x}d\omega,$$

we obtain

$$h(x) = (Tf)(x) = \frac{1}{2\pi} \int F(\omega)T(e^{-i\omega x})d\omega,$$

so that

$$h(x) = \frac{1}{2\pi} \int F(\omega)G(\omega)e^{-i\omega x}d\omega;$$

4. since we also have

$$h(x) = \frac{1}{2\pi} \int H(\omega)e^{-i\omega x}d\omega,$$

we can conclude that $H(\omega) = F(\omega)G(\omega)$;

5. if we define $g(x)$ to be $(T\delta)(x)$, then

$$g(x - y) = (T\delta)(x - y).$$

Writing

$$f(x) = \int f(y)\delta(x - y)dy,$$

we get

$$h(x) = (Tf)(x) = \int f(y)(T\delta)(x - y)dy = \int f(y)g(x - y)dy,$$

so that $h$ is the convolution of $f$ and $g$;

6. $g(x)$ is the inverse Fourier transform of $G(\omega)$.

## 6.11   A Question

Previously, we allowed the operator $T$ to move inside the integral. We know, however, that this is not always permissible. The differentiation operator $T = D$, with $D(f) = f'$, cannot always be moved inside the integral; as we learn in advanced calculus, we cannot always differentiate under the integral sign. This raises the interesting issue of how to represent the differentiation operator as a shift-invariant linear filter. In particular, what is the impulse-response function? The answer will involve the problem of differentiating the delta function, the Green's Function method for representing the inversion of linear differential operators, and generalized functions or distributions.

## 6.12   Band-Limiting

Suppose that $G(\omega) = \chi_\Omega(\omega)$. Then, if $F(\omega)$ is the Fourier transform of the input function, the Fourier transform of the output function $h(t)$ will be

$$H(\omega) = \begin{cases} F(\omega), \text{ if } |\omega| \leq \Omega ; \\ 0, \text{ if } |\omega| > \Omega . \end{cases}$$

The effect of the filter is to leave values $F(\omega)$ unchanged, if $|\omega| \leq \Omega$, and to replace $F(\omega)$ with zero, if $|\omega| > \Omega$. This is called *band-limiting*. Since the inverse Fourier transform of $G(\omega)$ is

$$g(t) = \frac{\sin(\Omega t)}{\pi t},$$

the band-limiting system can be described using convolution:

$$h(t) = \int f(s) \frac{\sin(\Omega(t - s))}{\pi(t - s)} ds.$$

# Chapter 7

# Infinite Sequences and Discrete Filters

## 7.1  Chapter Summary

Many textbooks on signal processing present filters in the context of infinite sequences. Although infinite sequences are no more realistic than functions $f(t)$ defined for all times $t$, they do simplify somewhat the discussion of filtering, particularly when it comes to the impulse response and to random signals. Systems that have as input and output infinite sequences are called *discrete* systems.

## 7.2  Shifting

We denote by $f = \{f_n\}_{n=-\infty}^{\infty}$ an infinite sequence. For a fixed integer $k$, the system that accepts $f$ as input and produces as output the shifted sequence $h = \{h_n = f_{n-k}\}$ is denoted $S_k$; therefore, we write $h = S_k f$.

## 7.3   Shift-Invariant Discrete Linear Systems

A discrete system $T$ is *linear* if

$$T(af^1 + bf^2) = aT(f^1) + bT(f^2),$$

for any infinite sequences $f^1$ and $f^2$ and scalars $a$ and $b$. As previously, a system $T$ is *shift-invariant* if $TS_k = S_kT$. This means that if input $f$ has output $h$, then input $S_kf$ has output $S_kh$; shifting the input by $k$ just shifts the output by $k$.

## 7.4   The Delta Sequence

The *delta sequence* $\delta = \{\delta_n\}$ has $\delta_0 = 1$ and $\delta_n = 0$, for $n$ not equal to zero. Then $S_k(\delta)$ is the sequence $S_k(\delta) = \{\delta_{n-k}\}$. For any sequence $f$ we have

$$f_n = \sum_{m=-\infty}^{\infty} f_m\delta_{n-m} = \sum_{m=-\infty}^{\infty} \delta_m f_{n-m}.$$

This means that we can write the sequence $f$ as an infinite sum of the sequences $S_m\delta$:

$$f = \sum_{m=-\infty}^{\infty} f_m S_m(\delta). \tag{7.1}$$

As in the continuous case, we use the delta sequence to understand better how a shift-invariant discrete linear system $T$ works.

## 7.5   The Discrete Impulse Response

We let $\delta$ be the input to the shift-invariant discrete linear system $T$, and denote the output sequence by $g = T(\delta)$. Now, for any input sequence

$f$ with $h = T(f)$, we write $f$ using Equation (7.1), so that

$$
\begin{aligned}
h &= T(f) = T\left(\sum_{m=-\infty}^{\infty} f_m S_m \delta\right) = \sum_{m=-\infty}^{\infty} f_m T S_m(\delta) \\
&= \sum_{m=-\infty}^{\infty} f_m S_m T(\delta) = \sum_{m=-\infty}^{\infty} f_m S_m(g).
\end{aligned}
$$

Therefore, we have

$$
h_n = \sum_{m=-\infty}^{\infty} f_m g_{n-m}, \tag{7.2}
$$

for each $n$. Equation (7.2) is the definition of *discrete convolution* or the *convolution of sequences*. This tells us that the output sequence $h = T(f)$ is the convolution of the input sequence $f$ with the impulse-response sequence $g$; that is, $h = T(f) = f * g$.

---

## 7.6 The Discrete Transfer Function

Associated with each $\omega$ in the interval $[0, 2\pi)$ we have the sequence $e_\omega = \{e^{-in\omega}\}_{n=-\infty}^{\infty}$; the minus sign in the exponent is just for notational convenience later. What happens when we let $f = e_\omega$ be the input to the system $T$? The output sequence $h$ will be the convolution of the sequence $e_\omega$ with the sequence $g$; that is,

$$
h_n = \sum_{m=-\infty}^{\infty} e^{-im\omega} g_{n-m} = \sum_{m=-\infty}^{\infty} g_m e^{-i(n-m)\omega} = e^{-in\omega} \sum_{m=-\infty}^{\infty} g_m e^{im\omega}.
$$

Defining

$$
G(\omega) = \sum_{m=-\infty}^{\infty} g_m e^{im\omega} \tag{7.3}
$$

for $0 \le \omega < 2\pi$, we can write

$$
h_n = e^{-in\omega} G(\omega),
$$

or

$$
h = T(e_\omega) = G(\omega) e_\omega.
$$

This tells us that when $e_\omega$ is the input, the output is a multiple of the input; the "frequency" $\omega$ has not changed, but the multiplication by $G(\omega)$ can alter the amplitude and phase of the complex-exponential sequence.

Notice that Equation (7.3) is the definition of the Fourier series associated with the sequence $g$ viewed as a sequence of Fourier coefficients. It follows that, once we have the function $G(\omega)$, we can recapture the original $g_n$ from the formula for Fourier coefficients:

$$g_n = \frac{1}{2\pi} \int_0^{2\pi} G(\omega)e^{-in\omega} d\omega.$$

## 7.7   Using Fourier Series

For any sequence $f = \{f_n\}$, we can define the function

$$F(\omega) = \sum_{n=-\infty}^{\infty} f_n e^{in\omega},$$

for $\omega$ in the interval $[0, 2\pi)$. Then each $f_n$ is a Fourier coefficient of $F(\omega)$ and we have

$$f_n = \frac{1}{2\pi} \int_0^{2\pi} F(\omega)e^{-in\omega} d\omega.$$

It follows that we can write

$$f = \frac{1}{2\pi} \int_0^{2\pi} F(\omega)e_\omega d\omega. \tag{7.4}$$

We interpret this as saying that the sequence $f$ is a superposition of the individual sequences $e_\omega$, with coefficients $F(\omega)$.

## 7.8   The Multiplication Theorem for Convolution

Now consider $f$ as the input to the system $T$, with $h = T(f)$ as output. Using Equation (7.4), we can write

$$h = T(f) = T\left(\frac{1}{2\pi} \int_0^{2\pi} F(\omega)e_\omega d\omega\right)$$

$$= \frac{1}{2\pi} \int_0^{2\pi} F(\omega)T(e_\omega) d\omega = \frac{1}{2\pi} \int_0^{2\pi} F(\omega)G(\omega)e_\omega d\omega.$$

But, applying Equation (7.4) to $h$, we have

$$h = \frac{1}{2\pi} \int_0^{2\pi} H(\omega) e_\omega d\omega.$$

It follows that $H(\omega) = F(\omega)G(\omega)$, which is analogous to what we found in the case of continuous systems. This tells us that the system $T$ works by multiplying the function $F(\omega)$ associated with the input by the transfer function $G(\omega)$, to get the function $H(\omega)$ associated with the output $h = T(f)$. In the next section we give an example.

---

## 7.9 The Three-Point Moving Average

We consider now the linear, shift-invariant system $T$ that performs the *three-point moving average* operation on any input sequence. Let $f$ be any input sequence. Then the output sequence is $h$ with

$$h_n = \frac{1}{3}(f_{n-1} + f_n + f_{n+1}).$$

The impulse-response sequence is $g$ with $g_{-1} = g_0 = g_1 = \frac{1}{3}$, and $g_n = 0$, otherwise.

To illustrate, for the input sequence with $f_n = 1$ for all $n$, the output is $h_n = 1$ for all $n$. For the input sequence

$$f = \{..., 3, 0, 0, 3, 0, 0, ...\},$$

the output $h$ is again the sequence $h_n = 1$ for all $n$. If our input is the difference of the previous two input sequences, that is, the input is $\{..., 2, -1, -1, 2, -1, -1, ...\}$, then the output is the sequence with all entries equal to zero.

The transfer function $G(\omega)$ is

$$G(\omega) = \frac{1}{3}(e^{i\omega} + 1 + e^{-i\omega}) = \frac{1}{3}(1 + 2\cos\omega).$$

The function $G(\omega)$ has a zero when $\cos\omega = -\frac{1}{2}$, or when $\omega = \frac{2\pi}{3}$ or $\omega = \frac{4\pi}{3}$. Notice that the sequence given by

$$f_n = \left(e^{i\frac{2\pi}{3}n} + e^{-i\frac{2\pi}{3}n}\right) = 2\cos\frac{2\pi}{3}n$$

is the sequence $\{..., 2, -1, -1, 2, -1, -1, ...\}$, which, as we have just seen, has as its output the zero sequence. We can say that the reason the output is zero is that the transfer function has a zero at $\omega = \frac{2\pi}{3}$ and at $\omega = \frac{4\pi}{3} = \frac{-2\pi}{3}$. Those complex-exponential components of the input sequence that correspond to values of $\omega$ where $G(\omega) = 0$ will be removed in the output. This is a useful role that filtering can play; we can *null out* an undesired complex-exponential component of an input signal by designing $G(\omega)$ to have a root at its frequency $\omega$.

## 7.10  Autocorrelation

If we take the input to our convolution filter to be the sequence $f$ related to the impulse-response sequence by

$$ f_n = \overline{g}_{-n}, $$

then the output sequence is $h$ with entries

$$ h_n = \sum_{k=-\infty}^{+\infty} g_k \overline{g_{k-n}} $$

and $H(\omega) = |G(\omega)|^2$. The sequence $h$ is called the *autocorrelation sequence* for $g$ and $|G(\omega)|^2$ is the *power spectrum* of $g$.

Autocorrelation sequences have special properties not shared with ordinary sequences, as the exercise below shows. The Cauchy Inequality is valid for infinite sequences: with the length of $g$ defined by

$$ \|g\| = \Big( \sum_{n=-\infty}^{+\infty} |g_n|^2 \Big)^{1/2} $$

and the inner product of any sequences $f$ and $g$ given by

$$ \langle f, g \rangle = \sum_{n=-\infty}^{+\infty} f_n \overline{g_n}, $$

we have

$$ |\langle f, g \rangle| \leq \|f\| \, \|g\|, $$

with equality if and only if $g$ is a constant multiple of $f$.

**Ex. 7.1** *Let $h$ be the autocorrelation sequence for $g$. Show that $h_{-n} = \overline{h_n}$ and $h_0 \geq |h_n|$ for all $n$.*

## 7.11   Stable Systems

An infinite sequence $f = \{f_n\}$ is called *bounded* if there is a constant $A > 0$ such that $|f_n| \le A$, for all $n$. The shift-invariant linear system with impulse-response sequence $g = T(\delta)$ is said to be *stable* [120] if the output sequence $h = \{h_n\}$ is bounded whenever the input sequence $f = \{f_n\}$ is. In Exercise 7.2 below we ask the reader to prove that, in order for the system to be stable, it is both necessary and sufficient that

$$\sum_{n=-\infty}^{\infty} |g_n| < +\infty.$$

Given a doubly infinite sequence, $g = \{g_n\}_{n=-\infty}^{+\infty}$, we associate with $g$ its *z-transform*, the function of the complex variable $z$ given by

$$G(z) = \sum_{n=-\infty}^{+\infty} g_n z^{-n}.$$

Doubly infinite series of this form are called *Laurent series* and occur in the representation of functions analytic in an annulus. Note that if we take $z = e^{-i\omega}$ then $G(z)$ becomes $G(\omega)$ as defined by Equation (7.3). The z-transform is a somewhat more flexible tool in that we are not restricted to those sequences $g$ for which the z-transform is defined for $z = e^{-i\omega}$.

**Ex. 7.2** *Show that the shift-invariant linear system with impulse-response sequence $g$ is stable if and only if*

$$\sum_{n=-\infty}^{+\infty} |g_n| < +\infty.$$

*Hint: If, on the contrary,*

$$\sum_{n=-\infty}^{+\infty} |g_n| = +\infty,$$

*consider as input the bounded sequence $f$ with*

$$f_n = \overline{g_{-n}}/|g_{-n}|$$

*and show that the output $h_0 = +\infty$.*

**Ex. 7.3** *Consider the linear system determined by the sequence $g_0 = 2$, $g_n = (\frac{1}{2})^{|n|}$, for $n \ne 0$. Show that this system is stable. Calculate the z-transform of $\{g_n\}$ and determine its region of convergence.*

## 7.12   Causal Filters

The shift-invariant linear system with impulse-response sequence $g$ is said to be a *causal system* if the sequence $\{g_n\}$ is itself *causal*; that is, $g_n = 0$ for $n < 0$. For causal systems the value of the output at $n$, that is, $h_n$, depends only on those input values $f_m$ for $m \leq n$. When the input is a time series, this says that the value of the output at any given time depends only on the value of the inputs up to that time, and not on future values of the input sequence. A number of important filters, such as band-limiting filters, are not causal and have to be approximated by causal filters to operate in real time.

**Ex. 7.4** *Show that the function $G(z) = (z - z_0)^{-1}$ is the z-transform of a causal sequence $g$, where $z_0$ is a fixed complex number. What is the region of convergence? Show that the resulting linear system is stable if and only if $|z_0| < 1$.*

# Chapter 8

## Convolution and the Vector DFT

## 8.1 Chapter Summary

Convolution is an important concept in signal processing and occurs in several distinct contexts. The simplest example of convolution is the nonperiodic convolution of finite vectors, which is what we do to the coefficients when we multiply two polynomials together. In Chapters 6 and 7 we considered the convolution of functions of a continuous variable and of infinite sequences. The reader may also recall an earlier encounter with convolution in a course on differential equations. In this chapter we shall discuss *nonperiodic convolution* and *periodic convolution* of vectors, with particular emphasis on the role of the vector DFT and the FFT algorithm.

## 8.2   Nonperiodic Convolution

Recall the algebra problem of multiplying one polynomial by another. Suppose

$$A(x) = a_0 + a_1 x + \ldots + a_M x^M$$

and

$$B(x) = b_0 + b_1 x + \ldots + b_N x^N.$$

Let $C(x) = A(x)B(x)$. With

$$C(x) = c_0 + c_1 x + \ldots + c_{M+N} x^{M+N},$$

each of the coefficients $c_j$, $j = 0, \ldots, M+N$, can be expressed in terms of the $a_m$ and $b_n$ (an easy exercise!). The vector $\mathbf{c} = (c_0, \ldots, c_{M+N})$ is called the *nonperiodic convolution* of the vectors $\mathbf{a} = (a_0, \ldots, a_M)$ and $\mathbf{b} = (b_0, \ldots, b_N)$. Nonperiodic convolution can be viewed as a particular case of periodic convolution, as we shall see.

## 8.3   The DFT as a Polynomial

Given the complex numbers $f_0, f_1, \ldots, f_{N-1}$, we form the vector $\mathbf{f} = (f_0, f_1, \ldots, f_{N-1})^T$. The DFT of the vector $\mathbf{f}$ is the function

$$DFT_{\mathbf{f}}(\omega) = \sum_{n=0}^{N-1} f_n e^{in\omega},$$

defined for $\omega$ in the interval $[0, 2\pi)$. Because $e^{in\omega} = (e^{i\omega})^n$, we can write the DFT as a polynomial

$$DFT_{\mathbf{f}}(\omega) = \sum_{n=0}^{N-1} f_n (e^{i\omega})^n.$$

If we have a second vector, say $\mathbf{d} = (d_0, d_1, \ldots, d_{N-1})^T$, then we define $DFT_{\mathbf{d}}(\omega)$ similarly. When we multiply $DFT_{\mathbf{f}}(\omega)$ by $DFT_{\mathbf{d}}(\omega)$, we are multiplying two polynomials together, so the result is a sum of powers of the form

$$c_0 + c_1 e^{i\omega} + c_2 (e^{i\omega})^2 + \ldots + c_{2N-2}(e^{i\omega})^{2N-2}, \tag{8.1}$$

for

$$c_j = f_0 d_j + f_1 d_{j-1} + \dots + f_j d_0.$$

This is *nonperiodic convolution* again. In the next section, we consider what happens when, instead of using arbitrary values of $\omega$, we consider only the $N$ special values $\omega_k = \frac{2\pi}{N}k$, $k = 0, 1, \dots, N-1$. Because of the periodicity of the complex exponential function, we have

$$(e^{i\omega_k})^{N+j} = (e^{i\omega_k})^j,$$

for each $k$. As a result, all the powers higher than $N-1$ that showed up in the previous multiplication in Equation (8.1) now become equal to lower powers, and the product now only has $N$ terms, instead of the $2N-1$ terms we got previously. When we calculate the coefficients of these powers, we find that we get more than we got when we did the nonperiodic convolution. Now what we get is called *periodic convolution*.

---

## 8.4 The Vector DFT and Periodic Convolution

As we just discussed, nonperiodic convolution is another way of looking at the multiplication of two polynomials. This relationship between convolution on the one hand and multiplication on the other is a fundamental aspect of convolution. Whenever we have a convolution we should ask what related mathematical objects are being multiplied. We ask this question now with regard to periodic convolution; the answer turns out to be the *vector discrete Fourier transform* (vDFT).

### 8.4.1 The Vector DFT

Let $\mathbf{f} = (f_0, f_1, \dots, f_{N-1})^T$ be a column vector whose entries are $N$ arbitrary complex numbers. For $k = 0, 1, \dots, N-1$, we let

$$F_k = \sum_{n=0}^{N-1} f_n e^{2\pi i k n/N} = DFT_{\mathbf{f}}(\omega_k). \tag{8.2}$$

Then we let $\mathbf{F} = (F_0, F_1, \dots, F_{N-1})^T$ be the column vector with the $N$ complex entries $F_k$. The vector $\mathbf{F}$ is called the *vector discrete Fourier transform* of the vector $\mathbf{f}$, and we denote it by $\mathbf{F} = vDFT_{\mathbf{f}}$.

The entries of the vector $\mathbf{F} = vDFT_{\mathbf{f}}$ are $N$ equi-spaced values of the function $DFT_{\mathbf{f}}(\omega)$. If the Fourier transform $F(\omega)$ is zero for $\omega$ outside the interval $[0, 2\pi]$, and $f_n = f(n)$, for $n = 0, 1, \dots, N-1$, then the entries of the vector $\mathbf{F}$ are $N$ estimated values of $F(\omega)$.

**Ex. 8.1** *Let $f_n$ be real, for each $n$. Show that $F_{N-k} = \overline{F_k}$, for each $k$.*

As we can see from Equation (8.2), there are $N$ multiplications involved in the calculation of each $F_k$, and there are $N$ values of $k$, so it would seem that, in order to calculate the vector DFT of $\mathbf{f}$, we need $N^2$ multiplications. In many applications, $N$ is quite large and calculating the vector $\mathbf{F}$ using the definition would be unrealistically time-consuming. The *fast Fourier transform* algorithm (FFT), to be discussed later, gives a quick way to calculate the vector $\mathbf{F}$ from the vector $\mathbf{f}$. The FFT, usually credited to Cooley and Tukey, was discovered in the mid-1960's and revolutionized signal and image processing.

### 8.4.2    Periodic Convolution

Given the $N$ by 1 vectors $\mathbf{f}$ and $\mathbf{d}$ with complex entries $f_n$ and $d_n$, respectively, we define a third $N$ by 1 vector $\mathbf{f} * \mathbf{d}$, the *periodic convolution* of $\mathbf{f}$ and $\mathbf{d}$, to have the entries

$$(\mathbf{f} * \mathbf{d})_n = f_0 d_n + f_1 d_{n-1} + ... + f_n d_0 + f_{n+1} d_{N-1} + ... + f_{N-1} d_{n+1}, \quad (8.3)$$

for $n = 0, 1, ..., N - 1$.

Notice that the term on the right side of Equation (8.3) is the sum of all products of entries, one from $\mathbf{f}$ and one from $\mathbf{d}$, where the sum of their respective indices is either $n$ or $n + N$. Periodic convolution is illustrated in Figure 8.1. The first exercise relates the periodic convolution to the vector DFT.

In the exercises that follow we investigate properties of the vector DFT and relate it to periodic convolution. It is not an exaggeration to say that these two exercises are the most important ones in signal processing. The first exercise establishes for finite vectors and periodic convolution a version of the multiplication theorems we saw earlier for continuous and discrete convolution.

**Ex. 8.2** *Let $\mathbf{F} = vDFT_{\mathbf{f}}$ and $\mathbf{D} = vDFT_{\mathbf{d}}$. Define a third vector $\mathbf{E}$ having for its kth entry $E_k = F_k D_k$, for $k = 0, ..., N - 1$. Show that $\mathbf{E}$ is the vDFT of the vector $\mathbf{f} * \mathbf{d}$.*

The vector $vDFT_{\mathbf{f}}$ can be obtained from the vector $\mathbf{f}$ by means of matrix multiplication by a certain matrix $G$, called the *DFT matrix*. The matrix $G$ has an inverse that is easily computed and can be used to go from $\mathbf{F} = vDFT_{\mathbf{f}}$ back to the original $\mathbf{f}$. The details are in Exercise 8.3.

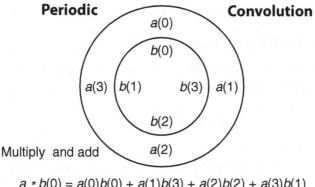

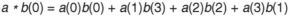

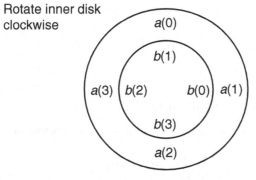

**FIGURE 8.1**: Periodic convolution of vectors $a = (a(0), a(1), a(2), a(3))$ and $b = (b(0), b(1), b(2), b(3))$.

**Ex. 8.3** *Let G be the N by N matrix whose entries are*

$$G_{jk} = e^{i(j-1)(k-1)2\pi/N}.$$

*The matrix G is sometimes called the DFT matrix. Show that the inverse of G is $G^{-1} = \frac{1}{N}G^{\dagger}$, where $G^{\dagger}$ is the conjugate transpose of the matrix G. Then $\mathbf{f} * \mathbf{d} = G^{-1}\mathbf{E} = \frac{1}{N}G^{\dagger}\mathbf{E}$.*

Every time I have taught this subject I have told my students that, if they learn nothing else in the course, they should understand the previous two exercises, which are fundamental in signal processing.

## 8.5   The vDFT of Sampled Data

For a doubly infinite sequence $\{f_n| -\infty < n < \infty\}$, the function of $F(\gamma)$ given by the infinite series

$$F(\gamma) = \sum_{n=-\infty}^{\infty} f_n e^{in\gamma} \qquad (8.4)$$

is sometimes called the *discrete-time Fourier transform* (DTFT) of the sequence, and the $f_n$ are called its *Fourier coefficients*. The function $F(\gamma)$ is $2\pi$-periodic, so we restrict our attention to the interval $0 \leq \gamma \leq 2\pi$. If we start with a function $F(\gamma)$, for $0 \leq \gamma \leq 2\pi$, we can find the Fourier coefficients by

$$f_n = \frac{1}{2\pi} \int_0^{2\pi} F(\gamma) e^{-i\gamma n} d\gamma. \qquad (8.5)$$

### 8.5.1   Superposition of Sinusoids

Equation (8.5) suggests a model for a function of a continuous variable $x$:

$$f(x) = \frac{1}{2\pi} \int_0^{2\pi} F(\gamma) e^{-i\gamma x} d\gamma.$$

The values $f_n$ then can be viewed as $f_n = f(n)$, that is, the $f_n$ are sampled values of the function $f(x)$, sampled at the points $x = n$. The function $F(\gamma)$ is now said to be the *spectrum* of the function $f(x)$. The function $f(x)$ is then viewed as a superposition of infinitely many simple functions, namely the complex exponentials or *sinusoidal functions* $e^{-i\gamma x}$, for values of $\gamma$ that lie in the interval $[0, 2\pi]$. The relative contribution of each $e^{-i\gamma x}$ to $f(x)$ is given by the complex number $\frac{1}{2\pi} F(\gamma)$.

### 8.5.2   Rescaling

In the model just discussed, we sampled the function $f(x)$ at the points $x = n$. In applications, the variable $x$ can have many meanings. In particular, $x$ is often time, denoted by the variable $t$. Then the variable $\gamma$ will be related to frequency. Depending on the application, the frequencies involved in the function $f(t)$ may be quite large numbers, or quite small ones; there is no reason to assume that they will all be in the interval $[0, 2\pi]$. For this reason, we have to modify our formulas.

Suppose that the function $g(t)$ is known to involve only frequencies in the interval $[0, \frac{2\pi}{\Delta}]$. Define $f(x) = g(x\Delta)$, so that

$$g(t) = f(t/\Delta) = \frac{1}{2\pi} \int_0^{2\pi} F(\gamma)e^{-i\gamma t/\Delta} d\gamma.$$

Introducing the variable $\omega = \gamma/\Delta$, and writing $G(\omega) = \Delta F(\omega\Delta)$, we get

$$g(t) = \frac{1}{2\pi} \int_0^{\frac{2\pi}{\Delta}} G(\omega)e^{-i\omega t} d\omega.$$

Now the typical problem is to estimate $G(\omega)$ from measurements of $g(t)$. Note that, using Equation (8.4), the function $G(\omega)$ can be written as follows:

$$G(\omega) = \Delta F(\omega\Delta) = \Delta \sum_{n=-\infty}^{\infty} f_n e^{in\omega\Delta},$$

so that

$$G(\omega) = \Delta \sum_{n=-\infty}^{\infty} g(n\Delta)e^{i(n\Delta)\omega}. \tag{8.6}$$

Note that this is what Shannon's Sampling Theorem tells us, and shows that the functions $G(\omega)$ and $g(t)$ can be completely recovered from the *infinite* sequence of samples $\{g(n\Delta)\}$, whenever $G(\omega)$ is zero outside an interval of total length $\frac{2\pi}{\Delta}$.

### 8.5.3 The Aliasing Problem

In the previous subsection, we assumed that we knew that the only frequencies involved in $g(t)$ were in the interval $[0, \frac{2\pi}{\Delta}]$, and that $\Delta$ was our sampling spacing. Notice that, given our data $g(n\Delta)$, it is impossible for us to distinguish a frequency $\omega$ from $\omega + \frac{2\pi k}{\Delta}$, for any integer $k$: for any integers $k$ and $n$ we have

$$e^{i(\omega + \frac{2\pi k}{\Delta})n\Delta} = e^{i\omega n\Delta}e^{2\pi ikn}.$$

### 8.5.4 The Discrete Fourier Transform

In practice, we will have only finitely many measurements $g(n\Delta)$; even these will typically be noisy, but we shall overlook this for now. Suppose our data is $g(n\Delta)$, for $n = 0, 1, ..., N-1$. For notational simplicity, we let $f_n = g(n\Delta)$. It seems reasonable, in this case, to base our estimate $\hat{G}(\omega)$ of $G(\omega)$ on Equation (8.6) and write

$$\hat{G}(\omega) = \Delta \sum_{n=0}^{N-1} g(n\Delta)e^{i(n\Delta)\omega}. \tag{8.7}$$

We shall call $\hat{G}(\omega)$ the DFT estimate of the function $G(\omega)$ and write

$$G_{DFT}(\omega) = \hat{G}(\omega);$$

it will be clear from the context that the DFT uses samples of $g(t)$ and estimates $G(\omega)$.

### 8.5.5   Calculating Values of the DFT

Suppose that we want to evaluate this estimate of $G(\omega)$ at the $N - 1$ points $\omega_k = \frac{2\pi k}{N\Delta}$, for $k = 0, 1, ..., N - 1$. Then we have

$$\hat{G}(\omega_k) = \Delta \sum_{n=0}^{N-1} g(n\Delta)e^{i(n\Delta)\frac{2\pi k}{N\Delta}} = \sum_{n=0}^{N-1} \Delta g(n\Delta)e^{2\pi ikn/N}.$$

Notice that this is the vector DFT entry $F_k$ for the choices $f_n = \Delta g(n\Delta)$.

To summarize, given the samples $g(n\Delta)$, for $n = 0, 1, ..., N - 1$, we can get the $N$ values $\hat{G}(\frac{2\pi k}{N\Delta})$ by taking the vector DFT of the vector $\mathbf{f} = (\Delta g(0), \Delta g(\Delta), ..., \Delta g((N - 1)\Delta))^T$. We would normally use the FFT algorithm to perform these calculations.

### 8.5.6   Zero-Padding

Suppose we simply want to graph the DFT estimate $G_{DFT}(\omega) = \hat{G}(\omega)$ on some uniform grid in the interval $[0, \frac{2\pi}{\Delta}]$, but want to use more than $N$ points in the grid. The FFT algorithm always gives us back a vector with the same number of entries as the one we begin with, so if we want to get, say, $M > N$ points in the grid, we need to give the FFT algorithm a vector with $M$ entries. We do this by *zero-padding*, that is, by taking as our input to the FFT algorithm the $M$ by 1 column vector

$$\mathbf{f} = (\Delta g(0), \Delta g(\Delta), ..., \Delta g((N - 1)\Delta), 0, 0, ..., 0)^T.$$

The resulting vector DFT $\mathbf{F}$ then has the entries

$$F_k = \Delta \sum_{n=0}^{N-1} g(n\Delta)e^{2\pi ikn/M},$$

for $k = 0, 1, ..., M - 1$; therefore, we have $F_k = \hat{G}(2\pi k/M)$.

### 8.5.7   What the vDFT Achieves

It is important to note that the values $F_k$ we calculate by applying the FFT algorithm to the sampled data $g(n\Delta)$ are not values of the function

$G(\omega)$, but of the estimate, $\hat{G}(\omega)$. Zero-padding allows us to use the FFT to see more of the values of $\hat{G}(\omega)$. It does not improve resolution, but simply shows us what is already present in the function $\hat{G}(\omega)$, which we may not have seen without the zero-padding. The FFT algorithm is most efficient when $N$ is a power of two, so it is common practice to zero-pad **f** using as $M$ the smallest power of two not less than $N$.

### 8.5.8   Terminology

In the signal processing literature no special name is given to what we call here $G_{DFT}(\omega)$, and the vector DFT of the data vector is called the DFT of the data. This is unfortunate, because the function of the continuous variable given in Equation (8.7) is the more fundamental entity, the vector DFT being merely the evaluation of that function at $N$ equi-spaced points. If we should wish to evaluate the $G_{DFT}(\omega)$ at $M > N$ equi-spaced points, say, for example, for the purpose of graphing the function, we would *zero-pad* the data vector, as we just discussed. The resulting vector DFT is not the same vector as the one obtained prior to zero-padding; it is not even the same size. But both of these vectors have, as their entries, values of the same function, $G_{DFT}(\omega)$.

---

## 8.6   Understanding the Vector DFT

Let $g(t)$ be the signal we are interested in. We sample the signal at the points $t = n\Delta$, for $n = 0, 1, ..., N-1$, to get our data values, which we label $f_n = g(n\Delta)$. To illustrate the significance of the vector DFT, we consider the simplest case, in which the signal $g(t)$ we are sampling is a single sinusoid.

Suppose that $g(t)$ is a complex exponential function with frequency the negative of $\omega_m = 2\pi m/N\Delta$; the reason for the negative is a technical one that we can safely ignore at this stage. Then

$$g(t) = e^{-i(2\pi m/N\Delta)t},$$

for some nonnegative integer $0 \le m \le N-1$. Our data is then

$$f_n = \Delta g(n\Delta) = \Delta e^{-i(2\pi m/N\Delta)n\Delta} = \Delta e^{-2\pi i m n/N}.$$

Now we calculate the components $F_k$ of the vector DFT. We have

$$F_k = \sum_{n=0}^{N-1} f_n e^{2\pi i k n/N} = \Delta \sum_{n=0}^{N-1} e^{2\pi i (k-m)/N}.$$

If $k = m$, then $F_m = N\Delta$, while, according to Equation 4.5, $F_k = 0$, for $k$ not equal to $m$. Let's try this on a more complicated signal.

Suppose now that our signal has the form

$$f(t) = \sum_{m=0}^{N-1} A_m e^{-2\pi imt/N\Delta}. \tag{8.8}$$

The data vector is now

$$f_n = \Delta \sum_{m=0}^{N-1} A_m e^{-2\pi imn/N}.$$

The entry $F_m$ of the vector DFT is now the sum of the values it would have if the signal had consisted only of the single sinusoid $e^{-i(2\pi m/N\Delta)t}$. As we just saw, all but one of these values would be zero, and so $F_m = N\Delta A_m$, and this holds for each $m = 0, 1, ..., N - 1$.

Summarizing, when the signal $f(t)$ is a sum of $N$ sinusoids, with the frequencies $\omega_k = 2\pi k/N\Delta$, for $k = 0, 1, ..., N-1$, and we sample at $t = n\Delta$, for $n = 0, 1, ..., N - 1$, the entries $F_k$ of the vector DFT are precisely $N\Delta$ times the corresponding amplitudes $A_k$. For this particular situation, calculating the vector DFT gives us the amplitudes of the different sinusoidal components of $f(t)$. We must remember, however, that this applies only to the case in which $f(t)$ has the form in Equation (8.8). In general, the entries of the vector DFT are to be understood as approximations, in the sense discussed above.

As mentioned previously, nonperiodic convolution is really a special case of periodic convolution. Extend the $M + 1$ by 1 vector $a$ to an $M + N + 1$ by 1 vector by appending $N$ zero entries; similarly, extend the vector $b$ to an $M + N + 1$ by 1 vector by appending zeros. The vector $c$ is now the periodic convolution of these extended vectors. Therefore, since we have an efficient algorithm for performing periodic convolution, namely the Fast Fourier Transform algorithm (FFT), we have a fast way to do the periodic (and thereby nonperiodic) convolution and polynomial multiplication.

---

## 8.7   The Fast Fourier Transform (FFT)

A fundamental problem in signal processing is to estimate finitely many values of the function $F(\omega)$ from finitely many values of its (inverse) Fourier transform, $f(t)$. As we have seen, the DFT arises in several ways in that estimation effort. The *Fast Fourier transform* (FFT), discovered in 1965 by Cooley and Tukey, is an important and efficient algorithm for calculating the vector DFT [58]. John Tukey has been quoted as saying that his main contribution to this discovery was the firm and often voiced belief that such an algorithm must exist.

## 8.7.1 Evaluating a Polynomial

To illustrate the main idea underlying the FFT, consider the problem of evaluating a real polynomial $P(x)$ at a point, say $x = c$. Let the polynomial be

$$P(x) = a_0 + a_1 x + a_2 x^2 + \ldots + a_{2K} x^{2K},$$

where $a_{2K}$ might be zero. Performing the evaluation efficiently by Horner's method,

$$P(c) = (((a_{2K}c + a_{2K-1})c + a_{2K-2})c + a_{2K-3})c + \ldots,$$

requires $2K$ multiplications, so the complexity is on the order of the degree of the polynomial being evaluated. But suppose we also want $P(-c)$. We can write

$$P(x) = (a_0 + a_2 x^2 + \ldots + a_{2K} x^{2K}) + x(a_1 + a_3 x^2 + \ldots + a_{2K-1} x^{2K-2})$$

or

$$P(x) = Q(x^2) + x R(x^2).$$

Therefore, we have $P(c) = Q(c^2) + cR(c^2)$ and $P(-c) = Q(c^2) - cR(c^2)$. If we evaluate $P(c)$ by evaluating $Q(c^2)$ and $R(c^2)$ separately, one more multiplication gives us $P(-c)$ as well. The FFT is based on repeated use of this idea, which turns out to be more powerful when we are using complex exponentials, because of their periodicity.

## 8.7.2 The DFT and Vector DFT

Suppose that the data are the samples $\{f(n\Delta), n = 1, \ldots, N\}$, where $\Delta > 0$ is the sampling increment or sampling spacing. The DFT estimate of $F(\omega)$ is the function $F_{DFT}(\omega)$, defined for $\omega$ in $[-\pi/\Delta, \pi/\Delta]$, and given by

$$F_{DFT}(\omega) = \Delta \sum_{n=1}^{N} f(n\Delta) e^{in\Delta\omega}.$$

The DFT estimate $F_{DFT}(\omega)$ is data consistent; its inverse Fourier-transform value at $t = n\Delta$ is $f(n\Delta)$ for $n = 1, \ldots, N$. The DFT is sometimes used in a slightly more general context in which the coefficients are not necessarily viewed as samples of a function $f(t)$.

Given the complex $N$-dimensional column vector $\mathbf{f} = (f_0, f_1, ..., f_{N-1})^T$, define the $DFT$ of vector $\mathbf{f}$ to be the function $DFT_{\mathbf{f}}(\omega)$, defined for $\omega$ in $[0, 2\pi)$, given by

$$DFT_{\mathbf{f}}(\omega) = \sum_{n=0}^{N-1} f_n e^{in\omega}.$$

Let $\mathbf{F}$ be the complex $N$-dimensional vector $\mathbf{F} = (F_0, F_1, ..., F_{N-1})^T$, where $F_k = DFT_{\mathbf{f}}(2\pi k/N), k = 0, 1, ..., N-1$. So the vector $\mathbf{F}$ consists of $N$ values of the function $DFT_{\mathbf{f}}$, taken at $N$ equi-spaced points $2\pi/N$ apart in $[0, 2\pi)$. From the formula for $DFT_{\mathbf{f}}$ we have, for $k = 0, 1, ..., N-1$,

$$F_k = F(2\pi k/N) = \sum_{n=0}^{N-1} f_n e^{2\pi i n k/N}. \tag{8.9}$$

To calculate a single $F_k$ requires $N$ multiplications; it would seem that to calculate all $N$ of them would require $N^2$ multiplications. However, using the FFT algorithm, we can calculate vector $\mathbf{F}$ in approximately $N \log_2(N)$ multiplications.

### 8.7.3 Exploiting Redundancy

Suppose that $N = 2M$ is even. We can rewrite Equation (8.9) as follows:

$$F_k = \sum_{m=0}^{M-1} f_{2m} e^{2\pi i (2m)k/N} + \sum_{m=0}^{M-1} f_{2m+1} e^{2\pi i (2m+1)k/N},$$

or, equivalently,

$$F_k = \sum_{m=0}^{M-1} f_{2m} e^{2\pi i m k/M} + e^{2\pi i k/N} \sum_{m=0}^{M-1} f_{2m+1} e^{2\pi i m k/M}. \tag{8.10}$$

Note that if $0 \le k \le M - 1$ then

$$F_{k+M} = \sum_{m=0}^{M-1} f_{2m} e^{2\pi i m k/M} - e^{2\pi i k/N} \sum_{m=0}^{M-1} f_{2m+1} e^{2\pi i m k/M}, \tag{8.11}$$

so there is no additional computational cost in calculating the second half of the entries of $\mathbf{F}$, once we have calculated the first half. The FFT is the algorithm that results when we take full advantage of the savings obtainable by splitting a DFT calculation into two similar calculations, each half the size.

We assume now that $N = 2^L$. Notice that if we use Equations (8.10) and (8.11) to calculate vector $\mathbf{F}$, the problem reduces to the calculation of two similar DFT evaluations, both involving half as many entries, followed by one multiplication for each of the $k$ between 0 and $M - 1$. We can split these in half as well. The FFT algorithm involves repeated splitting of the calculations of DFTs at each step into two similar DFTs, but with half the number of entries, followed by as many multiplications as there are entries in either one of these smaller DFTs. We use recursion to calculate the cost $C(N)$ of computing $\mathbf{F}$ using this FFT method. From Equation (8.10) we see that $C(N) = 2C(N/2) + (N/2)$. Applying the same reasoning to get $C(N/2) = 2C(N/4) + (N/4)$, we obtain

$$C(N) = 2C(N/2) + (N/2) = 4C(N/4) + 2(N/2) = \dots$$

$$= 2^L C(N/2^L) + L(N/2) = N + L(N/2).$$

Therefore, the cost required to calculate $\mathbf{F}$ is approximately $N \log_2 N$.

From our earlier discussion of discrete linear filters and convolution, we see that the FFT can be used to calculate the periodic convolution (or even the nonperiodic convolution) of finite length vectors.

Finally, let's return to the original context of estimating the Fourier transform $F(\omega)$ of function $f(t)$ from finitely many samples of $f(t)$. If we have $N$ equi-spaced samples, we can use them to form the vector $\mathbf{f}$ and perform the FFT algorithm to get vector $\mathbf{F}$ consisting of $N$ values of the DFT estimate of $F(\omega)$. It may happen that we wish to calculate more than $N$ values of the DFT estimate, perhaps to produce a smooth looking graph. We can still use the FFT, but we must trick it into thinking we have more data than the $N$ samples we really have. We do this by *zero-padding*. Instead of creating the $N$-dimensional vector $\mathbf{f}$, we make a longer vector by appending, say, $J$ zeros to the data, to make a vector that has dimension $N + J$. The DFT estimate is still the same function of $\omega$, since we have only included new zero coefficients as fake data; but, the FFT thinks we have $N + J$ data values, so it returns $N + J$ values of the DFT, at $N + J$ equi-spaced values of $\omega$ in $[0, 2\pi)$.

## 8.7.4 The Two-Dimensional Case

Suppose now that we have the data $\{f(m\Delta_x, n\Delta_y)\}$, for $m = 1, \dots, M$ and $n = 1, \dots, N$, where $\Delta_x > 0$ and $\Delta_y > 0$ are the sample spacings in the $x$ and $y$ directions, respectively. The DFT of this data is the function $F_{DFT}(\alpha, \beta)$ defined by

$$F_{DFT}(\alpha, \beta) = \Delta_x \Delta_y \sum_{m=1}^{M} \sum_{n=1}^{N} f(m\Delta_x, n\Delta_y) e^{i(\alpha m \Delta_x + \beta n \Delta_y)},$$

for $|\alpha| \leq \pi/\Delta_x$ and $|\beta| \leq \pi/\Delta_y$. The two-dimensional FFT produces $MN$ values of $F_{DFT}(\alpha, \beta)$ on a rectangular grid of $M$ equi-spaced values of $\alpha$

and $N$ equi-spaced values of $\beta$. This calculation proceeds as follows. First, for each fixed value of $n$, a FFT of the $M$ data points $\{f(m\Delta_x, n\Delta_y)\}, m = 1, ..., M$ is calculated, producing a function, say $G(\alpha_m, n\Delta_y)$, of $M$ equi-spaced values of $\alpha$ and the $N$ equi-spaced values $n\Delta_y$. Then, for each of the $M$ equi-spaced values of $\alpha$, the FFT is applied to the $N$ values $G(\alpha_m, n\Delta_y), n = 1, ..., N$, to produce the final result.

# Chapter 9

## Plane-Wave Propagation

## 9.1 Chapter Summary

We have seen how the Fourier transform arises naturally as we analyze the signals received in the far field from an array of transmitters or reflectors. In this chapter we describe the role played by the wave equation in remote sensing, focusing on plane-wave solutions. We shall consider this

133

topic in more detail in Chapter 24. We restrict our attention here to single-frequency, or narrow-band, signals. We begin with a simple illustration of some of the issues we deal with in greater detail later in this chapter.

---

## 9.2 The Bobbing Boats

Imagine a large swimming pool in which there are several toy boats arrayed in a straight line. Although we use Figure 9.1 for a slightly different purpose elsewhere, for now we can imagine that the black dots in that figure represent our toy boats. Far across the pool, someone is slapping the water repeatedly, generating waves that proceed outward, in essentially concentric circles, across the pool. By the time the waves reach the boats, the circular shape has flattened out so that the wavefronts are essentially straight lines. The straight lines in Figure 9.1 can represent these wavefronts.

As the wavefronts reach the boats, the boats bob up and down. If the lines of the wavefronts were oriented parallel to the line of the boats, then the boats would bob up and down in unison. When the wavefronts come in at some angle, as shown in the figure, the boats will bob up and down *out of sync* with one another, generally. By measuring the time it takes for the peak to travel from one boat to the next, we can estimate the angle of arrival of the wavefronts. This leads to two questions:

1. Is it possible to get the boats to bob up and down in unison, even though the wavefronts arrive at an angle, as shown in the figure?

2. Is it possible for wavefronts corresponding to two different angles of arrival to affect the boats in the same way, so that we cannot tell which of the two angles is the real one?

We need a bit of mathematical notation. We let the distance from each boat to the ones on both sides be a constant distance $\Delta$. We assume that the water is slapped $f$ times per second, so $f$ is the *frequency*, in units of cycles per second. As the wavefronts move out across the pool, the distance from one peak to the next is called the *wavelength*, denoted $\lambda$. The product $\lambda f$ is the *speed of propagation* $c$; so $\lambda f = c$. As the frequency changes, so does the wavelength, while the speed of propagation, which depends solely on the depth of the pool, remains constant. The angle $\theta$ measures the tilt between the line of the wavefronts and the line of the boats, so that $\theta = 0$ indicates that these wavefront lines are parallel to the line of the boats, while $\theta = \frac{\pi}{2}$ indicates that the wavefront lines are perpendicular to the line of the boats.

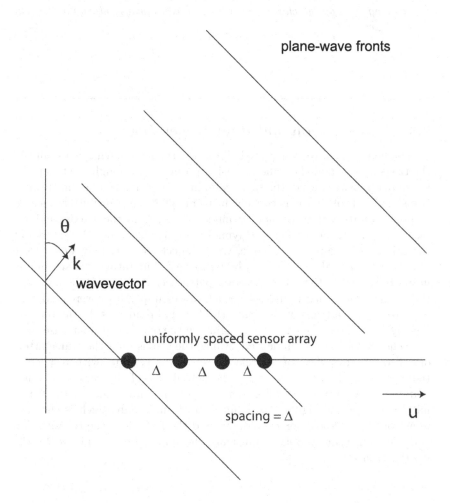

**FIGURE 9.1**: A uniform line array sensing a plane-wave field.

**Ex. 9.1** *Let the angle $\theta$ be arbitrary, but fixed, and let $\Delta$ be fixed. Can we select the frequency $f$ in such a way that we can make all the boats bob up and down in unison?*

**Ex. 9.2** *Suppose now that the frequency $f$ is fixed, but we are free to alter the spacing $\Delta$. Can we choose $\Delta$ so that we can always determine the true angle of arrival?*

---

## 9.3    Transmission and Remote Sensing

For pedagogical reasons, we shall discuss separately what we shall call the transmission and the remote-sensing problems, although the two problems are opposite sides of the same coin, in a sense. In the one-dimensional transmission problem, it is convenient to imagine the transmitters located at points $(x, 0)$ within a bounded interval $[-A, A]$ of the $x$-axis, and the measurements taken at points $P$ lying on a circle of radius $D$, centered at the origin. The radius $D$ is large, with respect to $A$. It may well be the case that no actual sensing is to be performed, but rather, we are simply interested in what the received signal pattern is at points $P$ distant from the transmitters. Such would be the case, for example, if we were analyzing or constructing a transmission pattern of radio broadcasts. In the remote-sensing problem, in contrast, we imagine, in the one-dimensional case, that our sensors occupy a bounded interval of the $x$-axis, and the transmitters or reflectors are points of a circle whose radius is large, with respect to the size of the bounded interval. The actual size of the radius does not matter and we are interested in determining the amplitudes of the transmitted or reflected signals, as a function of angle only. Such is the case in astronomy, far-field sonar or radar, and the like. Both the transmission and remote-sensing problems illustrate the important role played by the Fourier transform.

---

## 9.4    The Transmission Problem

We identify two distinct transmission problems: the direct problem and the inverse problem. In the direct transmission problem, we wish to determine the far-field pattern, given the complex amplitudes of the transmitted signals. In the inverse transmission problem, the array of transmitters or reflectors is the object of interest; we are given, or we measure, the far-field pattern and wish to determine the amplitudes. For simplicity, we consider only single-frequency signals.

We suppose that each point $x$ in the interval $[-A, A]$ transmits the signal $f(x)e^{i\omega t}$, where $f(x)$ is the complex amplitude of the signal and $\omega > 0$ is the common fixed frequency of the signals. Let $D > 0$ be large, with respect to $A$, and consider the signal received at each point $P$ given in polar coordinates by $P = (D, \theta)$. The distance from $(x, 0)$ to $P$ is approximately $D - x\cos\theta$, so that, at time $t$, the point $P$ receives from $(x, 0)$ the signal $f(x)e^{i\omega(t-(D-x\cos\theta)/c)}$, where $c$ is the propagation speed. Therefore, the combined signal received at $P$ is

$$B(P, t) = e^{i\omega t}e^{-i\omega D/c}\int_{-A}^{A} f(x)e^{ix\frac{\omega \cos\theta}{c}}\,dx.$$

The integral term, which gives the far-field pattern of the transmission, is

$$F(\frac{\omega \cos\theta}{c}) = \int_{-A}^{A} f(x)e^{ix\frac{\omega \cos\theta}{c}}\,dx,$$

where $F(\gamma)$ is the Fourier transform of $f(x)$, given by

$$F(\gamma) = \int_{-A}^{A} f(x)e^{ix\gamma}\,dx.$$

How $F(\frac{\omega \cos\theta}{c})$ behaves, as a function of $\theta$, as we change $A$ and $\omega$, is discussed in some detail in the chapter on direct transmission.

Consider, for example, the function $f(x) = 1$, for $|x| \le A$, and $f(x) = 0$, otherwise. The Fourier transform of $f(x)$ is

$$F(\gamma) = 2A\mathrm{sinc}(A\gamma),$$

where $\mathrm{sinc}(t)$ is defined to be

$$\mathrm{sinc}(t) = \frac{\sin(t)}{t},$$

for $t \ne 0$, and $\mathrm{sinc}(0) = 1$. Then $F(\frac{\omega \cos\theta}{c}) = 2A$ when $\cos\theta = 0$, so when $\theta = \frac{\pi}{2}$ and $\theta = \frac{3\pi}{2}$. We will have $F(\frac{\omega \cos\theta}{c}) = 0$ when $A\frac{\omega \cos\theta}{c} = \pi$, or $\cos\theta = \frac{\pi c}{A\omega}$. Therefore, the transmission pattern has no nulls if $\frac{\pi c}{A\omega} > 1$. In order for the transmission pattern to have nulls, we need $A > \frac{\lambda}{2}$, where $\lambda = \frac{2\pi c}{\omega}$ is the wavelength. This rather counterintuitive fact, namely that we need more signals transmitted in order to receive less at certain locations, illustrates the phenomenon of destructive interference.

## 9.5   Reciprocity

For certain remote-sensing applications, such as sonar and radar array processing and astronomy, it is convenient to switch the roles of sender and receiver. Imagine that superimposed plane-wave fields are sensed at points within some bounded region of the interior of the sphere, having been transmitted or reflected from the points $P$ on the surface of a sphere whose radius $D$ is large with respect to the bounded region. The *reciprocity principle* tells us that the same mathematical relation holds between points $P$ and $(x, 0)$, regardless of which is the sender and which the receiver. Consequently, the data obtained at the points $(x, 0)$ are then values of the inverse Fourier transform of the function describing the amplitude of the signal sent from each point $P$.

## 9.6   Remote Sensing

A basic problem in remote sensing is to determine the nature of a distant object by measuring signals transmitted by or reflected from that object. If the object of interest is sufficiently remote, that is, is in the far field, the data we obtain by sampling the propagating spatio-temporal field is related, approximately, to what we want by *Fourier transformation*. The problem is then to estimate a function from finitely many (usually noisy) values of its *Fourier transform*. The application we consider here is a common one of remote-sensing of transmitted or reflected waves propagating from distant sources. Examples include optical imaging of planets and asteroids using reflected sunlight, radio-astronomy imaging of distant sources of radio waves, active and passive sonar, and radar imaging.

## 9.7   The Wave Equation

In many areas of remote sensing, what we measure are the fluctuations in time of an electromagnetic or acoustic field. Such fields are described mathematically as solutions of certain partial differential equations, such as the *wave equation*. A function $u(x, y, z, t)$ is said to satisfy the *three-*

*dimensional wave equation* if

$$u_{tt} = c^2(u_{xx} + u_{yy} + u_{zz}) = c^2\nabla^2 u,$$

where $u_{tt}$ denotes the second partial derivative of $u$ with respect to the time variable $t$ twice and $c > 0$ is the (constant) speed of propagation. More complicated versions of the wave equation permit the speed of propagation $c$ to vary with the spatial variables $x, y, z$, but we shall not consider that here.

We use the method of *separation of variables* at this point, to get some idea about the nature of solutions of the wave equation. Assume, for the moment, that the solution $u(t, x, y, z)$ has the simple form

$$u(t, x, y, z) = g(t)f(x, y, z).$$

Inserting this separated form into the wave equation, we get

$$g''(t)f(x, y, z) = c^2 g(t)\nabla^2 f(x, y, z)$$

or

$$g''(t)/g(t) = c^2\nabla^2 f(x, y, z)/f(x, y, z).$$

The function on the left is independent of the spatial variables, while the one on the right is independent of the time variable; consequently, they must both equal the same constant, which we denote $-\omega^2$. From this we have two separate equations,

$$g''(t) + \omega^2 g(t) = 0, \tag{9.1}$$

and

$$\nabla^2 f(x, y, z) + \frac{\omega^2}{c^2} f(x, y, z) = 0. \tag{9.2}$$

Equation (9.2) is the *Helmholtz equation*.

Equation (9.1) has for its solutions the functions $g(t) = \cos(\omega t)$ and $\sin(\omega t)$, or, in complex form, the complex exponential functions $g(t) = e^{i\omega t}$ and $g(t) = e^{-i\omega t}$. Functions $u(t, x, y, z) = g(t)f(x, y, z)$ with such time dependence are called *time-harmonic* solutions.

In three-dimensional spherical coordinates with $r = \sqrt{x^2 + y^2 + z^2}$ a radial function $u(r, t)$ satisfies the wave equation if

$$u_{tt} = c^2\left(u_{rr} + \frac{2}{r}u_r\right).$$

**Ex. 9.3** *Show that the radial function $u(r, t) = \frac{1}{r}h(r - ct)$ satisfies the wave equation for any twice differentiable function $h$.*

## 9.8 Plane-Wave Solutions

Suppose that, beginning at time $t = 0$, there is a localized disturbance. As time passes, that disturbance spreads out spherically. When the radius of the sphere is very large, the surface of the sphere appears planar, to an observer on that surface, who is said then to be in the far field. This motivates the study of solutions of the wave equation that are constant on planes; the so-called *plane-wave solutions*.

**Ex. 9.4** *Let* $\mathbf{s} = (x, y, z)$ *and* $u(\mathbf{s}, t) = u(x, y, z, t) = e^{i\omega t} e^{i\mathbf{k} \cdot \mathbf{s}}$. *Show that* $u$ *satisfies the wave equation* $u_{tt} = c^2 \nabla^2 u$ *for any real vector* $\mathbf{k}$, *so long as* $||\mathbf{k}||^2 = \omega^2/c^2$. *This solution is a plane wave associated with frequency* $\omega$ *and wavevector* $\mathbf{k}$; *at any fixed time the function* $u(\mathbf{s}, t)$ *is constant on any plane in three-dimensional space having* $\mathbf{k}$ *as a normal vector.*

In radar and sonar, the field $u(\mathbf{s}, t)$ being sampled is usually viewed as a discrete or continuous superposition of plane-wave solutions with various amplitudes, frequencies, and wavevectors. We sample the field at various spatial locations $\mathbf{s}$, for various times $t$. Here we simplify the situation a bit by assuming that all the plane-wave solutions are associated with the same frequency, $\omega$. If not, we can perform an FFT on the functions of time received at each sensor location $\mathbf{s}$ and keep only the value associated with the desired frequency $\omega$.

## 9.9 Superposition and the Fourier Transform

In the continuous superposition model, the field is a superposition of plane-wave solutions

$$u(\mathbf{s}, t) = e^{i\omega t} \int F(\mathbf{k}) e^{i\mathbf{k} \cdot \mathbf{s}} d\mathbf{k}.$$

Our measurements at the sensor locations $\mathbf{s}$ give us the values

$$f(\mathbf{s}) = \int F(\mathbf{k}) e^{i\mathbf{k} \cdot \mathbf{s}} d\mathbf{k}. \tag{9.3}$$

The data are then Fourier transform values of the complex function $F(\mathbf{k})$; $F(\mathbf{k})$ is defined for all three-dimensional real vectors $\mathbf{k}$, but is zero, in theory, at least, for those $\mathbf{k}$ whose squared length $||\mathbf{k}||^2$ is not equal to $\omega^2/c^2$.

Our goal is then to estimate $F(\mathbf{k})$ from measured values of its Fourier transform. Since each $\mathbf{k}$ is a normal vector for its plane-wave field component, determining the value of $F(\mathbf{k})$ will tell us the strength of the plane-wave component coming from the direction $\mathbf{k}$.

### 9.9.1 The Spherical Model

We can imagine that the sources of the plane-wave fields are the points $P$ that lie on the surface of a large sphere centered at the origin. For each $P$, the ray from the origin to $P$ is parallel to some wavevector $\mathbf{k}$. The function $F(\mathbf{k})$ can then be viewed as a function $F(P)$ of the points $P$. Our measurements will be taken at points $\mathbf{s}$ inside this sphere. The radius of the sphere is assumed to be orders of magnitude larger than the distance between sensors. The situation is that of astronomical observation of the heavens using ground-based antennas. The sources of the optical or electromagnetic signals reaching the antennas are viewed as lying on a large sphere surrounding the earth. Distance to the sources is not considered now, and all we are interested in are the amplitudes $F(\mathbf{k})$ of the fields associated with each direction $\mathbf{k}$.

## 9.10 Sensor Arrays

In some applications the sensor locations are essentially arbitrary, while in others their locations are carefully chosen. Sometimes, the sensors are collinear, as in sonar towed arrays. Figure 9.1 illustrates a line array.

### 9.10.1 The Two-Dimensional Array

Suppose now that the sensors are in locations $\mathbf{s} = (x, y, 0)$, for various $x$ and $y$; then we have a *planar array* of sensors. Then the dot product $\mathbf{s} \cdot \mathbf{k}$ that occurs in Equation (9.3) is

$$\mathbf{s} \cdot \mathbf{k} = xk_1 + yk_2;$$

we cannot *see* the third component, $k_3$. However, since we know the size of the vector $\mathbf{k}$, we can determine $|k_3|$. The only ambiguity that remains is that we cannot distinguish sources on the upper hemisphere from those on the lower one. In most cases, such as astronomy, it is obvious in which hemisphere the sources lie, so the ambiguity is resolved.

The function $F(\mathbf{k})$ can then be viewed as $F(k_1, k_2)$, a function of the two variables $k_1$ and $k_2$. Our measurements give us values of $f(x, y)$, the two-dimensional Fourier transform of $F(k_1, k_2)$. Because of the limitation $||\mathbf{k}|| = \frac{\omega}{c}$, the function $F(k_1, k_2)$ has bounded support. Consequently, its Fourier transform cannot have bounded support. As a result, we can never have all the values of $f(x, y)$, and so cannot hope to reconstruct $F(k_1, k_2)$ exactly, even for noise-free data.

## 9.10.2   The One-Dimensional Array

If the sensors are located at points $\mathbf{s}$ having the form $\mathbf{s} = (x, 0, 0)$, then we have a *line array* of sensors. The dot product in Equation (9.3) becomes

$$\mathbf{s} \cdot \mathbf{k} = xk_1.$$

Now the ambiguity is greater than in the planar array case. Once we have $k_1$, we know that

$$k_2^2 + k_3^2 = \left(\frac{\omega}{c}\right)^2 - k_1^2,$$

which describes points $P$ lying on a circle on the surface of the distant sphere, with the vector $(k_1, 0, 0)$ pointing at the center of the circle. It is said then that we have a *cone of ambiguity*. One way to resolve the situation is to assume $k_3 = 0$; then $|k_2|$ can be determined and we have remaining only the ambiguity involving the sign of $k_2$. Once again, in many applications, this remaining ambiguity can be resolved by other means.

Once we have resolved any ambiguity, we can view the function $F(\mathbf{k})$ as $F(k_1)$, a function of the single variable $k_1$. Our measurements give us values of $f(x)$, the Fourier transform of $F(k_1)$. As in the two-dimensional case, the restriction on the size of the vectors $\mathbf{k}$ means that the function $F(k_1)$ has bounded support. Consequently, its Fourier transform, $f(x)$, cannot have bounded support. Therefore, we shall never have all of $f(x)$, and so cannot hope to reconstruct $F(k_1)$ exactly, even for noise-free data.

## 9.10.3   Limited Aperture

In both the one- and two-dimensional problems, the sensors will be placed within some bounded region, such as $|x| \le A$, $|y| \le B$ for the two-dimensional problem, or $|x| \le A$ for the one-dimensional case. These bounded regions are the *apertures* of the arrays. The larger these apertures are, in units of the wavelength, the better the resolution of the reconstructions.

In digital array processing there are only finitely many sensors, which then places added limitations on our ability to reconstruct the field amplitude function $F(\mathbf{k})$.

## 9.11   Sampling

In the one-dimensional case, the signal received at the point $(x, 0, 0)$ is essentially the inverse Fourier transform $f(x)$ of the function $F(k_1)$; for notational simplicity, we write $k = k_1$. The $F(k)$ supported on a bounded interval $|k| \leq \frac{\omega}{c}$, so $f(x)$ cannot have bounded support. As we noted earlier, to determine $F(k)$ exactly, we would need measurements of $f(x)$ on an unbounded set. But, which unbounded set?

Because the function $F(k)$ is zero outside the interval $[-\frac{\omega}{c}, \frac{\omega}{c}]$, the function $f(x)$ is *band-limited*. The *Nyquist spacing* in the variable $x$ is therefore

$$\Delta_x = \frac{\pi c}{\omega}.$$

The wavelength $\lambda$ associated with the frequency $\omega$ is defined to be

$$\lambda = \frac{2\pi c}{\omega},$$

so that

$$\Delta_x = \frac{\lambda}{2}.$$

The significance of the Nyquist spacing comes from *Shannon's Sampling Theorem*, which says that if we have the values $f(m\Delta_x)$, for all integers $m$, then we have enough information to recover $F(k)$ exactly. In practice, of course, this is never the case.

## 9.12   The Limited-Aperture Problem

In the remote-sensing problem, our measurements at points $(x, 0, 0)$ in the far field give us the values $f(x)$. Suppose now that we are able to take measurements only for limited values of $x$, say for $|x| \leq A$; then $2A$ is the *aperture* of our antenna or array of sensors. We describe this by saying that we have available measurements of $f(x)h(x)$, where $h(x) = \chi_A(x) = 1$, for $|x| \leq A$, and zero otherwise. So, in addition to describing blurring and low-pass filtering, the convolution-filter model can also be used to model the limited-aperture problem. As in the low-pass case, the limited-aperture problem can be attacked using extrapolation, but with the same sort of risks described for the low-pass case. A much different approach is to increase the aperture by physically moving the array of sensors, as in *synthetic aperture radar* (SAR).

Returning to the far-field remote-sensing model, if we have Fourier transform data only for $|x| \leq A$, then we have $f(x)$ for $|x| \leq A$. Using $h(x) = \chi_A(x)$ to describe the limited aperture of the system, the point-spread function is $H(\gamma) = 2A\mathrm{sinc}(\gamma A)$, the Fourier transform of $h(x)$. The first zeros of the numerator occur at $|\gamma| = \frac{\pi}{A}$, so the main lobe of the point-spread function has width $\frac{2\pi}{A}$. For this reason, the resolution of such a limited-aperture imaging system is said to be on the order of $\frac{1}{A}$. Since $|k| \leq \frac{\omega}{c}$, we can write $k = \frac{\omega}{c}\sin\theta$, where $\theta$ denotes the angle between the positive $y$-axis and the vector $\mathbf{k} = (k_1, k_2, 0)$; that is, $\theta$ points in the direction of the point $P$ associated with the wavevector $\mathbf{k}$. The resolution, as measured by the width of the main lobe of the point-spread function $H(\gamma)$, in units of $k$, is $\frac{2\pi}{A}$, but, the angular resolution will depend also on the frequency $\omega$. Since $k = \frac{2\pi}{\lambda}\sin\theta$, a distance of one unit in $k$ may correspond to a large change in $\theta$ when $\omega$ is large, but only to a relatively small change in $\theta$ when $\omega$ is small. For this reason, the aperture of the array is usually measured in units of the wavelength; an aperture of $A = 5$ meters may be acceptable if the frequency is high, so that the wavelength is small, but not if the radiation is in the one-meter-wavelength range.

---

## 9.13   Resolution

If $F(k) = \delta(k)$ and $h(x) = \chi_A(x)$ describes the aperture-limitation of the imaging system, then the point-spread function is $H(\gamma) = 2A\mathrm{sinc}(\gamma A)$. The maximum of $H(\gamma)$ still occurs at $\gamma = 0$, but the main lobe of $H(\gamma)$ extends from $-\frac{\pi}{A}$ to $\frac{\pi}{A}$; the point source has been spread out. If the point-source object shifts, so that $F(k) = \delta(k - a)$, then the reconstructed image of the object is $H(k - a)$, so the peak is still in the proper place. If we know *a priori* that the object is a single point source, but we do not know its location, the spreading of the point poses no problem; we simply look for the maximum in the reconstructed image. Problems arise when the object contains several point sources, or when we do not know *a priori* what we are looking at, or when the object contains no point sources, but is just a continuous distribution.

Suppose that $F(k) = \delta(k - a) + \delta(k - b)$; that is, the object consists of two point sources. Then Fourier transformation of the aperture-limited data leads to the reconstructed image

$$R(k) = 2A\Big(\mathrm{sinc}(A(k - a)) + \mathrm{sinc}(A(k - b))\Big).$$

If $|b - a|$ is large enough, $R(k)$ will have two distinct maxima, at approximately $k = a$ and $k = b$, respectively. For this to happen, we need $\pi/A$, half the width of the main lobe of the function $\text{sinc}(Ak)$, to be less than $|b - a|$. In other words, to resolve the two point sources a distance $|b - a|$ apart, we need $A \geq \pi/|b - a|$. However, if $|b - a|$ is too small, the distinct maxima merge into one, at $k = \frac{a+b}{2}$ and resolution will be lost. How small is too small will depend on both $A$ and $\omega$.

Suppose now that $F(k) = \delta(k - a)$, but we do not know *a priori* that the object is a single point source. We calculate

$$R(k) = H(k - a) = 2A\text{sinc}(A(k - a))$$

and use this function as our reconstructed image of the object, for all $k$. What we see when we look at $R(k)$ for some $k = b \neq a$ is $R(b)$, which is the same thing we see when the point source is at $k = b$ and we look at $k = a$. Point-spreading is, therefore, more than a cosmetic problem. When the object is a point source at $k = a$, but we do not know *a priori* that it is a point source, the spreading of the point causes us to believe that the object function $F(k)$ is nonzero at values of $k$ other than $k = a$. When we look at, say, $k = b$, we see a nonzero value that is caused by the presence of the point source at $k = a$.

Assume now that the object function $F(k)$ contains no point sources, but is simply an ordinary function of $k$. If the aperture $A$ is very small, then the function $H(k)$ is nearly constant over the entire extent of the object. The convolution of $F(k)$ and $H(k)$ is essentially the integral of $F(k)$, so the reconstructed object is $R(k) = \int F(k)dk$, for all $k$. Let's see what this means for the solar-emission problem discussed earlier.

### 9.13.1 The Solar-Emission Problem Revisited

The wavelength of the radiation is $\lambda = 1$ meter. Therefore, $\frac{\omega}{c} = 2\pi$, and $k$ in the interval $[-2\pi, 2\pi]$ corresponds to the angle $\theta$ in $[0, \pi]$. The sun has an angular diameter of 30 minutes of arc, which is about $10^{-2}$ radians. Therefore, the sun subtends the angles $\theta$ in $[\frac{\pi}{2} - (0.5) \cdot 10^{-2}, \frac{\pi}{2} + (0.5) \cdot 10^{-2}]$, which corresponds roughly to the variable $k$ in the interval $[-3 \cdot 10^{-2}, 3 \cdot 10^{-2}]$. Resolution of 3 minutes of arc means resolution in the variable $k$ of $3 \cdot 10^{-3}$. If the aperture is $2A$, then to achieve this resolution, we need

$$\frac{\pi}{A} \leq 3 \cdot 10^{-3},$$

or

$$A \geq \frac{\pi}{3} \cdot 10^3$$

meters, or $A$ not less than about 1000 meters.

The radio-wave signals emitted by the sun are focused, using a parabolic radio-telescope. The telescope is pointed at the center of the sun. Because the sun is a great distance from the earth and the subtended arc is small

(30 min.), the signals from each point on the sun's surface arrive at the parabola nearly head-on, that is, parallel to the line from the vertex to the focal point, and are reflected to the receiver located at the focal point of the parabola. The effect of the parabolic antenna is not to discriminate against signals coming from other directions, since there are none, but to effect a summation of the signals received at points $(x, 0, 0)$, for $|x| \leq A$, where $2A$ is the diameter of the parabola. When the aperture is large, the function $h(x)$ is nearly one for all $x$ and the signal received at the focal point is essentially

$$\int f(x)dx = F(0);$$

we are now able to distinguish between $F(0)$ and other values $F(k)$. When the aperture is small, $h(x)$ is essentially $\delta(x)$ and the signal received at the focal point is essentially

$$\int f(x)\delta(x)dx = f(0) = \int F(k)dk;$$

now all we get is the contribution from all the $k$, superimposed, and all resolution is lost.

Since the solar emission problem is clearly two-dimensional, and we need 3 min. resolution in both dimensions, it would seem that we would need a circular antenna with a diameter of about one kilometer, or a rectangular antenna roughly one kilometer on a side. Eventually, this problem was solved by converting it into essentially a tomography problem and applying the same techniques that are today used in CAT scan imaging.

### 9.13.2    Other Limitations on Resolution

In imaging regions of the earth from satellites in orbit there is a trade-off between resolution and the time available to image a given site. Satellites in *geostationary orbit*, such as weather and TV satellites, remain stationary, relative to a fixed position on the earth's surface, but to do so must orbit $22,000$ miles above the earth. If we tried to image the earth from that height, a telescope like the Hubble Space Telescope would have a resolution of about 21 feet, due to the unavoidable blurring caused by the optics of the lens itself. The Hubble orbits 353 miles above the earth, but because it looks out into space, not down to earth, it only needs to be high enough to avoid atmospheric distortions. Spy satellites operate in *low Earth orbit* (LEO), about 200 miles above the earth, and achieve a resolution of about 2 or 3 inches, at the cost of spending only about 1 or 2 minutes over their target. The satellites used in the GPS system maintain a *medium Earth orbit* (MEO) at a height of about $12,000$ miles, high enough to be seen over the horizon most of the time, but not so high as to require great power to send their signals.

In the February 2003 issue of *Harper's Magazine* there is an article on "scientific apocalypse" dealing with the search for near-earth asteroids. These objects are initially detected by passive optical observation, as small dots of reflected sunlight; once detected, they are then imaged by active radar to determine their size, shape, rotation and such. Some Russian astronomers are concerned about the near-earth asteroid Apophis 2004 MN4, which, they say, will pass within 30,000 km of earth in 2029, and come even closer in 2036. This is closer to earth than the satellites in geostationary orbit. As they say, "Stay tuned for further developments."

---

## 9.14 Discrete Data

A familiar topic in signal processing is the passage from functions of continuous variables to discrete sequences. This transition is achieved by *sampling*, that is, extracting values of the continuous-variable function at discrete points in its domain. Our example of far-field propagation can be used to explore some of the issues involved in sampling.

Imagine an infinite *uniform line array* of sensors formed by placing receivers at the points $(n\Delta, 0, 0)$, for some $\Delta > 0$ and all integers $n$. Then our data are the values $f(n\Delta)$. Because we defined $k = \frac{\omega}{c}\cos\theta$, it is clear that the function $F(k)$ is zero for $k$ outside the interval $[-\frac{\omega}{c}, \frac{\omega}{c}]$.

Our discrete array of sensors cannot distinguish between the signal arriving from $\theta$ and a signal with the same amplitude, coming from an angle $\alpha$ with

$$\frac{\omega}{c}\cos\alpha = \frac{\omega}{c}\cos\theta + \frac{2\pi}{\Delta}m,$$

$\Delta > 0$ so that

$$-\frac{\omega}{c} + \frac{2\pi}{\Delta} \geq \frac{\omega}{c},$$

or

$$\Delta \leq \frac{\pi c}{\omega} = \frac{\lambda}{2}.$$

The sensor spacing $\Delta_s = \frac{\lambda}{2}$ is the *Nyquist spacing*.

In the sunspot example, the object function $F(k)$ is zero for $k$ outside of an interval much smaller than $[-\frac{\omega}{c}, \frac{\omega}{c}]$. Knowing that $F(k) = 0$ for $|k| > K$, for some $0 < K < \frac{\omega}{c}$, we can accept ambiguities that confuse $\theta$ with another angle that lies outside the angular diameter of the object. Consequently, we can redefine the Nyquist spacing to be

$$\Delta_s = \frac{\pi}{K}.$$

This tells us that when we are imaging a distant object with a small angular diameter, the Nyquist spacing is greater than $\frac{\lambda}{2}$. If our sensor spacing has been chosen to be $\frac{\lambda}{2}$, then we have *oversampled*. In the oversampled case, band-limited extrapolation methods can be used to improve resolution.

### 9.14.1   Reconstruction from Samples

From the data gathered at our infinite array we have extracted the Fourier transform values $f(n\Delta)$, for all integers $n$. The obvious question is whether or not the data is sufficient to reconstruct $F(k)$. We know that, to avoid ambiguity, we must have $\Delta \leq \frac{\pi c}{\omega}$. The good news is that, provided this condition holds, $F(k)$ is uniquely determined by this data and formulas exist for reconstructing $F(k)$ from the data; this is the content of the *Shannon's Sampling Theorem*. Of course, this is only of theoretical interest, since we never have infinite data. Nevertheless, a considerable amount of traditional signal-processing exposition makes use of this infinite-sequence model. The real problem, of course, is that our data is always finite.

---

## 9.15   The Finite-Data Problem

Suppose that we build a *uniform line array* of sensors by placing receivers at the points $(n\Delta, 0, 0)$, for some $\Delta > 0$ and $n = -N, ..., N$. Then our data are the values $f(n\Delta)$, for $n = -N, ..., N$. Suppose, as previously, that the object of interest, the function $F(k)$, is nonzero only for values of $k$ in the interval $[-K, K]$, for some $0 < K < \frac{\omega}{c}$. Once again, we must have $\Delta \leq \frac{\pi c}{\omega}$ to avoid ambiguity; but this is not enough, now. The finite Fourier data is no longer sufficient to determine a unique $F(k)$. The best we can hope to do is to estimate the true $F(k)$, using both our measured Fourier data and whatever prior knowledge we may have about the function $F(k)$, such as where it is nonzero, if it consists of Dirac delta point sources, or if it is nonnegative. The data is also noisy, and that must be accounted for in the reconstruction process.

In certain applications, such as sonar array processing, the sensors are not necessarily arrayed at equal intervals along a line, or even at the grid points of a rectangle, but in an essentially arbitrary pattern in two, or even three, dimensions. In such cases, we have values of the Fourier transform of the object function, but at essentially arbitrary values of the variable. How best to reconstruct the object function in such cases is not obvious.

## 9.16 Functions of Several Variables

Fourier transformation applies, as well, to functions of several variables. As in the one-dimensional case, we can motivate the multi-dimensional Fourier transform using the far-field propagation model. As we noted earlier, the solar emission problem is inherently a two-dimensional problem.

### 9.16.1 A Two-Dimensional Far-Field Object

Assume that our sensors are located at points $\mathbf{s} = (x, y, 0)$ in the $x,y$-plane. As discussed previously, we assume that the function $F(\mathbf{k})$ can be viewed as a function $F(k_1, k_2)$. Since, in most applications, the distant object has a small angular diameter when viewed from a great distance – the sun's is only 30 minutes of arc – the function $F(k_1, k_2)$ will be supported on a small subset of vectors $(k_1, k_2)$.

### 9.16.2 Limited Apertures in Two Dimensions

Suppose we have the values of the Fourier transform, $f(x, y)$, for $|x| \leq A$ and $|y| \leq A$. We describe this limited-data problem using the function $h(x, y)$ that is one for $|x| \leq A$, and $|y| \leq A$, and zero, otherwise. Then the point-spread function is the Fourier transform of this $h(x, y)$, given by

$$H(\alpha, \beta) = 4AB\text{sinc}(A\alpha)\text{sinc}(B\beta).$$

The resolution in the horizontal ($x$) direction is on the order of $\frac{1}{A}$, and $\frac{1}{B}$ in the vertical, where, as in the one-dimensional case, aperture is best measured in units of wavelength.

Suppose our aperture is circular, with radius $A$. Then we have Fourier transform values $f(x, y)$ for $\sqrt{x^2 + y^2} \leq A$. Let $h(x, y)$ equal one, for $\sqrt{x^2 + y^2} \leq A$, and zero, otherwise. Then the point-spread function of this limited-aperture system is the Fourier transform of $h(x, y)$, given by $H(\alpha, \beta) = \frac{2\pi A}{r} J_1(rA)$, with $r = \sqrt{\alpha^2 + \beta^2}$. The resolution of this system is roughly the distance from the origin to the first null of the function $J_1(rA)$, which means that $rA = 4$, roughly.

For the solar emission problem, this says that we would need a circular aperture with radius approximately one kilometer to achieve 3 minutes of arc resolution. But this holds only if the antenna is stationary; a moving antenna is different! The solar emission problem was solved by using a rectangular antenna with a large $A$, but a small $B$, and exploiting the rotation of the earth. The resolution is then good in the horizontal, but bad in the vertical, so that the imaging system discriminates well between two distinct vertical lines, but cannot resolve sources within the same vertical line. Because $B$ is small, what we end up with is essentially the integral of the function $f(x, z)$ along each vertical line. By tilting the antenna, and waiting for the earth to rotate enough, we can get these integrals along any set of parallel lines. The problem then is to reconstruct $F(k_1, k_2)$ from such line integrals. This is also the main problem in tomography.

## 9.17   Broadband Signals

We have spent considerable time discussing the case of a distant point source or an extended object transmitting or reflecting a single-frequency signal. If the signal consists of many frequencies, the so-called broadband case, we can still analyze the received signals at the sensors in terms of time delays, but we cannot easily convert the delays to phase differences, and thereby make good use of the Fourier transform. One approach is to filter each received signal, to remove components at all but a single frequency, and then to proceed as previously discussed. In this way we can process one frequency at a time. The object now is described in terms of a function of both $\mathbf{k}$ and $\omega$, with $F(\mathbf{k}, \omega)$ the complex amplitude associated with the wave vector $\mathbf{k}$ and the frequency $\omega$. In the case of radar, the function $F(\mathbf{k}, \omega)$ tells us how the material at $P$ reflects the radio waves at the various frequencies $\omega$, and thereby gives information about the nature of the material making up the object near the point $P$.

There are times, of course, when we do not want to decompose a broadband signal into single-frequency components. A satellite reflecting a TV signal is a broadband point source. All we are interested in is receiving the broadband signal clearly, free of any other interfering sources. The direction of the satellite is known and the antenna is turned to face the satellite. Each location on the parabolic dish reflects the same signal. Because of its parabolic shape, the signals reflected off the dish and picked up at the focal point have exactly the same travel time from the satellite, so they combine coherently, to give us the desired TV signal.

# Chapter 10

## The Phase Problem

## 10.1 Chapter Summary

One of the main problems we consider in this book is the estimation of a function from finitely many values of its Fourier transform. In such cases, the data are complex numbers and the function to be estimated is a complex-valued function. As we mentioned previously, there are certain cases in which we have a *phase problem*, where it is not possible to measure the complex numbers, but only their magnitudes. Estimating the structure of a crystal from scattering data in x-ray crystallography and optical imaging through a turbulent atmosphere are two examples in which the phase problem arises. As you might imagine, reconstruction from magnitude-only data is more difficult than from the full complex data.

In this chapter we describe an algorithm for solving the phase problem that is based on the MDFT estimator discussed previously. This algorithm was originally introduced in [30]. The reader is invited to consult [30] for additional details and examples.

## 10.2    Reconstructing from Over-Sampled Complex FT Data

Let $f : [-\pi, \pi] \to \mathbb{C}$ have Fourier transform

$$F(\gamma) = \int_{-\pi}^{\pi} f(x)e^{i\gamma x}dx.$$

The Fourier series expansion for $f(x)$ is then

$$f(x) \approx \frac{1}{2\pi} \sum_{n=-\infty}^{\infty} F(n)e^{-inx}.$$

If we are able to obtain only the values $F(n)$ for $|n| \leq N$, then the DFT estimate of $f(x)$ is

$$f_{DFT}(x) = \frac{1}{2\pi} \sum_{n=-N}^{N} F(n)e^{-inx},$$

for $|x| \leq \pi$. We denote the data vector by $d = (F(-N), ..., F(N))^T$.

We assume now that $f(x) = 0$ for $x$ outside the interval $V = [-v, v]$, for some $v$ with $0 < v < \pi$.

**Ex. 10.1** *Let $S$ be the $2N + 1$ by $2N + 1$ matrix with entries*

$$S_{m,n} = \frac{\sin v(n - m)}{\pi(n - m)},$$

*for $m \neq n$, and $S_{m,m} = \frac{v}{\pi}$. Show that*

$$2\pi \int_{-v}^{v} |f_{DFT}(x)|^2 dx = d^\dagger Sd,$$

*and*

$$2\pi \int_{-\pi}^{\pi} |f_{DFT}(x)|^2 dx = d^\dagger d.$$

Therefore, the amount of DFT energy outside the interval $[-v, v]$ is

$$\int_{-\pi}^{\pi} |f_{DFT}(x)|^2 dx - \int_{-v}^{v} |DFT(x)|^2 dx = \frac{1}{2\pi}(d^\dagger d - d^\dagger Sd).$$

The proportion of DFT energy outside $V = [-v, v]$ is then

$$1 - \frac{d^\dagger Sd}{d^\dagger d} \geq 1 - \lambda_{max}(S),$$

where $\lambda_{max}(S)$ denotes the largest eigenvalue of the positive-definite symmetric matrix $S$.

When $v$ is close to $\pi$ or $N$ is large, $\lambda_{max}(S)$ is near one. The trace of the matrix $S$ is $\text{trace}(S) = (2N+1)\frac{v}{\pi}$, and, as $N$ approaches $+\infty$, roughly $(2N+1)\frac{v}{\pi}$ eigenvalues of $S$ have values that are approximately equal to one and the remainder have values approximately equal to zero. It is curious to note that, for large values of $N$, a plot of the eigenvalues of $S$, in decending order of size, resembles the graph of the right half of the function $\chi_V(x)$. This is one case in which an eigenspectrum and a power spectrum are related.

The lower bound on the proportion of energy outside $V$ will be attained if $d$ is replaced by an eigenvector of $S$ with eigenvalue $\lambda_{max}(S)$. Then the DFT will be maximally concentrated within $V$, but will be quite smooth and have little structure. When the function $f(x)$ has structure within the interval $V$ that we wish to reconstruct, we will need to employ eigenvectors of $S$ other than the ones associated with the largest eigenvalues of $S$. One way to do this is to use the MDFT estimator discussed previously.

The MDFT estimator of the function $f(x)$ is

$$f_{MDFT}(x) = \chi_V(x) \sum_{n=-N}^{N} b_n e^{-inx},$$

where the vector $b$ of coefficients is $b = \frac{1}{2\pi} S^{-1} d$. The energy of the function $f_{MDFT}(x)$ is then

$$\int_{-v}^{v} |f_{MDFT}(x)|^2 dx = 2\pi b^\dagger S b = \frac{1}{2\pi} d^\dagger S^{-1} d.$$

**Ex. 10.2** *Show that*
$$\frac{d^\dagger S^{-1} d}{d^\dagger d} \geq 1 \geq \frac{d^\dagger S d}{d^\dagger d}.$$

When the data are truly values of $F(n)$ and the function $f(x)$ has reasonable values and is actually supported on the interval $V = [-v, v]$, then the energy of the MDFT estimator will not be abnormally large. However, if the data values are not at least approximately equal to the values $F(n)$, or $f(x)$ is not supported on the interval $V$, then the MDFT energy will be quite large, indicating a mismatch between our data and our assumptions about $f(x)$. This behavior can actually be put to good use. In some cases, we may not know $V$, but do not want to overestimate it; we want $V$ to be as small as is allowable, but not smaller. We can take a decreasing sequence of intervals $V$ and stop when we see the MDFT energy begin to explode. We can also use this behavior to solve the phase problem.

## 10.3   The Phase Problem

We suppose now that $F(n) = |F(n)|e^{i\phi(n)}$, and we have only the magnitude data, $|F(n)|$, for $|n| \leq N$. If we take arbitrary phase angles $\theta(n)$ and create complex "data"

$$G(n) = |F(n)|e^{i\theta(n)},$$

for $|n| \leq N$, we can then pretend to have the complex FT data for $f(x)$ and compute the MDFT estimate. Fortunately for us, as the phase angles $\theta(n)$ begin to differ substantially from the true phase angles $\phi(n)$, the MDFT reacts to this mismatch and the MDFT energy increases dramatically. The idea is then to monitor the MDFT energy as we make choices of phase angles, attempting to find ones that are approximately correct. In the next section we present an iterative algorithm to implement this idea.

## 10.4   A Phase-Retrieval Algorithm

Let $\theta = (\theta(-N), ..., \theta(N))$ be an arbitrary selection of phase angles and

$$d(\theta) = (|F(-N)|e^{i\theta(-N)}, ..., |F(N)|e^{i\theta(N)})^T$$

our constructed "data" vector having the true magnitudes, but arbitrary phases. We shall also denote by $d(\theta)$ the infinite sequence whose only nonzero entries are the entries of the finite vector $d(\theta)$; the context will make clear which interpretation we are using. The energy in the resulting MDFT estimator is

$$E(\theta) = \frac{1}{2\pi}d(\theta)^{\dagger}S^{-1}d(\theta).$$

Our objective is to find a choice of angles $\theta(n)$ for which $E(\theta)$ is not unreasonably large, in the hope that the resulting MDFT will be a decent approximation of the true $f(x)$.

One approach would be to design an iterative algorithm that takes us from one phase vector $\theta^k$ to a new one, $\theta^{k+1}$, in such a way that $E(\theta^k) > E(\theta^{k+1})$. Perhaps a gradient-descent algorithm could be devised to do this. Instead, we have an iterative algorithm that, at least in our simulations, achieves much the same result, by a more indirect approach.

Let the Hilbert space $H$ be $L^2[-\pi, \pi]$ and $P_V$ the orthogonal projection of $H$ onto the subspace $L^2[-v, v]$. For any infinite sequence $G = \{G(n)\}$, denote by $\mathcal{F}^{-1}G$ the function

$$(\mathcal{F}^{-1}G)(x) = g(x) = \frac{1}{2\pi} \sum_{n=-\infty}^{\infty} G(n)e^{-inx},$$

for $|x| \leq \pi$. Then we write $G = \mathcal{F}g$. Define $(AG)(n) = G(n)$, for $|n| \leq N$, and $(AG)(n) = 0$, otherwise. Define $(DG)(n) = 0$, if $G(n) = 0$, $(DG)(n) = G(n)$, for $|n| > N$, and

$$(DG)(n) = \frac{|F(n)|}{|G(n)|}G(n),$$

otherwise. Then $DA = AD$ as operators.

We begin with an arbitrary phase vector $\theta^0$ and use it to define $g^0 = P_V \mathcal{F}^{-1} d(\theta^0)$. We let $\mathcal{F}g^0 = G^0$. Having found $g^k$ and $G^k = \mathcal{F}g^k$, we define $\theta^{k+1}$ by

$$d(\theta^{k+1}) = DAG^k.$$

The iterative step is then

$$g^{k+1} = P_V \mathcal{F}^{-1}[(I - A)\mathcal{F}g^k + AD\mathcal{F}g^k].$$

We can also write

$$g^{k+1} = P_V \mathcal{F}^{-1} d(\theta^{k+1}).$$

Note that

$$g^{k+1} - g^k = P_V \mathcal{F}^{-1}(DAG^k - AG^k) = P_V \mathcal{F}^{-1}c^k,$$

for

$$c^k = DAG^k - AG^k.$$

Therefore,

$$g^{k+1} = g^0 + \sum_{m=0}^{k} P_V \mathcal{F}^{-1}c^m = P_V \mathcal{F}^{-1}a^{k+1} \tag{10.1}$$

for

$$a^{k+1} = d(\theta^0) + \sum_{m=0}^{k} c^m.$$

It follows then that

$$a^{k+1} = a^k + DAG^k - AG^k.$$

From Equation (10.1) we see that each function $g^k$ has the form of an

MDFT estimator associated with the subset $V$ and the "data" $AG^k$. Therefore,

$$Sa^k = AG^k.$$

The iteration then becomes

$$a^{k+1} = a^k + DG^k - G^k = a^k + DSa^k - Sa^k.$$

We iterate using this updating step until convergence to some $a^\infty$ and then take

$$g^\infty = P_V \mathcal{F}^{-1} a^\infty$$

as our final estimate of $f(x)$. The energy at each step is

$$E(\theta^k) = d(\theta^k)^\dagger S^{-1} d(\theta^k),$$

so we can easily monitor the energy at each stage of the iteration.

## 10.5    Fienup's Method

Our algorithm has the iterative step

$$g^{k+1} = P_V \mathcal{F}^{-1}[(I - A)\mathcal{F}g^k + AD\mathcal{F}g^k],$$

where the operators $\mathcal{F}$ and $\mathcal{F}^{-1}$ relate infinite sequences to functions of a continuous variable. If we choose, instead, to view $g^k$ as a finite vector and these operators as relating finite vectors to one another via the FFT, we get Fienup's error-reduction method [76, 77]. In the error-reduction method what we call here the function $g^k(x)$, defined for $x$ in the interval $[-\pi, \pi]$, is discretized and replaced by a vector in $\mathbb{C}^J$, where $J > 2N + 1$. Similarly, the infinite sequence $\mathcal{F}g^k$ is replaced by a vector in $\mathbb{C}^J$ and the operator $\mathcal{F}$ is replaced by the FFT.

## 10.6    Does the Iteration Converge?

The operator $P_V$ is an orthogonal projection onto a subspace of $H$, and the operator

$$P = \mathcal{F}^{-1}(I - A)\mathcal{F} + \mathcal{F}^{-1}DA\mathcal{F}$$

is also a projection, but its range is not a convex set; therefore, the useful convergence theorems about composition of orthogonal projections onto convex sets do not apply here. All is not lost, however.

In [108] Levi and Stark define the *set-distance error*

$$J(g) = \|P_1 g - g\|_2 + \|P_2 g - g\|_2,$$

for projections $P_1$ and $P_2$, when one of the projections has nonconvex range. They show that, for the sequence generated by the iterative step $g^{k+1} = P_1 P_2 g^k$,

$$J(g^{k+1}) \le J(P_2 g^k) \le J(g^k).$$

In our case, with $P_1 = P_V$ and $P_2 = P$, we find that $g^{k+1}$ is at least as close to being consistent with the magnitude data as $g^k$ is, and $P g^{k+1}$ is at least as close to being supported on $V$ as $P g^k$ is.

# Chapter 11

## Transmission Tomography

## 11.1  Chapter Summary

Our topic is now transmission tomography. This chapter will provide a detailed description of how the data is gathered, the mathematical model of the scanning process, the problem to be solved, the various mathematical techniques needed to solve this problem, and the manner in which these techniques are applied, including filtering methods for inverting the two-dimensional Fourier transform.

According to the Central Slice Theorem, if we have all the line integrals through the attenuation function $f(x, y)$ then we have the two-dimensional Fourier transform of $f(x, y)$. To get $f(x, y)$ we need to invert the two-dimensional Fourier transform.

## 11.2  X-ray Transmission Tomography

Although transmission tomography is not limited to scanning living beings, we shall concentrate here on the use of x-ray tomography in medical diagnosis and the issues that concern us in that application. The mathematical formulation will, of course, apply more generally.

In x-ray tomography, x-rays are transmitted through the body along many lines. In some, but not all, cases, the lines will all lie in the same plane. The strength of the x-rays upon entering the body is assumed known, and the strength upon leaving the body is measured. This data can then be used to estimate the amount of attenuation the x-ray encountered along that line, which is taken to be the integral, along that line, of the attenuation function. On the basis of these line integrals, we estimate the attenuation function. This estimate is presented to the physician as one or more two-dimensional images.

## 11.3  The Exponential-Decay Model

As an x-ray beam passes through the body, it encounters various types of matter, such as soft tissue, bone, ligaments, air, each weakening the beam to a greater or lesser extent. If the intensity of the beam upon entry is $I_{in}$ and $I_{out}$ is its lower intensity after passing through the body, then

$$I_{out} = I_{in}e^{-\int_L f},$$

where $f = f(x,y) \geq 0$ is the *attenuation function* describing the two-dimensional distribution of matter within the slice of the body being scanned and $\int_L f$ is the integral of the function $f$ over the line $L$ along which the x-ray beam has passed. To see why this is the case, imagine the line $L$ parameterized by the variable $s$ and consider the intensity function $I(s)$ as a function of $s$. For small $\Delta s > 0$, the drop in intensity from the start to the end of the interval $[s, s + \Delta s]$ is approximately proportional to the intensity $I(s)$, to the attenuation $f(s)$ and to $\Delta s$, the length of the interval; that is,

$$I(s) - I(s + \Delta s) \approx f(s)I(s)\Delta s.$$

Dividing by $\Delta s$ and letting $\Delta s$ approach zero, we get

$$I'(s) = -f(s)I(s).$$

**Ex. 11.1** *Show that the solution to this differential equation is*

$$I(s) = I(0) \exp \left( - \int_{u=0}^{u=s} f(u)du \right).$$

*Hint: Use an integrating factor.*

From knowledge of $I_{\text{in}}$ and $I_{\text{out}}$, we can determine $\int_L f$. If we know $\int_L f$ for every line in the $x, y$-plane we can reconstruct the attenuation function $f$. In the real world we know line integrals only approximately and only for finitely many lines. The goal in x-ray transmission tomography is to estimate the attenuation function $f(x, y)$ in the slice, from finitely many noisy measurements of the line integrals. We usually have prior information about the values that $f(x, y)$ can take on. We also expect to find sharp boundaries separating regions where the function $f(x, y)$ varies only slightly. Therefore, we need algorithms capable of providing such images.

---

## 11.4 Difficulties to Be Overcome

There are several problems associated with this model. The paths taken by x-ray beams are not exactly straight lines; the beams tend to spread out. The x-rays are not monochromatic, and their various frequency components are attenuated at different rates, resulting in *beam hardening*, that is, changes in the spectrum of the beam as it passes through the object. The beams consist of photons obeying statistical laws, so our algorithms probably should be based on these laws. How we choose the line segments is determined by the nature of the problem; in certain cases we are somewhat limited in our choice of these segments. Patients move; they breathe, their hearts beat, and, occasionally, they shift position during the scan. Compensating for these motions is an important, and difficult, aspect of the image reconstruction process. Finally, to be practical in a clinical setting, the processing that leads to the reconstructed image must be completed in a short time, usually around fifteen minutes. This time constraint is what motivates viewing the three-dimensional attenuation function in terms of its two-dimensional slices.

As we shall see, the Fourier transform and the associated theory of convolution filters play important roles in the reconstruction of transmission tomographic images.

The data we actually obtain at the detectors are counts of detected photons. These counts are not the line integrals; they are random quantities whose means, or expected values, are related to the line integrals. The Fourier inversion methods for solving the problem ignore its statistical aspects; in contrast, other methods, such as likelihood maximization, are based on a statistical model that involves Poisson-distributed emissions.

---

## 11.5    Reconstruction from Line Integrals

We turn now to the underlying problem of reconstructing attenuation functions from line-integral data.

### 11.5.1    The Radon Transform

Our goal is to reconstruct the function $f(x, y) \geq 0$ from line-integral data. Let $\theta$ be a fixed angle in the interval $[0, \pi)$. Form the $t, s$-axis system with the positive $t$-axis making the angle $\theta$ with the positive $x$-axis, as shown in Figure 11.1. Each point $(x, y)$ in the original coordinate system has coordinates $(t, s)$ in the second system, where the $t$ and $s$ are given by

$$t = x \cos \theta + y \sin \theta,$$

and

$$s = -x \sin \theta + y \cos \theta.$$

If we have the new coordinates $(t, s)$ of a point, the old coordinates are $(x, y)$ given by

$$x = t \cos \theta - s \sin \theta,$$

and

$$y = t \sin \theta + s \cos \theta.$$

We can then write the function $f$ as a function of the variables $t$ and $s$. For each fixed value of $t$, we compute the integral

$$\int_L f(x, y) ds = \int f(t \cos \theta - s \sin \theta, t \sin \theta + s \cos \theta) ds$$

along the single line $L$ corresponding to the fixed values of $\theta$ and $t$. We repeat this process for every value of $t$ and then change the angle $\theta$ and repeat again. In this way we obtain the integrals of $f$ over every line $L$ in the plane. We denote by $r_f(\theta, t)$ the integral

$$r_f(\theta, t) = \int_L f(x, y) ds = \int f(t \cos \theta - s \sin \theta, t \sin \theta + s \cos \theta) ds.$$

The function $r_f(\theta, t)$ is called the *Radon transform* of $f$.

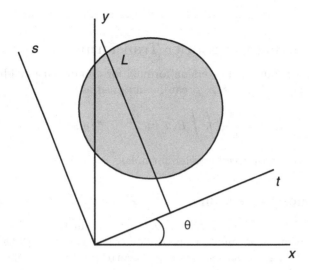

**FIGURE 11.1**: The Radon transform of $f$ at $(t, \theta)$ is the line integral of $f$ along line $L$.

## 11.5.2 The Central Slice Theorem

For fixed $\theta$ the function $r_f(\theta, t)$ is a function of the single real variable $t$; let $R_f(\theta, \omega)$ be its Fourier transform. Then

$$
\begin{aligned}
R_f(\theta, \omega) &= \int r_f(\theta, t) e^{i\omega t} dt \\
&= \int \int f(t \cos\theta - s \sin\theta, t \sin\theta + s \cos\theta) e^{i\omega t} ds dt \\
&= \int \int f(x, y) e^{i\omega(x \cos\theta + y \sin\theta)} dx dy \\
&= F(\omega \cos\theta, \omega \sin\theta),
\end{aligned}
$$

where $F(\omega \cos\theta, \omega \sin\theta)$ is the two-dimensional Fourier transform of the function $f(x, y)$, evaluated at the point $(\omega \cos\theta, \omega \sin\theta)$; this relationship is called the *Central Slice Theorem*. For fixed $\theta$, as we change the value of $\omega$, we obtain the values of the function $F$ along the points of the line making the angle $\theta$ with the horizontal axis. As $\theta$ varies in $[0, \pi)$, we get all the values of the function $F$. Once we have $F$, we can obtain $f$ using the formula for the two-dimensional inverse Fourier transform. We conclude that we are able to determine $f$ from its line integrals. As we shall see, inverting the Fourier transform can be implemented by combinations of frequency-domain filtering and back projection.

## 11.6     Inverting the Fourier Transform

The Fourier-transform inversion formula for two-dimensional functions tells us that the function $f(x, y)$ can be obtained as

$$f(x, y) = \frac{1}{4\pi^2} \int \int F(u, v) e^{-i(xu+yv)} du dv. \qquad (11.1)$$

We now derive alternative inversion formulas.

### 11.6.1     Back Projection

For $0 \leq \theta < \pi$ and all real $t$, let $h(\theta, t)$ be any function of the variables $\theta$ and $t$; for example, it could be the Radon transform. As with the Radon transform, we imagine that each pair $(\theta, t)$ corresponds to one line through the $x, y$-plane. For each fixed point $(x, y)$ we assign to this point the average, over all $\theta$, of the quantities $h(\theta, t)$ for every pair $(\theta, t)$ such that the point $(x, y)$ lies on the associated line. The summing process is integration and the *back-projection* function at $(x, y)$ is

$$BP_h(x, y) = \int_0^\pi h(\theta, x \cos \theta + y \sin \theta) d\theta.$$

The operation of back projection will play an important role in what follows in this chapter.

### 11.6.2     Ramp Filter, then Back Project

Expressing the double integral in Equation (11.1) in polar coordinates $(\omega, \theta)$, with $\omega \geq 0$, $u = \omega \cos \theta$, and $v = \omega \sin \theta$, we get

$$f(x, y) = \frac{1}{4\pi^2} \int_0^{2\pi} \int_0^\infty F(u, v) e^{-i(xu+yv)} \omega d\omega d\theta,$$

or

$$f(x, y) = \frac{1}{4\pi^2} \int_0^\pi \int_{-\infty}^\infty F(u, v) e^{-i(xu+yv)} |\omega| d\omega d\theta.$$

Now write

$$F(u, v) = F(\omega \cos \theta, \omega \sin \theta) = R_f(\theta, \omega),$$

where $R_f(\theta, \omega)$ is the FT with respect to $t$ of $r_f(\theta, t)$, so that

$$\int_{-\infty}^\infty F(u, v) e^{-i(xu+yv)} |\omega| d\omega = \int_{-\infty}^\infty R_f(\theta, \omega) |\omega| e^{-i\omega t} d\omega.$$

The function $g_f(\theta, t)$ defined for $t = x\cos\theta + y\sin\theta$ by

$$g_f(\theta, x\cos\theta + y\sin\theta) = \frac{1}{2\pi}\int_{-\infty}^{\infty} R_f(\theta, \omega)|\omega|e^{-i\omega t}d\omega \qquad (11.2)$$

is the result of a linear filtering of $r_f(\theta, t)$ using a *ramp filter* with transfer function $H(\omega) = |\omega|$. Then,

$$f(x, y) = \frac{1}{2\pi}BP_{g_f}(x, y) = \frac{1}{2\pi}\int_0^\pi g_f(\theta, x\cos\theta + y\sin\theta)d\theta$$

gives $f(x, y)$ as the result of a *back-projection operator*; for every fixed value of $(\theta, t)$ add $g_f(\theta, t)$ to the current value at the point $(x, y)$ for all $(x, y)$ lying on the straight line determined by $\theta$ and $t$ by $t = x\cos\theta + y\sin\theta$. The final value at a fixed point $(x, y)$ is then the average of all the values $g_f(\theta, t)$ for those $(\theta, t)$ for which $(x, y)$ is on the line $t = x\cos\theta + y\sin\theta$. It is therefore said that $f(x, y)$ can be obtained by *filtered back-projection* (FBP) of the line-integral data.

Knowing that $f(x, y)$ is related to the complete set of line integrals by filtered back-projection suggests that, when only finitely many line integrals are available, a similar ramp filtering and back-projection can be used to estimate $f(x, y)$; in the clinic this is the most widely used method for the reconstruction of tomographic images.

## 11.6.3 Back Project, then Ramp Filter

There is a second way to recover $f(x, y)$ using back projection and filtering, this time in the reverse order; that is, we back project the Radon transform and then ramp filter the resulting function of two variables. We begin with the back-projection operation, as applied to the function $h(\theta, t) = r_f(\theta, t)$.

We have

$$BP_{r_f}(x, y) = \int_0^\pi r_f(\theta, x\cos\theta + y\sin\theta)d\theta.$$

Replacing $r_f(\theta, t)$ with

$$r_f(\theta, t) = \frac{1}{2\pi}\int_{-\infty}^{\infty} R_f(\theta, \omega)e^{-i\omega t}d\omega,$$

and inserting

$$R_f(\theta, \omega) = F(\omega\cos\theta, \omega\sin\theta),$$

and

$$t = x\cos\theta + y\sin\theta,$$

we get

$$BP_{r_f}(x,y) = \int_0^\pi \left( \frac{1}{2\pi} \int_{-\infty}^\infty F(\omega\cos\theta, \omega\sin\theta) e^{-i(x\cos\theta + y\sin\theta)} d\omega \right) d\theta.$$

With $u = \omega\cos\theta$ and $v = \omega\sin\theta$, this becomes

$$\begin{aligned}
BP_{r_f}(x,y) &= \int_0^\pi \left( \frac{1}{2\pi} \int_{-\infty}^\infty \frac{F(u,v)}{\sqrt{u^2+v^2}} e^{-i(xu+yv)} |\omega| d\omega \right) d\theta, \\
&= \int_0^\pi \left( \frac{1}{2\pi} \int_{-\infty}^\infty G(u,v) e^{-i(xu+yv)} |\omega| d\omega \right) d\theta \\
&= \frac{1}{2\pi} \int_{-\infty}^\infty \int_{-\infty}^\infty G(u,v) e^{-i(xu+yv)} du\, dv.
\end{aligned}$$

This tells us that the back projection of $r_f(\theta, t)$ is the function $g(x,y)$ whose two-dimensional Fourier transform is

$$G(u,v) = \frac{1}{2\pi} F(u,v)/\sqrt{u^2+v^2}.$$

Therefore, we can obtain $f(x,y)$ from $r_f(\theta,t)$ by first back projecting $r_f(\theta,t)$ to get $g(x,y)$ and then filtering $g(x,y)$ by forming $G(u,v)$, multiplying by $\sqrt{u^2+v^2}$, and taking the inverse Fourier transform.

### 11.6.4　Radon's Inversion Formula

To get Radon's inversion formula, we need two basic properties of the Fourier transform. First, if $f(x)$ has Fourier transform $F(\gamma)$ then the derivative $f'(x)$ has Fourier transform $-i\gamma F(\gamma)$. Second, if $F(\gamma) = \text{sgn}(\gamma)$, the function that is $\frac{\gamma}{|\gamma|}$ for $\gamma \neq 0$, and equal to zero for $\gamma = 0$, then its inverse Fourier transform is $f(x) = \frac{1}{i\pi x}$.

Writing Equation (11.2) as

$$g_f(\theta, t) = \frac{1}{2\pi} \int_{-\infty}^\infty \omega R_f(\theta, \omega) \text{sgn}(\omega) e^{-i\omega t} d\omega,$$

we see that $g_f$ is the inverse Fourier transform of the product of the two functions $\omega R_f(\theta, \omega)$ and $\text{sgn}(\omega)$. Consequently, $g_f$ is the convolution of their individual inverse Fourier transforms, $i\frac{\partial}{\partial t} r_f(\theta, t)$ and $\frac{1}{i\pi t}$; that is,

$$g_f(\theta, t) = \frac{1}{\pi} \int_{-\infty}^\infty \frac{\partial}{\partial t} r_f(\theta, s) \frac{1}{t-s} ds,$$

which is the Hilbert transform of the function $\frac{\partial}{\partial t} r_f(\theta, t)$, with respect to the variable $t$. Radon's inversion formula is then

$$f(x,y) = \frac{1}{2\pi} \int_0^\pi HT \left( \frac{\partial}{\partial t} r_f(\theta, t) \right) d\theta.$$

## 11.7 From Theory to Practice

What we have just described is the theory. What happens in practice?

### 11.7.1 The Practical Problems

Of course, in reality we never have the Radon transform $r_f(\theta, t)$ for all values of its variables. Only finitely many angles $\theta$ are used, and, for each $\theta$, we will have (approximate) values of line integrals for only finitely many $t$. Therefore, taking the Fourier transform of $r_f(\theta, t)$, as a function of the single variable $t$, is not something we can actually do. At best, we can approximate $R_f(\theta, \omega)$ for finitely many $\theta$. From the Central Slice Theorem, we can then say that we have approximate values of $F(\omega \cos\theta, \omega \sin\theta)$, for finitely many $\theta$. This means that we have (approximate) Fourier transform values for $f(x, y)$ along finitely many lines through the origin, like the spokes of a wheel. The farther from the origin we get, the fewer values we have, so the *coverage* in Fourier space is quite uneven. The low-spatial-frequencies are much better estimated than higher ones, meaning that we have a low-pass version of the desired $f(x, y)$. The filtered-back-projection approaches we have just discussed both involve ramp filtering, in which the higher frequencies are increased, relative to the lower ones. This too can only be implemented approximately, since the data is noisy and careless ramp filtering will cause the reconstructed image to be unacceptably noisy.

### 11.7.2 A Practical Solution: Filtered Back Projection

We assume, to begin with, that we have finitely many line integrals, that is, we have values $r_f(\theta, t)$ for finitely many $\theta$ and finitely many $t$. For each fixed $\theta$ we estimate the Fourier transform, $R_f(\theta, \omega)$. This step can be performed in various ways, and we can freely choose the values of $\omega$ at which we perform the estimation. The FFT will almost certainly be involved in calculating the estimates of $R_f(\theta, \omega)$.

For each fixed $\theta$ we multiply our estimated values of $R_f(\theta, \omega)$ by $|\omega|$ and then use the FFT again to inverse Fourier transform, to achieve a ramp filtering of $r_f(\theta, t)$ as a function of $t$. Note, however, that when $|\omega|$ is large, we may multiply by a smaller quantity, to avoid enhancing noise. We do this for each angle $\theta$, to get a function of $(\theta, t)$, which we then back project to get our final image. This is ramp filtering, followed by back projection, as applied to the finite data we have.

It is also possible to mimic the second approach to inversion, that is, to back project onto the pixels each $r_f(\theta, t)$ that we have, and then to perform a ramp filtering of this two-dimensional array of numbers to obtain the final image. In this case, the two-dimensional ramp filtering involves many applications of the FFT.

There is a third approach. Invoking the Central Slice Theorem, we can say that we have finitely many approximate values of $F(u, v)$, the Fourier transform of the attenuation function $f(x, y)$, along finitely many lines through the origin. The first step is to use these values to estimate the values of $F(u, v)$ at the points of a rectangular grid. This step involves *interpolation* [157]. Once we have (approximate) values of $F(u, v)$ on a rectangular grid, we perform a two-dimensional FFT to obtain our final estimate of the (discretized) $f(x, y)$.

## 11.8 Some Practical Concerns

As computer power increases and scanners become more sophisticated, there is pressure to include more dimensionality in the scans. This means going beyond slice-by-slice tomography to fully three-dimensional images, or even including time as the fourth dimension, to image dynamically. This increase in dimensionality comes at a cost, however. Besides the increase in radiation to the patient, there are other drawbacks, such as longer acquisition time, storing large amounts of data, processing and analyzing this data, displaying the results, reading and understanding the higher-dimensional images, and so on.

## 11.9 Summary

We have seen how the problem of reconstructing a function from line integrals arises in transmission tomography. The Central Slice Theorem connects the line integrals and the Radon transform to the Fourier transform of the desired attenuation function. Various approaches to implementing the Fourier Inversion Formula lead to filtered-back-projection algorithms for the reconstruction. In x-ray tomography, as well as in PET, viewing the data as line integrals ignores the statistical aspects of the problem, and in SPECT, it ignores, as well, the important physical effects of attenuation. To incorporate more of the physics of the problem, iterative algorithms based on statistical models have been developed. We consider some of these algorithms in the books [41] and [42].

# Chapter 12

# Random Sequences

## 12.1 Chapter Summary

When we sample a function $f(x)$ we usually make some error, and the data we get is not precisely $f(n\Delta)$, but contains *additive noise*, that is, our data value is really $f(n\Delta)$ + noise. Noise is best viewed as random, so it becomes necessary to treat *random sequences* $f = \{f_n\}$ in which each $f_n$ is a random variable. The random variables $f_n$ and $f_m$ may or may not be statistically independent. In this chapter we survey several topics from probability and stochastic processes that are particularly important in signal processing.

---

## 12.2    What Is a Random Variable?

The simplest answer to the question "What is a random variable?" is "A random variable is a mathematical model". Imagine that we repeatedly drop a baseball from eye-level to the floor. Each time, the baseball behaves the same. If we were asked to describe this behavior with a mathematical model, we probably would choose to use a differential equation as our model. Ignoring everything except the force of gravity, we would write

$$h''(t) = -32$$

as the equation describing the downward acceleration due to gravity. Integrating, we have

$$h'(t) = -32t + h'(0)$$

as the velocity of the baseball at time $t \geq 0$, and integrating once more,

$$h(t) = -16t^2 + h'(0)t + h(0)$$

as the equation of position of the baseball at time $t \geq 0$, up to the moment when it hits the floor. Knowing $h(0)$, the distance from eye-level to the floor, and knowing that, since we dropped the ball, $h'(0) = 0$, we can determine how long it will take the baseball to hit the floor, and the speed with which it will hit. This analysis will apply every time we drop the baseball. There will, of course, be slight differences from one drop to the next, depending, perhaps, on how the ball was held, but these will be so small as to be insignificant.

Now imagine that, instead of a baseball, we drop a feather. A few repetitions are all that is necessary to convince us that the model used for the baseball no longer suffices. The factors that we safely ignored with regard to the baseball, such as air resistance, air currents, and how the object was held, now become important. The feather does not always land in the same place, it doesn't always take the same amount of time to reach the floor, and doesn't always land with the same velocity. It doesn't even fall in a straight vertical line. How can we possibly model such behavior? Must we try to describe accurately the air resistance encountered by the feather? The answer is that we use random variables as our model.

While we cannot predict exactly the place where the feather will land, and, of course, we must be careful to specify how we are to determine "the place" where it does land, we can learn, from a number of trials, where it tends to land, and we can postulate the probability that it will land within any given region of the floor. In this way, the place where the feather will land becomes a random variable with associated probability density function. Similarly, we can postulate the probability that the time for the fall will lie within any interval of elapsed time, making the elapsed time a random variable. Finally, we can postulate the probability that its velocity vector upon hitting the ground will lie within any given set of three-dimensional vectors, making the velocity a random vector. On the basis of these probabilistic models we can proceed to predict the outcome of the next drop.

It is important to remember that the random variable is the model that we set up prior to the dropping of the feather, not the outcome of any particular drop.

---

## 12.3  The Coin-Flip Random Sequence

The simplest example of a random sequence is the *coin-flip* sequence, which we denote by $c = \{c_n\}_{n=-\infty}^{\infty}$. We imagine that, at each "time" $n$, a coin is flipped, and $c_n = 1$ if the coin shows heads, and $c_n = -1$ if the coin shows tails. When we speak of this coin-flip sequence, we refer to this random model, not to any specific sequence of ones and minus ones; the random coin-flip sequence is not, therefore, a particular sequence, just as a random variable is not actually a specific number. Any particular sequence of ones and minus ones can be thought of as having resulted from such an infinite number of flips of the coin, and is called a *realization* of the random coin-flip sequence.

It will be convenient to allow for the coin to be *biased*, that is, for the probabilities of heads and tails to be unequal. We denote by $p$ the probability that heads occurs and $1 - p$ the probability of tails; the coin is called *unbiased* or *fair* if $p = 1/2$. To find the *expected value* of $c_n$, written $E(c_n)$, we multiply each possible value of $c_n$ by its probability and sum; that is,

$$E(c_n) = (+1)p + (-1)(1 - p) = 2p - 1.$$

If the coin is fair then $E(c_n) = 0$. The variance of the random variable $c_n$, measuring its tendency to deviate from its expected value, is $var(c_n) = E([c_n - E(c_n)]^2)$. We have

$$var(c_n) = [+1 - (2p - 1)]^2 p + [-1 - (2p - 1)]^2(1 - p) = 4p - 4p^2.$$

If the coin is fair then $var(c_n) = 1$. It is important to note that we do not change the coin at any time during the generation of a realization of the random sequence $c$; in particular, the $p$ does not depend on $n$.

Also, we assume that the random variables $c_n$ are statistically independent. This means that, for any $N$, any choice of "times" $n_1, ..., n_N$, and any values $m_1, ..., m_N$ in the set $\{-1, 1\}$, the probability that $c_{n_1} = m_1$, $c_{n_2} = m_2, ..., c_{n_N} = m_N$ is the product of the individual probabilities. For example, the probablity that $c_1 = -1$, $c_2 = +1$ and $c_4 = +1$ is $(1-p)p^2$.

---

## 12.4   Correlation

Let $u$ and $v$ be (possibly complex-valued) random variables with expected values $E(u)$ and $E(v)$, respectively. The covariance between $u$ and $v$ is defined to be

$$cov(u, v) = E\Big( (u - E(u))\overline{(v - E(v))}\Big),$$

and the cross-correlation between $u$ and $v$ is

$$corr(u, v) = E(u\bar{v}).$$

It is easily shown that $cov(u, v) = corr(u, v) - E(u)\overline{E(v)}$. When $u = v$ we get $cov(u, u) = var(u)$ and $corr(u, u) = E(|u|^2)$. If $E(u) = E(v) = 0$ then $cov(u, v) = corr(u, v)$. In statistics the "correlation coefficient" is the quantity $cov(u, v)$ divided by the standard deviations of $u$ and $v$.

When $u$ and $v$ are independent, we have

$$E(u\bar{v}) = E(u)E(\bar{v}),$$

and

$$E\Big( (u - E(u))\overline{(v - E(v))}\Big) = E(u - E(u)) E\Big(\overline{(v - E(v))}\Big) = 0.$$

To illustrate, let $u = c_n$ and $v = c_{n-m}$. Then, if the coin is fair, $E(c_n) = E(c_{n-m}) = 0$ and

$$cov(c_n, c_{n-m}) = corr(c_n, c_{n-m}) = E(c_n\overline{c_{n-m}}).$$

Because the $c_n$ are independent, $E(c_n\overline{c_{n-m}}) = 0$ for $m$ not equal to 0, and $E(|c_n|^2) = var(c_n) = 1$. Therefore

$$cov(c_n, c_{n-m}) = corr(c_n, c_{n-m}) = 0, \text{for } m \neq 0,$$

and
$$cov(c_n, c_n) = corr(c_n, c_n) = 1.$$

In the next section we shall use the random coin-flip sequence to generate a wide class of random sequences, obtained by viewing $c = \{c_n\}$ as the input into a shift-invariant discrete linear filter.

## 12.5 Filtering Random Sequences

Suppose, once again, that $T$ is a shift-invariant discrete linear filter with impulse-response sequence $g$. Now let us take as input, not a particular sequence, but the random coin-flip sequence $c$, with $p = 0.5$. The output will therefore not be a particular sequence either, but will be another random sequence, say $d$. Then, for each $n$ the random variable $d_n$ is

$$d_n = \sum_{m=-\infty}^{\infty} c_m g_{n-m} = \sum_{m=-\infty}^{\infty} g_m c_{n-m}. \tag{12.1}$$

We compute the correlation $corr(d_n, d_{n-m}) = E(d_n \overline{d_{n-m}})$. Using the convolution formula Equation (12.1), we find that

$$corr(d_n, d_{n-m}) = \sum_{k=-\infty}^{\infty} \sum_{j=-\infty}^{\infty} g_k \overline{g_j} corr(c_{n-k}, c_{n-m-j}).$$

Since
$$corr(c_{n-k}, c_{n-m-j}) = 0, \text{for } k \neq m + j,$$

we have
$$corr(d_n, d_{n-m}) = \sum_{k=-\infty}^{\infty} g_k \overline{g_{k-m}}. \tag{12.2}$$

The expression of the right side of Equation (12.2) is the definition of the *autocorrelation* of the non-random sequence $g$, denoted $\rho_g = \{\rho_g(m)\}$; that is,

$$\rho_g(m) = \sum_{k=-\infty}^{\infty} g_k \overline{g_{k-m}}. \tag{12.3}$$

It is important to note that the expected value of $d_n$ is

$$E(d_n) = \sum_{k=-\infty}^{\infty} g_k E(c_{n-k}) = 0$$

and the correlation $corr(d_n, d_{n-m})$ depends only on $m$; neither quantity depends on $n$ and the sequence $d$ is therefore called *wide-sense stationary*. Let's consider an example.

## 12.6   An Example

Take $g_0 = g_1 = 0.5$ and $g_k = 0$ otherwise. Then the system is the two-point moving-average, with

$$d_n = 0.5c_n + 0.5c_{n-1}.$$

In the case of the random-coin-flip sequence $c$ each $c_n$ is unrelated to all other $c_m$; the coin flips are independent. This is no longer the case for the $d_n$; one effect of the filter $g$ is to introduce correlation into the output. To illustrate, since $d_0$ and $d_1$ both depend, to some degree, on the value $c_0$, they are related. Using Equation (12.3) we have

$$corr(d_n, d_n) = \rho_g(0) = g_0 g_0 + g_1 g_1 = 0.25 + 0.25 = 0.5,$$

$$corr(d_n, d_{n+1}) = \rho_g(-1) = g_0 g_1 = 0.25,$$

$$corr(d_n, d_{n-1}) = \rho_g(+1) = g_1 g_0 = 0.25,$$

and

$$corr(d_n, d_{n-m}) = \rho_g(m) = 0, \text{otherwise}.$$

So we see that $d_n$ and $d_{n-m}$ are related, for $m = -1, 0, +1$, but not otherwise.

## 12.7   Correlation Functions and Power Spectra

As we have seen, any non-random sequence $g = \{g_n\}$ has its autocorrelation function defined, for each integer $m$, by

$$\rho_g(m) = \sum_{k=-\infty}^{\infty} g_k \overline{g_{k-m}}.$$

For a random sequence $d_n$ that is wide-sense stationary, its correlation function is defined to be

$$\rho_d(m) = E(d_n \overline{d_{n-m}}).$$

The *power spectrum* of $g$ is defined for $\omega$ in $[-\pi, \pi]$ by

$$R_g(\omega) = \sum_{m=-\infty}^{\infty} \rho_g(m)e^{im\omega}.$$

It is easy to see that

$$R_g(\omega) = |G(\omega)|^2,$$

where

$$G(\omega) = \sum_{n=-\infty}^{\infty} g_n e^{in\omega},$$

so that $R_g(\omega) \geq 0$. The power spectrum of the random sequence $d = \{d_n\}$ is defined as

$$R_d(\omega) = \sum_{m=-\infty}^{\infty} \rho_d(m)e^{im\omega}.$$

Although it is not immediately obvious, we also have $R_d(\omega) \geq 0$. One way to see this is to consider

$$D(\omega) = \sum_{n=-\infty}^{\infty} d_n e^{in\omega}$$

and to calculate

$$E(|D(\omega)|^2) = \sum_{m=-\infty}^{\infty} E(d_n \overline{d_{n-m}})e^{im\omega} = R_d(\omega).$$

Given any power spectrum $R_d(\omega) \geq 0$ we can construct $G(\omega)$ by selecting an arbitrary phase angle $\theta$ and letting

$$G(\omega) = \sqrt{R_d(\omega)}e^{i\theta}.$$

We then obtain the non-random sequence $g$ associated with $G(\omega)$ using

$$g_n = \frac{1}{2\pi} \int_{-\pi}^{\pi} G(\omega)e^{-in\omega}d\omega.$$

It follows that $\rho_g(m) = \rho_d(m)$ for each $m$ and $R_g(\omega) = R_d(\omega)$ for each $\omega$.

What we have discovered is that, when the input to the system is the random-coin-flip sequence $c$, the output sequence $d$ has a correlation function $\rho_d(m)$ that is equal to the autocorrelation of the sequence $g$. As we just saw, for any wide-sense stationary random sequence $d$ with expected value $E(d_n)$ constant and correlation function $corr(d_n, d_{n-m})$ independent of $n$, there is a shift-invariant discrete linear system $T$ with impulse-response sequence $g$, such that $\rho_g(m) = \rho_d(m)$ for each $m$. Therefore, any wide-sense stationary random sequence $d$ can be viewed as the output of a shift-invariant discrete linear system, when the input is the random-coin-flip sequence $c = \{c_n\}$.

## 12.8 The Dirac Delta in Frequency Space

Consider the "function" defined by the infinite sum

$$\delta(\omega) = \frac{1}{2\pi} \sum_{n=-\infty}^{\infty} e^{in\omega} = \frac{1}{2\pi} \sum_{n=-\infty}^{\infty} e^{-in\omega}. \qquad (12.4)$$

This is a Fourier series in which all the Fourier coefficients are one. The series doesn't converge in the usual sense, but still has some uses. In particular, look what happens when we take

$$F(\omega) = \sum_{n=-\infty}^{\infty} f(n)e^{-in\omega},$$

for $\pi \le \omega \le \pi$, and calculate

$$\int_{-\pi}^{\pi} F(\omega)\delta(\omega)d\omega = \sum_{n=-\infty}^{\infty} \frac{1}{2\pi} \int_{-\pi}^{\pi} F(\omega)e^{-in\omega}d\omega.$$

We have

$$\int_{-\pi}^{\pi} F(\omega)\delta(\omega)d\omega = \frac{1}{2\pi} \sum_{n=-\infty}^{\infty} f(n) = F(0),$$

where the $f(n)$ are the Fourier coefficients of $F(\omega)$. This means that $\delta(\omega)$ has the *sifting property*, just like we saw with the Dirac delta $\delta(x)$; that is why we call it $\delta(\omega)$. When we shift $\delta(\omega)$ to get $\delta(\omega - \alpha)$, we find that

$$\int_{-\pi}^{\pi} F(\omega)\delta(\omega - \alpha)d\omega = F(\alpha).$$

The "function" $\delta(\omega)$ is the Dirac delta for $\omega$ space.

## 12.9 Random Sinusoidal Sequences

Consider $A = |A|e^{i\theta}$, with amplitude $|A|$ a positive-valued random variable and phase angle $\theta$ a random variable taking values in the interval $[-\pi, \pi]$; then $A$ is a complex-valued random variable. For a fixed frequency $\omega_0$ we define a random sinusoidal sequence $s = \{s_n\}$ by $s_n = Ae^{-in\omega_0}$. We assume that $\theta$ has the uniform distribution over $[-\pi, \pi]$ so that the expected value of $s_n$ is zero. The correlation function for $s$ is

$$\rho_s(m) = E(s_n \overline{s_{n-m}}) = E(|A|^2)e^{-im\omega_0}$$

and the power spectrum of $s$ is

$$R_s(\omega) = E(|A|^2) \sum_{m=-\infty}^{\infty} e^{-im(\omega_0-\omega)},$$

so that, by Equation (12.4), we have

$$R_s(\omega) = 2\pi E(|A|^2)\delta(\omega - \omega_0).$$

We generalize this example to the case of multiple independent sinusoids. Suppose that, for $j = 1, ..., J$, we have fixed frequencies $\omega_j$ and independent complex-valued random variables $A_j$. We let our random sequence be defined by

$$s_n = \sum_{j=1}^{J} A_j e^{-in\omega_j}.$$

Then the correlation function for $s$ is

$$\rho_s(m) = \sum_{j=1}^{J} E(|A_j|^2)e^{-im\omega_j}$$

and the power spectrum for $s$ is

$$R_s(\omega) = 2\pi \sum_{j=1}^{J} E(|A_j|^2)\delta(\omega - \omega_j).$$

This is the commonly used model of independent sinusoids. The problem of *power spectrum estimation* is to determine the values $J$, the frequencies $\omega_j$ and the variances $E(|A_j|^2)$ from finitely many samples from one or more realizations of the random sequence $s$.

---

## 12.10 Random Noise Sequences

Let $q = \{q_n\}$ be an arbitrary wide-sense stationary discrete random sequence, with correlation function $\rho_q(m)$ and power spectrum $R_q(\omega)$. We say that $q$ is *white noise* if $\rho_q(m) = 0$ for $m$ not equal to zero, or, equivalently, if the power spectrum $R_q(\omega)$ is constant over the interval $[-\pi, \pi]$. The *independent sinusoids in additive white noise* model is a random sequence of the form

$$x_n = \sum_{j=1}^{J} A_j e^{-in\omega_j} + q_n.$$

The *signal power* is defined to be $\rho_s(0)$, which is the sum of the $E(|A_j|^2)$, while the noise power is $\rho_q(0)$. The *signal-to-noise ratio* (SNR) is the ratio of signal power to noise power.

## 12.11 Increasing the SNR

It is often the case that the SNR is quite low and it is desirable to process the data from $x$ to enhance this ratio. The data we have is typically finitely many values of one realization of $x$. We say we have $f_n$ for $n = 1, 2, ..., N$; we don't say we have $x_n$ because $x_n$ is the random variable, not one value of the random variable. One way to process the data is to estimate $\rho_x(m)$ for some small number of integers $m$ around zero, using, for example, the *lag products* estimate

$$\hat{\rho}_x(m) = \frac{1}{N-m} \sum_{n=1}^{N-m} f_n \overline{f_{n-m}},$$

for $m = 0, 1, ..., M < N$ and $\hat{\rho}_x(-m) = \overline{\hat{\rho}_x(m)}$. Because $\rho_q(m) = 0$ for $m$ not equal to zero, we will have $\hat{\rho}_x(m)$ approximating $\rho_s(m)$ for nonzero values of $m$, thereby reducing the effect of the noise. Therefore, our estimates of $\rho_s(m)$ are relatively noise-free for $m \neq 0$.

## 12.12 Colored Noise

The additive noise is said to be *correlated* or *non-white* if it is not the case that $\rho_x(m) = 0$ for all nonzero $m$. In this case the noise power spectrum is not constant, and so may be concentrated in certain regions of the interval $[-\pi, \pi]$.

The next few sections deal with applications of random sequences.

## 12.13 Spread-Spectrum Communication

In this section we return to the random-coin-flip model, this time allowing the coin to be biased, that is, $p$ need not be 0.5. Let $s = \{s_n\}$ be a random sequence, such as $s_n = Ae^{in\omega_0}$, with $E(s_n) = \mu$ and correlation function $\rho_s(m)$. Define a second random sequence $x$ by

$$x_n = s_n c_n.$$

The random sequence $x$ is generated from the random signal $s$ by randomly changing its signs. We can show that

$$E(x_n) = \mu(2p - 1)$$

and, for $m$ not equal to zero,

$$\rho_x(m) = \rho_s(m)(2p - 1)^2,$$

with

$$\rho_x(0) = \rho_s(0) + 4p(1 - p)\mu^2.$$

Therefore, if $p = 1$ or $p = 0$ we get $\rho_x(m) = \rho_s(m)$ for all $m$, but for $p = 0.5$ we get $\rho_x(m) = 0$ for $m$ not equal to zero. If the coin is unbiased, then the random sign changes convert the original signal $s$ into white noise. Generally, we have

$$R_x(\omega) = (2p - 1)^2 R_s(\omega) + (1 - (2p - 1)^2)(\mu^2 + \rho_s(0)),$$

which says that the power spectrum of $x$ is a combination of the signal power spectrum and a white-noise power spectrum, approaching the white-noise power spectrum as $p$ approaches 0.5. If the original signal power spectrum is concentrated within a small interval, then the effect of the random sign changes is to spread that spectrum. Once we know what the particular realization of the random sequence $c$ is that has been used, we can recapture the original signal from $s_n = x_n c_n$. The use of such a spread spectrum permits the sending of multiple narrow-band signals, without confusion, as well as protecting against any narrow-band additive interference.

---

## 12.14 Stochastic Difference Equations

The ordinary first-order differential equation $y'(t) + ay(t) = f(t)$, with initial condition $y(0) = 0$, has for its solution $y(t) = e^{-at} \int_0^t e^{as} f(s) ds$. One way to look at such differential equations is to consider $f(t)$ to be the input to a system having $y(t)$ as its output. The system determines which terms will occur on the left side of the differential equation. In many applications the input $f(t)$ is viewed as random noise and the output is then a continuous-time random process. Here we want to consider the discrete analog of such differential equations.

We replace the first derivative with the first difference, $y_{n+1} - y_n$ and we replace the input with the random-coin-flip sequence $c = \{c_n\}$, to obtain the random difference equation

$$y_{n+1} - y_n + ay_n = c_n.$$

With $b = 1 - a$ and $0 < b < 1$ we have

$$y_{n+1} - by_n = c_n. \tag{12.5}$$

The solution is $y = \{y_n\}$ given by

$$y_n = b^{n-1} \sum_{k=-\infty}^{n-1} b^{-k} c_k. \tag{12.6}$$

Comparing this with the solution of the differential equation, we see that the term $b^{n-1}$ plays the role of $e^{-at} = (e^{-a})^t$, so that $b = 1 - a$ is substituting for $e^{-a}$. The infinite sum replaces the infinite integral, with $b^{-k} c_k$ replacing the integrand $e^{as} f(s)$.

The solution sequence $y$ given by Equation (12.6) is a wide-sense stationary random sequence and its correlation function is

$$\rho_y(m) = b^m / (1 - b^2).$$

Since

$$b^{n-1} \sum_{k=-\infty}^{n-1} b^{-k} = \frac{1}{1-b}$$

the random sequence $(1 - b)y_n = ay_n$ is an infinite *moving-average* random sequence formed from the random sequence $c$.

We can derive the solution in Equation (12.6) using *z-transforms*. We write

$$Y(z) = \sum_{n=-\infty}^{\infty} y_n z^{-n},$$

and

$$C(z) = \sum_{n=-\infty}^{\infty} c_n z^{-n}.$$

From Equation (12.5) we have

$$zY(z) - bY(z) = C(z),$$

or

$$Y(z) = C(z)(z - b)^{-1}.$$

Expanding in a geometric series, we get

$$Y(z) = C(z)z^{-1}\Big(1 + bz^{-1} + b^2z^{-2} + ...\Big),$$

from which the solution given in Equation (12.6) follows immediately.

---

## 12.15   Random Vectors and Correlation Matrices

In estimation and detection theory, the task is to distinguish *signal vectors* from *noise vectors*. In order to perform such a task, we need to know how signal vectors differ from noise vectors. Most frequently, what we have is statistical information. The signal vectors of interest, which we denote by $\mathbf{s} = (s_1, ..., s_N)^T$, typically exhibit some patterns of behavior among their entries. For example, a constant signal, such as $\mathbf{s} = (1, 1, ..., 1)^T$, has all its entries identical. A sinusoidal signal, such as $\mathbf{s} = (1, -1, 1, -1, ..., 1, -1)^T$, exhibits a periodicity in its entries. If the signal is a vectorization of a two-dimensional image, then the patterns will be more difficult to describe, but will be there, nevertheless. In contrast, a typical noise vector, denoted $\mathbf{q} = (q_1, ..., q_N)^T$, may have entries that are statistically unrelated to each other, as in white noise. Of course, what is signal and what is noise depends on the context; unwanted interference in radio may be viewed as noise, even though it may be a weather report or a song.

To deal with these notions mathematically, we adopt statistical models. The entries of $\mathbf{s}$ and $\mathbf{q}$ are taken to be random variables, so that $\mathbf{s}$ and $\mathbf{q}$ are random vectors. Often we assume that the mean values, $E(\mathbf{s})$ and $E(\mathbf{q})$, are both equal to the zero vector. Then patterns that may exist among the entries of these vectors are described in terms of *correlations*. The *noise covariance matrix*, which we denote by $Q$, has for its entries $Q_{mn} = E\Big((q_m - E(q_m))\overline{(q_n - E(q_n))}\Big)$, for $m, n = 1, ..., N$. The signal covariance matrix is defined similarly. If $E(q_n) = 0$ and $E(|q_n|^2) = 1$ for each $n$, then $Q$ is the *noise correlation matrix*. Such matrices $Q$ are Hermitian and nonnegative definite, that is, $\mathbf{x}^\dagger Q\mathbf{x}$ is nonnegative, for every vector $\mathbf{x}$. If $Q$ is a positive multiple of the identity matrix, then the noise vector $\mathbf{q}$ is said to be a *white noise* random vector.

## 12.16 The Prediction Problem

An important problem in signal processing is the estimation of the next term in a sequence of numbers from knowledge of the previous values. This is called the *prediction problem*. The numbers might be the values at closing of a certain stock market index; knowing what has happened up to today, can we predict, with some accuracy, tomorrow's closing value? The numbers might describe the position in space of a missile; knowing where it has been for the past few minutes, can we predict where it will be for the next few? The numbers might be the noon-time temperature in New York City on successive days; can we predict tomorrow's temperature from our knowledge of the temperatures on previous days? It is helpful, in weather prediction and elsewhere, to use not only the previous values of the sequence of interest, but those of related sequences; the recent temperatures in Pittsburgh might be helpful in predicting tomorrow's weather in New York City. In this chapter we begin a discussion of the prediction problem.

## 12.17 Prediction Through Interpolation

Suppose that our data are the real numbers $x_1, ..., x_m$, corresponding to times $t = 1, ..., m$. Our goal is to estimate $x_{m+1}$. One way to do this is by interpolation.

A function $f(t)$ is said to *interpolate* the data if $f(n) = x_n$ for $n = 1, ..., m$. Having found such an interpolating function, we can take as our prediction of $x_{m+1}$ the number $\hat{x}_{m+1} = f(m + 1)$. Of course, there are infinitely many choices for the interpolating function $f(t)$. In our discussion of Fourier transform estimation, we considered methods of interpolation that incorporated prior knowledge about the function being sampled, such as that it was band-limited. In the absence of such additional information polynomial interpolation is one obvious choice.

Polynomial interpolation involves selecting as the function $f(t)$ the polynomial of least degree that interpolates the data. Given $m$ data points, we seek a polynomial of degree $m - 1$. Lagrange's method is a well-known procedure for solving this problem.

For $k = 1, ..., m$, let $L_k(t)$ be the unique polynomial of degree $m - 1$ with the properties $L_k(k) = 1$ and $L_k(n) = 0$ for $n = 1, ..., m$ and $n \neq k$.

We can write each $L_k(t)$ explicitly, since we know its zeros:

$$L_k(t) = \frac{(t-1)\cdots(t-(k-1))(t-(k+1))\cdots(t-m)}{(k-1)\cdots(k-(k-1))(k-(k+1))\cdots(k-m)}.$$

Then the polynomial

$$P_m(t) = \sum_{k=1}^{m} x_k L_k(t)$$

is the interpolating polynomial we seek.

**Ex. 12.1** *Show that for $m = 1$ the predicted value of $x_2$ is $\hat{x}_2 = x_1$, so that*

$$\hat{x}_2 - x_1 = 0.$$

*This is the "Tomorrow will be like today" prediction.*

**Ex. 12.2** *Show that for $m = 2$ the predicted value of $x_3$ is $\hat{x}_3 = 2x_2 - x_1$, or $\hat{x}_3 - x_2 = (x_2 - x_1)$ so that*

$$\hat{x}_3 - 2x_2 + x_1 = 0.$$

*This prediction amounts to assuming the change from today to tomorrow will be the same as the change from yesterday to today; that is, we assume a constant slope.*

**Ex. 12.3** *Show that for $m = 3$ the predicted value of $x_4$ is $\hat{x}_4 = 3x_3 - 3x_2 + x_1$, so that*

$$\hat{x}_4 - 3x_3 + 3x_2 - x_1 = 0.$$

**Ex. 12.4** *The coefficients in the previous exercises fit a pattern. Using this pattern, determine the predicted value of $x_5$ for the case of $m = 4$. In general, what will be the predicted value of $x_{m+1}$ based on the $m$ previous values?*

The concept of divided difference plays a significant role in interpolation, as we shall see.

---

## 12.18   Divided Differences

The zeroth *divided difference* of a function $f(t)$ with respect to the point $t_0$ is $f[t_0] = f(t_0)$. The first divided difference with respect to the points $t_0$ and $t_1$ is

$$f[t_0, t_1] = \frac{f(t_1) - f(t_0)}{t_1 - t_0}.$$

The $m$th divided difference with respect to the points $t_0, ..., t_m$ is

$$f[t_0, ..., t_m] = \frac{f[t_1, ..., t_m] - f[t_0, ..., t_{m-1}]}{t_m - t_0}.$$

These quantities are discrete analogs of the derivatives of a function. Indeed, if $f(t)$ is a polynomial of degree at most $m - 1$ then the $m$th divided difference is zero, for any points $t_0, ..., t_m$.

When the points $t_0, ..., t_m$ are consecutive integers the divided differences take on a special form. Suppose $t_0 = 1, t_1 = 2, ..., t_m = m + 1$. Then,

$$f[t_0, t_1] = f(2) - f(1);$$

$$f[t_0, t_1, t_2] = \frac{1}{2}(f(3) - 2f(2) + f(1));$$

$$f[t_0, t_1, t_2, t_3] = \frac{1}{6}(f(4) - 3f(3) + 3f(2) - f(1))$$

and so on, with each successive divided difference involving the coefficients in the expansion of the binomial $(a - b)^k$.

For each fixed value of $m \geq 1$ and $1 \leq n \leq m$, we have $f(n) = x_n$ and $f(m + 1) = \hat{x}_{m+1}$. According to the previous exercises, for $m = 1$ we can write

$$\hat{x}_2 - x_1 = 0,$$

which says that the first divided difference is zero; that is, $f[1, 2] = 0$. For $m = 2$ we have

$$[\hat{x}_3 - x_2] - [x_2 - x_1] = 0,$$

or $f[1, 2, 3] = 0$, so the second divided difference is zero. For $m = 3$

$$[[\hat{x}_4 - x_3] - [x_3 - x_2]] - [[x_3 - x_2] - [x_2 - x_1]] = 0,$$

which says that the third divided difference, $f[1, 2, 3, 4]$, is zero. The interpolation is achieved by assuming that the $m$ data points as well as the point to be interpolated lie on a polynomial of degree at most $m - 1$. Under this assumption the $m$th divided difference with respect to the points $1, 2, ..., m + 1$ would be zero. The interpolated value can then be calculated by setting the $m$th divided difference equal to zero, but replacing $x_{m+1}$ with the estimate $\hat{x}_{m+1}$.

The coefficients that occur in these various predictors are those in the expansion of the binomial $(a - b)^m$. To investigate this matter further, we define the *first difference operator* on an arbitrary sequence $x = \{x_n\}$ to be the operator $D$ such that $y = Dx$, where $y = \{y_n\}$ is the sequence with entries $y_n = x_n - x_{n-1}$. Notice that the operator $D$ can be written as $D = I - S$, where $I$ is the identity operator and $S$ is the shift operator; that is, $Sx = z$ where $z = \{z_n\}$ is the sequence with entries $z_n = x_{n-1}$.

The $k$th difference operator is $D^k = (I - S)^k$; expanding this product in terms of powers of $S$ leads to the binomial coefficients that we saw earlier.

This method of predicting using the interpolating polynomial of degree $m - 1$ will be perfectly accurate if the sequence $\{x_n\}$ is formed by taking values from a polynomial of degree $m-1$ or less. Typically, our data contains noise and interpolating the data exactly, while theoretically possible, is not wise or useful.

The prediction method used here is linear in the sense that our predicted value is a linear combination of the data values and the coefficients we use do not involve the data. Another approach, linear predictive coding, is somewhat different.

---

## 12.19  Linear Predictive Coding

Suppose once again that we have the data $x_1, ..., x_m$ and we want to predict $x_{m+1}$. Instead of using a linear combination of all the values $x_1, ..., x_m$, we choose to use as our prediction of $x_{m+1}$ a linear combination of $x_{m-p}, x_{m-p+1}, ..., x_m$, where $p$ is a positive integer much smaller than $m$. So, our prediction has the form

$$\hat{x}_{m+1} = a_0 x_m + a_1 x_{m-1} + ... + a_p x_{m-p}.$$

To find the best coefficients $a_0, ..., a_p$ to use, we imagine trying out each possible choice of coefficients, using them to predict data values we already know. Specifically, for each set of coefficients $\{a_0, ..., a_p\}$, we form the predictions

$$\hat{x}_{p+2} = a_0 x_{p+1} + a_1 x_p + a_2 x_{p-1} + ... + a_p x_1,$$

$$\hat{x}_{p+3} = a_0 x_{p+2} + a_1 x_{p+1} + a_2 x_p + ... + a_p x_2,$$

and so on, down to

$$\hat{x}_m = a_0 x_{m-1} + a_1 x_{m-2} + ... + a_p x_{m-(p+1)}.$$

Since we already know what the true values are, we can compare the predicted values with the true ones and then find the choice of coefficients that minimizes the average squared error. This amounts to finding the least-squares solution of the system of equations obtained by replacing the predictions with the true values on the left side of the previous equations:

$$\begin{bmatrix} x_{p+1} & x_p & \cdots & x_1 \\ x_{p+2} & x_{p+1} & \cdots & x_2 \\ \cdot & & & \\ \cdot & & & \\ \cdot & & & \\ x_m & x_{m-1} & \cdots & x_{m-p-1} \end{bmatrix} \begin{bmatrix} a_0 \\ a_1 \\ \cdot \\ \cdot \\ \cdot \\ a_p \end{bmatrix} = \begin{bmatrix} x_{p+2} \\ x_{p+3} \\ \cdot \\ \cdot \\ \cdot \\ x_m \end{bmatrix},$$

which we write as $G\mathbf{a} = \mathbf{b}$. Since $m$ is typically larger than $p$, this system is overdetermined. The least-squares solution is

$$\mathbf{a} = (G^\dagger G)^{-1} G^\dagger \mathbf{b}.$$

The resulting set of coefficients is then used to make a linear combination of the values $x_m, ..., x_{m-p}$, which is then our predicted value. But note that although a linear combination of data forms the predicted value, the coefficients are determined from the data values themselves, so the overall method is nonlinear.

This method of prediction forms the basis of a data-compression technique known as *linear predictive coding* (LPC). In many applications a long sequence of numbers has a certain amount of local redundancy, and many of the values can be well predicted from a small number of previous ones, using the method just described. Instead of transmitting the entire sequence of numbers, only some of the numbers, along with the coefficients and occasional outliers, are sent.

The entry in the $k$th row, $n$th column of the matrix $G^\dagger G$ is

$$(G^\dagger G)_{kn} = \sum_{j=1}^{m-p} x_{p+1-k+j}\overline{x_{p+1-n+j}}.$$

If we view the data as values of a stationary random process, then the quantity $\frac{1}{m-p}(G^\dagger G)_{kn}$ is an estimate of the autocorrelation value $r_x(n-k)$. Similarly, the $k$th entry of the vector $G^\dagger \mathbf{b}$ is

$$(G^\dagger \mathbf{b})_k = \sum_{j=1}^{m-p} x_{p+1-k+j}\overline{x_{p+1+j}},$$

and $\frac{1}{m-p}(G^\dagger \mathbf{b})_k$ is an estimate of $r_x(-k)$, for $k = 1, ..., p+1$. This brings us to the problem of predicting the next value for a (possibly nonstationary) random process.

## 12.20 Discrete Random Processes

The most common model used in signal processing is that of a sum of complex exponential functions plus noise. The noise is viewed as a sequence of random variables, and the signal components also may involve random parameters, such as random amplitudes and phase angles. Such models are best studied as particular cases of *discrete random processes*.

A discrete random process is an infinite sequence $\{X_n\}_{n=-\infty}^{+\infty}$ in which each $X_n$ is a complex-valued random variable. The *autocorrelation function* associated with the random process is defined for all index values $m$ and $n$ by $r_x(m,n) = E(X_m\overline{X_n})$, where $E(\cdot)$ is the expectation or expected value operator. For $m = n$ we get $r(n,n) = E(|X_n|^2)$. Generally, we have

$$\text{variance}(X_n) = E(|X_n - E(X_n)|^2) = E(|X_n|^2) - |E(X_n)|^2.$$

### 12.20.1 Wide-Sense Stationary Processes

We say that the random process is *wide-sense stationary* if $E(X_n)$ is independent of $n$ and $r_x(m,n)$ is a function only of the difference, $m - n$. Since $E(X_n)$ does not depend on $n$, it is common to assume that this constant mean has been subtracted, so that $E(X_n) = 0$. Then variance$(X_n) = E(|X_n|^2)$, which is independent of $n$ as well. For the remainder of this chapter all random processes will be wide-sense stationary.

For wide-sense stationary processes the autocorrelation function becomes $r_x(k) = E(X_{n+k}\overline{X_n})$, so that $r_x(0)$ is the constant variance of the $X_n$. The *power spectrum* $R_x(\omega)$ of the random process is defined using the values $r_x(k)$ as its Fourier coeffcients:

$$R_x(\omega) = \sum_{k=-\infty}^{+\infty} r_x(k)e^{ik\omega},$$

for all $\omega$ in the interval $[-\pi, \pi]$. It can be proved that the power spectrum is a nonnegative function of the form $R_x(\omega) = |G(\omega)|^2$ and the autocorrelation sequence $\{r_x(k)\}$ satisfies the equations

$$r_x(k) = \sum_{n=-\infty}^{+\infty} g_{k+n}\overline{g_n},$$

for

$$G(\omega) = \sum_{n=-\infty}^{+\infty} g(n)e^{in\omega}.$$

In practice we will have actual values $X_n = x_n$, for only finitely many of the $X_n$, say for $n = 1, ..., m$. These can be used to estimate the values $r_x(k)$, at least for values of $k$ between, say, $-M/5$ and $M/5$. For example, we could estimate $r_x(k)$ by averaging all the products of the form $x_{k+m}\overline{x_m}$ that we can compute from the data. Clearly, as $k$ gets farther away from zero we have fewer such products, so our average is a less accurate estimate.

Once we have $r_x(k)$, $|k| \leq N$, we form the $N+1$ by $N+1$ *autocorrelation matrix R* having the entries $R_{m,n} = r_x(m-n)$. This autocorrelation matrix is what is used in the design of optimal filtering.

The matrix $R$ is *Hermitian*, that is, $R_{n,m} = \overline{R_{m,n}}$, so that $R^\dagger = R$. An $M$ by $M$ Hermitian matrix $H$ is said to be *nonnegative definite* if, for all complex column vectors $\mathbf{a} = (a_1, ..., a_M)^T$, the *quadratic form* $\mathbf{a}^\dagger H \mathbf{a}$ is a nonnegative number and *positive definite* if such a quadratic form is always positive, when $\mathbf{a}$ is not zero.

**Ex. 12.5** *Show that the autocorrelation matrix $R$ is nonnegative definite. Under what conditions can $R$ fail to be positive definite? Hint: Let*

$$A(\omega) = \sum_{n=1}^{N+1} a_n e^{in\omega}$$

*and express the integral*

$$\int |A(\omega)|^2 R(\omega) d\omega$$

*in terms of the $a_n$ and the $R_{m,n}$.*

In Chapter 13 we shall consider the *maximum entropy* method for estimating the power spectrum from finitely many values of $r_x(k)$.

## 12.20.2   Autoregressive Processes

We noted previously that the case of a discrete-time signal with additive random noise provides a good example of a discrete random process; there are others. One particularly important type is the *autoregressive* (AR) process, which is closely related to ordinary linear differential equations.

When a smooth periodic function has noise added the new function is rough. Imagine, though, a fairly weighty pendulum of a clock, moving smoothly and periodically. Now imagine that a young child is throwing small stones at the bob of the pendulum. The movement of the pendulum is no longer periodic, but it is not rough. The pendulum is moving randomly in response to the random external disturbance, but not as if a random noise component has been added to its motion. To model such random processes we need to extend the notion of an ordinary differential equation. That leads us to the AR processes.

Recall that an ordinary linear $M$th order differential equation with constant coefficients has the form

$$x^{(M)}(t) + c_1 x^{(M-1)}(t) + c_2 x^{(M-2)}(t) + ... + c_{M-1} x'(t) + c_M x(t) = f(t),$$

where $x^{(m)}(t)$ denotes the $m$th derivative of the function $x(t)$ and the $c_m$ are constants. In many applications the variable $t$ is time and the function $f(t)$ is an external effect driving the linear system, with system response given by the unknown function $x(t)$. How the system responds to a variety of external drivers is of great interest. It is sometimes convenient to replace this continuous formulation with a discrete analog called a *difference equation*.

In switching from differential equations to difference equations, we discretize the time variable and replace the driving function $f(t)$ with $f_n$, $x(t)$ with $x_n$, the first derivative at time $t$, $x'(t)$, with the first difference, $x_n - x_{n-1}$, the second derivative $x''(t)$ with the second difference, $(x_n - x_{n-1}) - (x_{n-1} - x_{n-2})$, and so on. The differential equation is then replaced by the difference equation

$$x_n - a_1 x_{n-1} - a_2 x_{n-2} - ... - a_M x_{n-M} = f_n \tag{12.7}$$

for some constants $a_m$; the negative signs are a technical convenience only.

We now assume that the driving function is a discrete random process $\{f_n\}$, so that the system response becomes a discrete random process, $\{X_n\}$. If we assume that the driver $f_n$ is a mean-zero white noise process that is independent of the $\{X_n\}$, then the process $\{X_n\}$ is called an *autoregressive* (AR) process. What the system does at time $n$ depends partly on what it has done at the $M$ discrete times prior to time $n$, as well as partly on what the external disturbance $f_n$ is at time $n$. Our goal is usually to determine the constants $a_m$; this is *system identification*. Our data is typically some number of consecutive measurements of the $X_n$.

Multiplying both sides of Equation (12.7) by $\overline{X_{n-k}}$, for some $k > 0$ and taking the expected value, we obtain

$$E(X_n \overline{X_{n-k}}) - ... - a_M E(X_{n-M} \overline{X_{n-k}}) = 0,$$

or

$$r_x(k) - a_1 r_x(k-1) - ... - a_M r_x(k-M) = 0.$$

Taking $k = 0$, we get

$$r_x(0) - a_1 r_x(-1) - ... - a_M r_x(-M) = E(|f_n|^2) = \text{var}(f_n).$$

To find the $a_m$ we use the data to estimate $r_x(k)$ at least for $k = 0, 1, ..., M$. Then, we use these estimates in the previous linear equations, solving them for the $a_m$.

## 12.20.3 Linear Systems with Random Input

In our discussion of discrete linear filters, also called time-invariant linear systems, we noted that it is common to consider as the input to such a system a discrete random process, $\{X_n\}$. The output is then another random process $\{Y_n\}$ given by

$$Y_n = \sum_{m=-\infty}^{+\infty} g_m X_{n-m},$$

for each $n$.

**Ex. 12.6** *Show that if the input process is wide-sense stationary then so is the output. Show that the power spectrum $R_y(\omega)$ of the output is*

$$R_y(\omega) = |G(\omega)|^2 R_x(\omega).$$

---

## 12.21   Stochastic Prediction

In time series analysis, stochastic prediction methods are studied. In that case the numbers $x_n$ are viewed as values of a discrete random process $\{X_n\}$. The coefficients are determined by considering the statistical description of how the random variable $X_{m+1}$ is related to the previous $X_n$. The prediction of $X_{m+1}$ is a linear combination of the random variables $X_n$, $n = 1, ..., m$,

$$\hat{X}_{m+1} = a_0 X_m + a_1 X_{m-1} + ... + a_{m-1} X_1,$$

with the coefficients determined using the orthogonality principle. Consequently, the coefficients satisfy the system of linear equations

$$E(X_{m+1}\overline{X_k}) = a_0 E(X_m \overline{X_k}) + ... + a_{m-1} E(X_1 \overline{X_k}),$$

for $k = 1, 2, ..., m$. The expected values in these equations are the autocorrelations associated with the random process.

### 12.21.1   Prediction for an Autoregressive Process

Suppose that the random process $\{X_n\}$ is an $M$th order AR process, so that

$$X_n - a_1 X_{n-1} - ... - a_M X_{n-M} = f_n,$$

where $\{f_n\}$ is a mean-zero white noise process, independent of the $\{X_n\}$.

**Ex. 12.7** *Use our earlier discussion of the relationship between the autocorrelation values $r_x(k)$ and the coefficients $a_m$ to show that the best* *linear predictor for the random variable $X_n$ in terms of the values of $X_{n-1}, ..., X_{n-M}$ is*

$$\hat{X}_n = a_1 X_{n-1} + ... + a_M X_{n-M}$$

*and the mean-squared error is*

$$E(|\hat{X}_n - X_n|^2) = \text{var}(f_n).$$

In fact, it can be shown that, because the process is an $M$th order AR process, this is the best linear predictor of $X_n$ in terms of the entire history of the process.

# Chapter 13

# Nonlinear Methods

## 13.1 Chapter Summary

It is common to speak of classical, as opposed to modern, signal processing methods. In this chapter we describe briefly the distinction. Then we discuss entropy maximization, eigenvector methods, and related nonlinear methods in signal processing. We first encounter infinite series expansions for functions in calculus when we study Maclaurin and Taylor series. Fourier series are usually first met in different contexts, such as partial differential equations and boundary value problems. Laurent expansions come later when we study functions of a complex variable. There are, nevertheless, important connections among these different types of infinite series expansions that we consider in this chapter.

## 13.2 The Classical Methods

In [48] Candy locates the beginning of the classical period of spectral estimation in Schuster's use of Fourier techniques in 1898 to analyze sunspot data [138]. The role of Fourier techniques grew with the discovery, by Wiener in the USA and Khintchine in the USSR, of the relation between the power spectrum and the autocorrelation function. Much of Wiener's important work on control and communication remained classified and became known only with the publication of his classic text *Time Series* in 1949 [162]. The book by Blackman and Tukey, *Measurement of Power Spectra* [10], provides perhaps the best description of the classical methods. With the discovery of the FFT by Cooley and Tukey in 1965, all the pieces were in place for the rapid development of this DFT-based approach to spectral estimation.

## 13.3 Modern Signal Processing and Entropy

Until about the middle of the 1970s most signal processing depended almost exclusively on the DFT, as implemented using the FFT. Algorithms such as the Gerchberg-Papoulis bandlimited extrapolation method were performed as iterative operations on finite vectors, using the FFT at every step. Linear filters and related windowing methods involving the FFT were also used to enhance the resolution of the reconstructed objects. The proper design of these filters was an area of interest to quite a number of researchers, John Tukey among them. Then, around the end of that decade, interest in entropy maximization began to grow, as researchers began to wonder if high-resolution methods developed for seismic oil exploration could be applied successfully in other areas.

John Burg had developed his *maximum entropy method* (MEM) while working in the oil industry in the 1960s. He then went to Stanford as a mature graduate student and received his doctorate in 1975 for a thesis based largely on his earlier work on MEM [21]. This thesis and a handful of earlier presentations at meetings [19, 20] fueled the interest in entropy.

It was not only the effectiveness of Burg's techniques that attracted the attention of members of the signal-processing community. The classical methods seemed to some to be *ad hoc*, and they sought a more intellectually satisfying basis for spectral estimation. Classical methods start with the time series data, say $x_n$, for $n = 1, ..., N$. In the direct approach, slightly simplified, the data is *windowed*; that is, $x_n$ is replaced with $x_n w_n$ for some choice of constants $w_n$. Then, the vDFT is computed, using the FFT, and the squared magnitudes of the entries of the vDFT provide the desired estimate of the power spectrum. In the more indirect approach, autocorrelation values $r_x(m)$ are first estimated, for $m = 0, 1, ..., M$, where $M$ is some fraction of the data length $N$. Then, these estimates of $r_x(m)$ are windowed and the vDFT calculated, again using the FFT.

What some people objected to was the use of these windows. After all, the measured data was $x_n$, not $x_n w_n$, so why corrupt the data at the first step? The classical methods produced answers that depended to some extent on which window function one used; there had to be a better way. Entropy maximization was the answer to their prayers.

In 1981 the first of several international workshops on entropy maximization was held at the University of Wyoming, bringing together most of the people working in this area. The books [145] and [146] contain the papers presented at those workshops. As one can see from reading those papers, the general theme is that a new day has dawned.

## 13.4 Related Methods

It was soon recognized that maximum entropy methods were closely related to model-based techniques that had been part of statistical time series for decades. This realization led to a broader use of *autoregressive* (AR) and *autoregressive, moving average* (ARMA) models for spectral estimation [129], as well as of eigenvector methods, such as Pisarenko's method [126]. What Candy describes as the modern approach to spectral estimation is one based on explicit parametric models, in contrast to the classical non-parametric approach. The book edited by Don Childers [53] is a collection of journal articles that captures the state-of-the-art at the end of the 1970s.

In a sense the transition from the classical ways to the modern methods solved little; the choice of models is as *ad hoc* as the choice of windows was before. On the other hand, we do have a wider collection of techniques from which to choose and we can examine these techniques to see when they perform well and when they do not. We do not expect one approach to work in all cases. High-speed computation permits the use of more complicated parametric models tailored to the physics of a given situation.

Our estimates are intended to be used for some purpose. In medical imaging a doctor is going to make a diagnosis based in part on what the image reveals. How good the image needs to be depends on the purpose for which it is made. Judging the quality of a reconstructed image based on somewhat subjective criteria, such as how useful it is to a doctor, is a problem that is not yet solved. Human-observer studies are one way to obtain this nonmathematical evaluation of reconstruction and estimation methods. The next step beyond that is to develop computer software that judges the images or spectra as a human would.

---

## 13.5 Entropy Maximization

The problem of estimating the nonnegative function $R(\omega)$, for $|\omega| \leq \pi$, from the finitely many Fourier coefficients

$$r(n) = \int_{-\pi}^{\pi} R(\omega) \exp(-in\omega) d\omega/2\pi, \, n = -N, ..., N$$

is an *under-determined problem*, meaning that the data alone is insufficient to determine a unique answer. In such situations we must select one solution out of the infinitely many that are mathematically possible. The obvious questions we need to answer are: What criteria do we use in this selection? How do we find algorithms that meet our chosen criteria? In this chapter we look at some of the answers people have offered and at one particular algorithm, Burg's *maximum entropy method* (MEM) [19, 20].

## 13.6 Estimating Nonnegative Functions

The values $r(n)$ are autocorrelation-function values associated with a random process having $R(\omega)$ for its power spectrum. In many applications, such as seismic remote sensing, these autocorrelation values are estimates obtained from relatively few samples of the underlying random process, so that $N$ is not large. The DFT estimate,

$$R_{DFT}(\omega) = \sum_{n=-N}^{N} r(n) \exp(in\omega),$$

is real-valued and consistent with the data, but is not necessarily nonnegative. For small values of $N$, the DFT may not be sufficiently resolving to be useful. This suggests that one criterion we can use to perform our selection process is to require that the method provide better resolution than the DFT for relatively small values of $N$, when reconstructing power spectra that consist mainly of delta functions.

## 13.7 Philosophical Issues

Generally speaking, we would expect to do a better job of estimating a function from data pertaining to that function if we also possess additional prior information about the function to be estimated and are able to employ estimation techniques that make use of that additional information. There is the danger, however, that we may end up with an answer that is influenced more by our prior guesses than by the actual measured data. Striking a balance between including prior knowledge and letting the data speak for itself is a noble goal; how to achieve that is the question. At this stage, we begin to suspect that the problem is as much philosophical as it is mathematical.

We are essentially looking for principles of induction that enable us to extrapolate from what we have measured to what we have not. Unwilling to turn the problem over entirely to the philosophers, a number of mathematicians and physicists have sought mathematical solutions to this inference problem, framed in terms of what the *most likely* answer is, or which answer involves the smallest amount of additional prior information [60]. This is not, of course, a new issue; it has been argued for centuries with regard to the use of what we now call Bayesian statistics; *objective* Bayesians allow the use of prior information, but only if it is the "right" prior information. The interested reader should consult the books [145] and [146], containing papers by Ed Jaynes, Roy Frieden, and others originally presented at workshops on this topic held in the early 1980s.

The maximum entropy method is a general approach to such problems that includes Burg's algorithm as a particular case. It is argued that by maximizing entropy we are, in some sense, being maximally noncommittal about what we do not know and thereby introducing a minimum of prior knowledge (some would say prior guesswork) into the solution. In the case of Burg's MEM, a somewhat more mathematical argument is available.

Let $\{X_n\}_{n=-\infty}^{\infty}$ be a stationary random process with autocorrelation sequence $r(m)$ and power spectrum $R(\omega)$, $|\omega| \leq \pi$. The prediction problem is the following: Suppose we have measured the values, at "times" prior to $n$, of one realization of the process and we want to predict the value of the process at time $n$. On average, how much error do we expect to make in predicting $X_n$ from knowledge of the infinite past? The answer, according to Szegö's Theorem [93], is

$$\exp \left( \int_{-\pi}^{\pi} \log R(\omega) d\omega \right) ;$$

the integral

$$\int_{-\pi}^{\pi} \log R(\omega) d\omega$$

is the *Burg entropy* of the random process [129]. Processes that are very predictable have low entropy, while those that are quite unpredictable, or, like white noise, completely unpredictable, have high entropy; to make entropies comparable, we assume a fixed value of $r(0)$. Given the data $r(n)$, $|n| \leq N$, Burg's method selects that power spectrum consistent with these autocorrelation values that corresponds to the most unpredictable random process.

Other similar procedures are also based on selection through optimization. We have seen the minimum norm approach to finding a solution to an underdetermined system of linear equations, and the minimum expected squared error approach in statistical filtering, and later we shall see the maximum likelihood method used in detection. We must keep in mind that, however comforting it may be to feel that we are on solid philosophical ground (if such exists) in choosing our selection criteria, if the method does not work well, we must use something else. As we shall see, the MEM, like every other reasonable method, works well sometimes and not so well other times. There is certainly philosophical precedent for considering the consequences of our choices, as Blaise Pascal's famous wager about the existence of God nicely illustrates. As an attentive reader of the books [145] and [146] will surely note, there is a certain theological tone to some of the arguments offered in support of entropy maximization. One group of authors (reference omitted) went so far as to declare that entropy maximization was what one did if one cared what happened to one's data.

The objective of Burg's MEM for estimating a power spectrum is to seek better resolution by combining nonnegativity and data-consistency in a single closed-form estimate. The MEM is remarkable in that it is the only closed-form (that is, noniterative) estimation method that is guaranteed to produce an estimate that is both nonnegative and consistent with the autocorrelation samples. Later we shall consider a more general method, the inverse PDFT (IPDFT), that is both data-consistent and positive in most cases.

## 13.8 The Autocorrelation Sequence $\{r(n)\}$

We begin our discussion with important properties of the sequence $\{r(n)\}$. Because $R(\omega) \geq 0$, the values $r(n)$ are often called *autocorrelation values*.

Since $R(\omega) \geq 0$, it follows immediately that $r(0) \geq 0$. In addition, $r(0) \geq |r(n)|$ for all $n$:

$$|r(n)| = \left| \int_{-\pi}^{\pi} R(\omega) \exp(-in\omega) d\omega/2\pi \right|$$

$$\leq \int_{-\pi}^{\pi} R(\omega) |\exp(-in\omega)| d\omega/2\pi = r(0).$$

In fact, if $r(0) = |r(n)| > 0$ for some $n > 0$, then $R$ is a sum of at most $n + 1$ delta functions with nonnegative amplitudes. To see this, suppose that $r(n) = |r(n)| \exp(i\theta) = r(0) \exp(i\theta)$. Then,

$$\int_{-\pi}^{\pi} R(\omega) |1 - \exp(i(\theta + n\omega))|^2 d\omega/2\pi$$

$$= \int_{-\pi}^{\pi} R(\omega)(1 - \exp(i(\theta + n\omega)))(1 - \exp(-i(\theta + n\omega))) d\omega/2\pi$$

$$= \int_{-\pi}^{\pi} R(\omega)[2 - \exp(i(\theta + n\omega)) - \exp(-i(\theta + n\omega))] d\omega/2\pi$$

$$= 2r(0) - \exp(i\theta)\overline{r(n)} - \exp(-i\theta)r(n) = 2r(0) - r(0) - r(0) = 0.$$

Therefore, $R(\omega) > 0$ only at the values of $\omega$ where $|1 - \exp(i(\theta + n\omega))|^2 = 0$; that is, only at $\omega = n^{-1}(2\pi k - \theta)$ for some integer $k$. Since $|\omega| \leq \pi$, there are only finitely many such $k$.

This result is important in any discussion of resolution limits. It is natural to feel that if we have only the Fourier coefficients $r(n)$ for $|n| \leq N$ then we have only the low frequency information about the function $R(\omega)$. How is it possible to achieve higher resolution? Notice, however, that in the case just considered, the infinite sequence of Fourier coefficients is periodic. Of course, we do not know this *a priori*, necessarily. The fact that $|r(N)| = r(0)$ does not, *by itself*, tell us that $R(\omega)$ consists solely of delta functions and that the sequence of Fourier coefficients is periodic. But, under the added assumption that $R(\omega) \geq 0$, it does! When we put in this prior information about $R(\omega)$ we find that the data now tells us more than it did before. This is a good example of the point made in the Introduction; to get information out we need to put information in.

In discussing the Burg MEM estimate, we shall need to refer to the concept of *minimum-phase* vectors. We consider that briefly now.

---

## 13.9 Minimum-Phase Vectors

We say that the finite column vector with complex entries $(a_0, a_1, ..., a_N)^T$ is a *minimum-phase* vector if the complex polynomial

$$A(z) = a_0 + a_1 z + ... + a_N z^N$$

has the property that $A(z) = 0$ implies that $|z| > 1$; that is, all roots of $A(z)$ are outside the unit circle. Consequently, the function $B(z)$ given by $B(z) = 1/A(z)$ is analytic in a disk centered at the origin and including the unit circle. Therefore, we can write

$$B(z) = b_0 + b_1 z + b_2 z^2 + ...,$$

and taking $z = \exp(i\omega)$, we get

$$B(\exp(i\omega)) = b_0 + b_1 \exp(i\omega) + b_2 \exp(2i\omega) + ... .$$

The point here is that $B(\exp(i\omega))$ is a one-sided trigonometric series, with only terms corresponding to $\exp(in\omega)$ for nonnegative $n$.

---

## 13.10 Burg's MEM

The approach is to estimate $R(\omega)$ by the function $S(\omega) > 0$ that maximizes the so-called Burg entropy, $\int_{-\pi}^{\pi} \log S(\omega) d\omega$, subject to the data constraints.

The Euler–Lagrange equation from the calculus of variations allows us to conclude that $S(\omega)$ has the form

$$S(\omega) = 1/H(\omega)$$

for

$$H(\omega) = \sum_{n=-N}^{N} h_n e^{in\omega} > 0.$$

From the Fejér–Riesz Theorem 13.2 we know that $H(\omega) = |A(e^{i\omega})|^2$ for minimum phase $A(z)$. As we now show, the coefficients $a_n$ satisfy a system of linear equations formed using the data $r(n)$.

Given the data $r(n), |n| \leq N$, we form the *autocorrelation matrix* $R$ with entries $R_{mn} = r(m-n)$, for $-N \leq m, n \leq N$. Let $\delta$ be the column vector $\delta = (1, 0, ..., 0)^T$. Let $a = (a_0, a_1, ..., a_N)^T$ be the solution of the system $Ra = \delta$. Then, Burg's MEM estimate is the function $S(\omega) = R_{MEM}(\omega)$ given by

$$R_{MEM}(\omega) = a_0/|A(\exp(i\omega))|^2, |\omega| \leq \pi.$$

Once we show that $a_0 \geq 0$, it will be obvious that $R_{MEM}(\omega) \geq 0$. We also must show that $R_{MEM}$ is data-consistent; that is,

$$r(n) = \int_{-\pi}^{\pi} R_{MEM}(\omega) \exp(-in\omega) d\omega/2\pi =, \ n = -N, ..., N.$$

Let us write $R_{MEM}(\omega)$ as a Fourier series; that is,

$$R_{MEM}(\omega) = \sum_{n=-\infty}^{+\infty} q(n) \exp(in\omega), |\omega| \leq \pi.$$

From the form of $R_{MEM}(\omega)$, we have

$$R_{MEM}(\omega)\overline{A(\exp(i\omega))} = a_0 B(\exp(i\omega)). \tag{13.1}$$

Suppose, as we shall see shortly, that $A(z)$ has all its roots outside the unit circle, so $B(\exp(i\omega))$ is a one-sided trigonometric series, with only terms corresponding to $\exp(in\omega)$ for nonnegative $n$. Then, multiplying on the left side of Equation (13.1), and equating coefficients corresponding to $n = 0, -1, -2, ...$, we find that, provided $q(n) = r(n)$, for $|n| \leq N$, we must have $Ra = \delta$. Notice that these are precisely the same equations we solve in calculating the coefficients of an AR process. For that reason the MEM is sometimes called an autoregressive method for spectral estimation.

### 13.10.1    The Minimum-Phase Property

We now show that if $Ra = \delta$ then $A(z)$ has all its roots outside the unit circle. Let $r \exp(i\theta)$ be a root of $A(z)$. Then, write

$$A(z) = (z - r \exp(i\theta))C(z),$$

where

$$C(z) = c_0 + c_1 z + c_2 z^2 + ... + c_{N-1} z^{N-1}.$$

The vector $a = (a_0, a_1, ..., a_N)^T$ can be written as $a = -r \exp(i\theta)c + d$, where $c = (c_0, c_1, ..., c_{N-1}, 0)^T$ and $d = (0, c_0, c_1, ..., c_{N-1})^T$. So, $\delta = Ra = -r \exp(i\theta)Rc + Rd$ and

$$0 = d^\dagger \delta = -r \exp(i\theta)d^\dagger Rc + d^\dagger Rd,$$

so that

$$r \exp(i\theta)d^\dagger Rc = d^\dagger Rd.$$

From the Cauchy Inequality we know that

$$|d^\dagger Rc|^2 \le (d^\dagger Rd)(c^\dagger Rc) = (d^\dagger Rd)^2, \qquad (13.2)$$

where the last equality comes from the special form of the matrix $R$ and the similarity between $c$ and $d$.

With

$$D(\omega) = c_0 e^{i\omega} + c_1 e^{2i\omega} ... + c_{N-1} e^{iN\omega}$$

and

$$C(\omega) = c_0 + c_1 e^{i\omega} + ... + c_{N-1} e^{i(N-1)\omega},$$

we can easily show that

$$d^\dagger Rd = c^\dagger Rc = \frac{1}{2\pi} \int_{-\pi}^{\pi} R(\omega)|D(\omega)|^2 d\omega$$

and

$$d^\dagger Rc = \frac{1}{2\pi} \int_{-\pi}^{\pi} R(\omega)\overline{D(\omega)}C(\omega)d\omega.$$

If there is equality in the Cauchy Inequality (13.2), then $r = 1$ and we would have

$$\exp(i\theta)\frac{1}{2\pi} \int_{-\pi}^{\pi} R(\omega)\overline{D(\omega)}C(\omega)d\omega = \frac{1}{2\pi} \int_{-\pi}^{\pi} R(\omega)|D(\omega)|^2 d\omega.$$

From the Cauchy Inequality for integrals, we can conclude that

$$\exp(i\theta)\overline{D(\omega)}C(\omega) = |D(\omega)|^2$$

for all $\omega$ for which $R(\omega) > 0$. But,

$$\exp(i\omega)C(\omega) = D(\omega).$$

Therefore, we cannot have $r = 1$ unless $R(\omega)$ consists of a single delta function; that is, $R(\omega) = \delta(\omega - \theta)$. In all other cases we have

$$|d^\dagger Rc|^2 < |r|^2 |d^\dagger Rc|^2,$$

from which we conclude that $|r| > 1$.

## 13.10.2 Solving $Ra = \delta$ Using Levinson's Algorithm

Because the matrix $R$ is Toeplitz, that is, constant on diagonals, and positive definite, there is a fast algorithm for solving $Ra = \delta$ for $a$. Instead of a single $R$, we let $R_M$ be the matrix defined for $M = 0, 1, ..., N$ by

$$R_M = \begin{bmatrix} r(0) & r(-1) & ... & r(-M) \\ r(1) & r(0) & ... & r(-M+1) \\ & \cdot & & \\ & \cdot & & \\ & \cdot & & \\ r(M) & r(M-1) & ... & r(0) \end{bmatrix}$$

so that $R = R_N$. We also let $\delta^M$ be the $(M+1)$-dimensional column vector $\delta^M = (1, 0, ..., 0)^T$. We want to find the column vector $a^M = (a_0^M, a_1^M, ..., a_M^M)^T$ that satisfies the equation $R_M a^M = \delta^M$. The point of Levinson's algorithm is to calculate $a^{M+1}$ quickly from $a^M$.

For fixed $M$ find constants $\alpha$ and $\beta$ so that

$$\delta^M = R_M \left\{ \alpha \begin{bmatrix} a_0^{M-1} \\ a_1^{M-1} \\ \cdot \\ \cdot \\ \cdot \\ a_{M-1}^{M-1} \\ 0 \end{bmatrix} + \beta \begin{bmatrix} 0 \\ \overline{a}_{M-1}^{M-1} \\ \overline{a}_{M-2}^{M-1} \\ \cdot \\ \cdot \\ \cdot \\ \overline{a}_0^{M-1} \end{bmatrix} \right\}$$

$$= \left\{ \alpha \begin{bmatrix} 1 \\ 0 \\ \cdot \\ \cdot \\ \cdot \\ 0 \\ \gamma^M \end{bmatrix} + \beta \begin{bmatrix} \overline{\gamma}^M \\ 0 \\ \cdot \\ \cdot \\ \cdot \\ 0 \\ 1 \end{bmatrix} \right\},$$

where

$$\gamma^M = r(M)a_0^{M-1} + r(M-1)a_1^{M-1} + ... + r(1)a_{M-1}^{M-1}.$$

We then have

$$\alpha + \beta \overline{\gamma^M} = 1, \; \alpha \gamma^M + \beta = 0$$

or

$$\beta = -\alpha \gamma^M, \; \alpha - \alpha |\gamma^M|^2 = 1,$$

so

$$\alpha = 1/(1 - |\gamma^M|^2), \; \beta = -\gamma^M/(1 - |\gamma^M|^2).$$

Therefore, the algorithm begins with $M = 0$, $R_0 = [r(0)]$, $a_0^0 = r(0)^{-1}$. At each step calculate the $\gamma^M$, solve for $\alpha$ and $\beta$ and form the next $a^M$.

The MEM resolves better than the DFT when the true power spectrum being reconstructed is a sum of delta functions plus a flat background. When the background itself is not flat, performance of the MEM degrades rapidly; the MEM tends to interpret any nonflat background in terms of additional delta functions. In the next chapter we consider an extension of the MEM, called the indirect PDFT (IPDFT), that corrects this flaw.

Why Burg's MEM and the IPDFT are able to resolve closely spaced sinusoidal components better than the DFT is best answered by studying the eigenvalues and eigenvectors of the matrix $R$; we turn to this topic in Chapter 14.

---

## 13.11    A Sufficient Condition for Positive-Definiteness

If the function

$$R(\omega) = \sum_{n=-\infty}^{\infty} r(n) e^{in\omega}$$

is nonnegative on the interval $[-\pi, \pi]$, then the matrices $R_M$ are nonnegative definite for every $M$. Theorems by Herglotz and by Bochner go in the reverse direction [2]. Katznelson [99] gives the following result.

**Theorem 13.1** *Let $\{f(n)\}_{n=-\infty}^{\infty}$ be a sequence of nonnegative real numbers converging to zero, with $f(-n) = f(n)$ for each $n$. If, for each $n > 0$, we have*

$$(f(n-1) - f(n)) - (f(n) - f(n+1)) > 0,$$

*then there is a nonnegative function $R(\omega)$ on the interval $[-\pi, \pi]$ with $f(n) = r(n)$ for each $n$.*

The following figures illustrate the behavior of the MEM. In Figures 13.1, 13.2, and 13.3, the true object has two delta functions at $0.95\pi$ and $1.05\pi$. The data is $f(n)$ for $|n| \le 10$. The DFT cannot resolve the two spikes. The

SNR is high in Figure 13.1, and the MEM easily resolves them. In Figure 13.2 the SNR is much lower and MEM no longer resolves the spikes.

**Ex. 13.1** *In Figure 13.3 the SNR is much higher than in Figure 13.1. Explain why the graph looks as it does.*

In Figure 13.4 the true object is a box supported between $0.75\pi$ and $1.25\pi$. Here $N = 10$, again. The MEM does a poor job reconstructing the box. This weakness in MEM will become a problem in the last two figures, in which the true object consists of the box with the two spikes added. In Figure 13.5 we have $N = 10$, while, in Figure 13.6, $N = 25$.

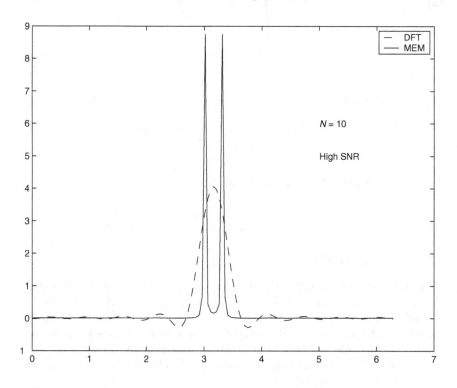

**FIGURE 13.1**: The DFT and MEM, $N = 10$, high SNR.

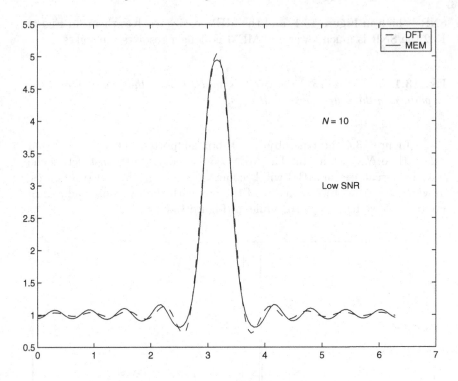

**FIGURE 13.2**: The DFT and MEM, $N = 10$, low SNR.

## 13.12 The IPDFT

Experience with Burg's MEM shows that it is capable of resolving closely spaced delta functions better than the DFT, provided that the background is flat. When the background is not flat, MEM tends to interpret the non-flat background as additional delta functions to be resolved. In this chapter we consider an extension of MEM based on the PDFT that can resolve in the presence of non-flat background. This method is called the *indirect* PDFT (IPDFT) [26].

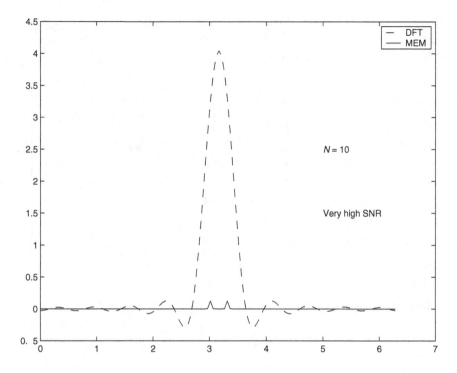

**FIGURE 13.3**: The DFT and MEM, $N = 10$, very high SNR. What happened?

## 13.13  The Need for Prior Information in Nonlinear Estimation

As we saw previously, the PDFT is a linear method for incorporating prior knowledge into the estimation of the Fourier transform. Burg's MEM is a nonlinear method for estimating a non-negative Fourier transform.

The IPDFT applies to the reconstruction of one-dimensional power spectra, but the main idea can be used to generate high-resolution methods for multi-dimensional spectra as well. The IPDFT method is suggested by considering the MEM equations $Ra = \delta$ as a particular case of the equations that arise in Wiener filter approximation. As in the previous chapter, we assume that we have the autocorrelation values $r(n)$ for $|n| \leq N$, from

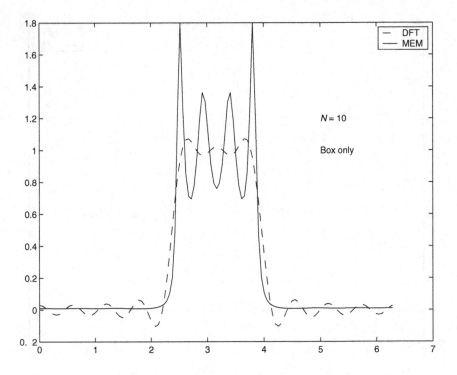

**FIGURE 13.4**: MEM and DFT for a box object; $N = 10$.

which we wish to estimate the power spectrum

$$R(\omega) = \sum_{n=-\infty}^{+\infty} r(n)e^{in\omega}, \ |\omega| \leq \pi.$$

## 13.14  What Wiener Filtering Suggests

In Chapter 20 on Wiener filter approximation, we show that the best finite length filter approximation of the Wiener filter $H(\omega)$ is obtained by minimizing the integral in Equation (20.3)

$$\int_{-\pi}^{\pi} \left| H(\omega) - \sum_{k=-K}^{L} f_k e^{ik\omega} \right|^2 (R_s(\omega) + R_u(\omega))d\omega.$$

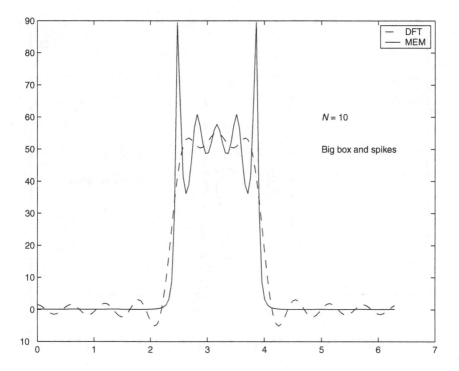

**FIGURE 13.5**: The DFT and MEM: two spikes on a large box; $N = 10$.

The optimal coefficients then must satisfy Equation (20.4):

$$r_s(m) = \sum_{k=-K}^{L} f_k(r_s(m-k) + r_u(m-k)), \qquad (13.3)$$

for $-K \leq m \leq L$.

Consider the case in which the power spectrum we wish to estimate consists of a signal component that is the sum of delta functions and a noise component that is white noise. If we construct a finite-length Wiener filter that filters out the signal component and leaves only the noise, then that filter should be able to zero out the delta-function components. By finding the locations of those zeros, we can find the supports of the delta functions. So the approach is to reverse the roles of signal and noise, viewing the signal as the component called $u$ and the noise as the component called $s$ in the discussion of the Wiener filter. The autocorrelation function $r_s(n)$ corresponds to the white noise now and so $r_s(n) = 0$ for $n \neq 0$. The terms $r_s(n) + r_u(n)$ are the data values $r(n)$, for $|n| \leq N$. Taking $K = 0$ and

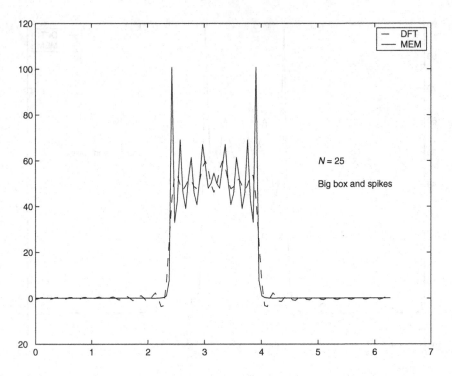

**FIGURE 13.6**: The DFT and MEM: two spikes on a large box; $N = 25$.

$L = N$ in Equation (13.3), we obtain

$$\sum_{k=0}^{N} f_k r(m - k) = 0,$$

for $m = 1, 2, ..., N$ and

$$\sum_{k=0}^{N} f_k r(0 - k) = r(0),$$

which is precisely that same system $R\mathbf{a} = \delta$ that occurs in MEM.

This approach reveals that the vector $\mathbf{a} = (a_0, ..., a_N)^T$ we find in MEM can be viewed as a finite-length approximation of the Wiener filter designed to remove the delta-function component and to leave the remaining flat white-noise component untouched. The polynomial

$$A(\omega) = \sum_{n=0}^{N} a_n e^{in\omega}$$

will then have zeros near the supports of the delta functions. What happens to MEM when the background is not flat is that the filter tries to eliminate any component that is not white noise and so places the zeros of $A(\omega)$ in the wrong places.

---

## 13.15   Using a Prior Estimate

Suppose we take $P(\omega) \geq 0$ to be our estimate of the background component of $R(\omega)$; that is, we believe that $R(\omega)$ equals a multiple of $P(\omega)$ plus a sum of delta functions. We now ask for the finite length approximation of the Wiener filter that removes the delta functions and leaves any background component that looks like $P(\omega)$ untouched. We then take $r_s(n) = p(n)$, where

$$P(\omega) = \sum_{n=-\infty}^{+\infty} p(n)e^{in\omega}, \ |\omega| \leq \pi.$$

The desired filter is $\mathbf{f} = (f_0, ..., f_N)^T$ satisfying the equations

$$p(m) = \sum_{k=0}^{N} f_k r(m-k). \tag{13.4}$$

Once we have found $\mathbf{f}$ we form the polynomial

$$F(\omega) = \sum_{k=0}^{N} f_k e^{ik\omega}, \ |\omega| \leq \pi.$$

The zeros of $F(\omega)$ should then be near the supports of the delta function components of the power spectrum $R(\omega)$, provided that our original estimate of the background is not too inaccurate.

In the PDFT it is important to select the prior estimate $P(\omega)$ nonzero wherever the function being reconstructed is nonzero; for the IPDFT the situation is different. Comparing Equation (13.4) with Equation (2.23), we see that in the IPDFT the true $R(\omega)$ is playing the role previously given to $P(\omega)$, while $P(\omega)$ is in the role previously played by the function we wished to estimate, which, in the IPDFT, is $R(\omega)$. It is important, therefore, that $R(\omega)$ not be zero where $P(\omega) \neq 0$; that is, we should choose the $P(\omega) = 0$ wherever $R(\omega) = 0$. Of course, we usually do not know the support of $R(\omega)$ a priori. The point is simply that it is better to make $P(\omega) = 0$ than to make it nonzero, if we have any doubt as to the value of $R(\omega)$.

## 13.16    Properties of the IPDFT

In our discussion of the MEM, we obtained an estimate for the function $R(\omega)$, not simply a way of locating the delta-function components. As we shall show, the IPDFT can also be used to estimate $R(\omega)$. Although the resulting estimate is not guaranteed to be nonnegative and data consistent, it usually is both of these.

For any function $G(\omega)$ on $[-\pi, \pi]$ with Fourier series

$$G(\omega) = \sum_{n=-\infty}^{\infty} g(n)e^{in\omega},$$

the *additive causal part* of the function $G(\omega)$ is

$$G_+(\omega) = \sum_{n=0}^{\infty} g(n)e^{in\omega}.$$

Any function such as $G_+$ that has Fourier coefficients that are zero for negative indices is called a *causal function*. The Equation (13.4) then says that the two causal functions $P_+$ and $(FR)_+$ have Fourier coefficients that agree for $m = 0, 1, ..., N$.

Because $F(\omega)$ is a finite causal trigonometric polynomial, we can write

$$(FR)_+(\omega) = R_+(\omega)F(\omega) + J(\omega),$$

where

$$J(\omega) = \sum_{m=0}^{N-1} \Big( \sum_{k=1}^{N-m} r(-k)f(m+k) \Big)e^{im\omega}.$$

Treating $P_+$ as approximately equal to $(FR)_+ = R_+F + J$, we obtain as an estimate of $R_+$ the function $Q = (P_+ - J)/F$. In order for this estimate of $R_+$ to be causal, it is sufficient that the function $1/F$ be causal. This means that the trigonometric polynomial $F(\omega)$ must be minimum phase; that is, all its roots lie outside the unit circle. In our discussion of the MEM, we saw that this is always the case for MEM. It is not always the case for the IPDFT, but it is usually the case in practice; in fact, it was difficult (but possible) to construct a counterexample. We then construct our IPDFT estimate of $R(\omega)$, which is

$$R_{IPDFT}(\omega) = 2\text{Re}(Q(\omega)) - r(0).$$

The IPDFT estimate is real-valued and, when $1/F$ is causal, guaranteed to be data consistent. Although this estimate is not guaranteed to be non-negative, it usually is.

We showed previously that the vector **a** that solves $R\mathbf{a} = \delta$ corresponds to a polynomial $A(z)$ having all its roots on or outside the unit circle; that is, it is minimum phase. The IPDFT involves the solution of the system $R\mathbf{f} = \mathbf{p}$, where $\mathbf{p} = (p(0), ..., p(N))^T$ is the vector of initial Fourier coefficients of another power spectrum, $P(\omega) \geq 0$ on $[-\pi, \pi]$. When $P(\omega)$ is constant, we get $\mathbf{p} = \delta$. For the IPDFT to be data-consistent, it is sufficient that the polynomial $F(z) = f_0 + ... + f_N z^N$ be minimum phase. Although this need not be the case, it is usually observed in practice.

**Ex. 13.2** *Find conditions on the power spectra $R(\omega)$ and $P(\omega)$ that cause $F(z)$ to be minimum phase. Warning: I have not solved this, so it is probably not an easy exercise.*

---

## 13.17 Illustrations

The figures below illustrate the IPDFT. The prior function in each case is the box object supported on the central fourth of the interval $[0, 2\pi]$. The value $r(0)$ has been increased slightly to regularize the matrix inversion. Figure 13.7 shows the behavior of the IPDFT when the object is only the box. Contrast this with the behavior of MEM in this case, as seen in Figure 13.4. Figures 13.8 and 13.9 show the abilty of the IPDFT to resolve the two spikes at $0.95\pi$ and $1.05\pi$ against the box background. Again, contrast this with the MEM reconstructions in Figures 13.5 and 13.6. To show that the IPDFT is actually indicating the presence of the spikes and not just rolling across the top of the box, we reconstruct two unequal spikes in Figure 13.10. Figure 13.11 shows how the IPDFT behaves when we increase the number of data points; now, $N = 25$ and the SNR is very low.

---

## 13.18 Fourier Series and Analytic Functions

Suppose that $f(z)$ is analytic in an annulus containing the unit circle $C = \{z \,|\, |z| = 1\}$. Then $f(z)$ has a Laurent series expansion

$$f(z) = \sum_{n=-\infty}^{\infty} f_n z^n$$

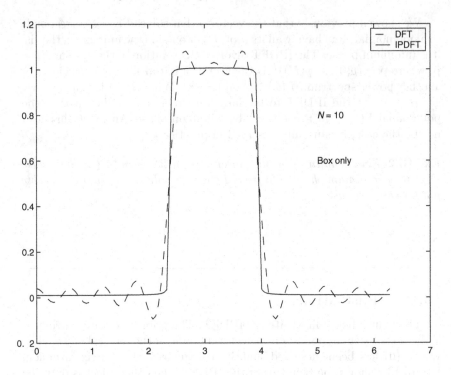

**FIGURE 13.7**: The DFT and IPDFT: box only, $N = 1$.

valid for $z$ within that annulus. Substituting $z = e^{i\theta}$, we get $f(e^{i\theta})$, also written as $b(\theta)$, defined for $\theta$ in the interval $[-\pi, \pi]$ by

$$b(\theta) = f(e^{i\theta}) = \sum_{n=-\infty}^{\infty} f_n e^{in\theta};$$

here the Fourier series for $b(\theta)$ is derived from the Laurent series for the analytic function $f(z)$. If $f(z)$ is actually analytic in $(1 + \epsilon)D$, where $D = \{z|\,|z| < 1\}$ is the open unit disk, then $f(z)$ has a Taylor series expansion and the Fourier series for $b(\theta)$ contains only terms corresponding to nonnegative $n$.

## 13.18.1  An Example

As an example, consider the rational function

$$f(z) = \frac{1}{z - \frac{1}{2}} - \frac{1}{z - 3} = -\frac{5}{2} \Big/ \left( z - \frac{1}{2} \right)(z - 3). \qquad (13.5)$$

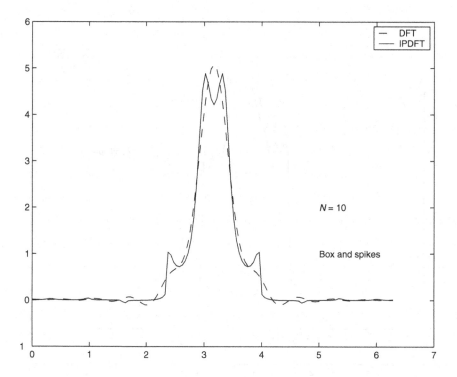

**FIGURE 13.8**: The DFT and IPDFT, box and two spikes, $N = 10$, high SNR.

In an annulus containing the unit circle this function has the Laurent series expansion

$$f(z) = \sum_{n=-\infty}^{-1} 2^{n+1} z^n + \sum_{n=0}^{\infty} \left(\frac{1}{3}\right)^{n+1} z^n;$$

replacing $z$ with $e^{i\theta}$, we obtain the Fourier series for the function $b(\theta) = f(e^{i\theta})$ defined for $\theta$ in the interval $[-\pi, \pi]$.

The function $F(z) = 1/f(z)$ is analytic for all complex $z$, but because it has a root inside the unit circle, its reciprocal, $f(z)$, is not analytic in a disk containing the unit circle. Consequently, the Fourier series for $b(\theta)$ is doubly infinite. We saw in the chapter on complex varables that the function $G(z) = \frac{z - \bar{a}}{1 - az}$ has $|G(e^{i\theta})| = 1$. With $a = 2$ and $H(z) = F(z)G(z)$, we have

$$H(z) = \frac{1}{5}(z - 3)(z - 2),$$

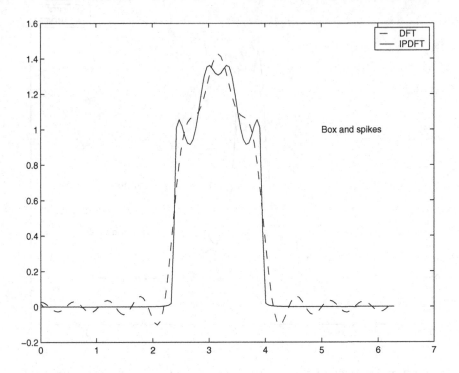

**FIGURE 13.9**: The DFT and IPDFT, box and two spikes, $N = 10$, moderate SNR.

and its reciprocal has the form

$$1/H(z) = \sum_{n=0}^{\infty} a_n z^n.$$

Because

$$G(e^{i\theta})/H(e^{i\theta}) = 1/F(e^{i\theta}),$$

it follows that

$$|1/H(e^{i\theta})| = |1/F(e^{i\theta})| = |b(\theta)|$$

and so

$$|b(\theta)| = \left| \sum_{n=0}^{\infty} a_n e^{in\theta} \right|.$$

Multiplication by $G(z)$ permits us to move a root from inside $C$ to outside $C$ without altering the magnitude of the function's values on $C$.

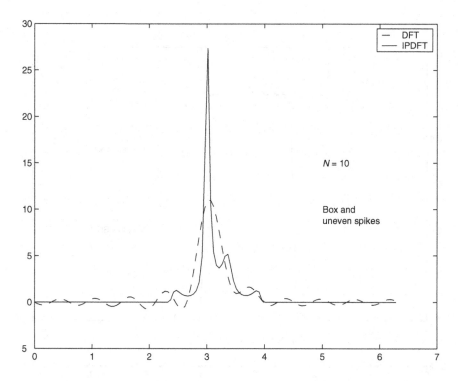

**FIGURE 13.10**: The DFT and IPDFT, box and unequal spikes, $N = 10$, high SNR.

The relationships between functions defined on $C$ and functions analytic (or harmonic) in $D$ form the core of *harmonic analysis* [93]. The factorization $F(z) = H(z)/G(z)$ above is a special case of the *inner-outer factorization* for functions in Hardy spaces; the function $H(z)$ is an *outer function*, and the functions $G(z)$ and $1/G(z)$ are *inner functions*.

## 13.18.2 Hyperfunctions

The rational function $f(z)$ given by Equation (13.5) is analytic in an annulus containing the unit circle in its interior. The annulus has width equal to 2.5, the distance between the roots $z = 0.5$ and $z = 3$. Within that annulus the function has a convergent Laurent expansion, and by setting $z = e^{i\theta}$ we get the Fourier series for the function $b(\theta)$ on $[-\pi, \pi]$. But not every function that has a convergent Fourier series is the restriction to the unit circle of a function analytic in an annulus containing the unit circle. To extend the notion that Fourier series are related to Laurent series, we

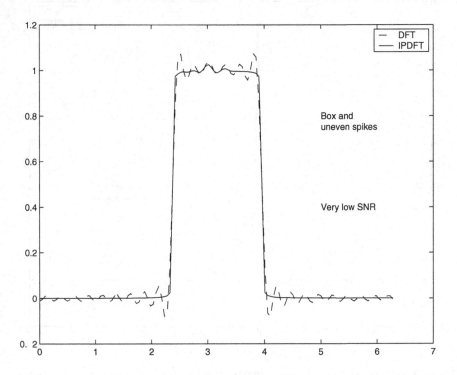

**FIGURE 13.11**: The DFT and IPDFT, box and unequal spikes, $N = 25$, very low SNR.

have to entertain the possibility of the width of the annulus shrinking to zero. This leads to the theory of *hyperfunctions*, introduced in 1958 by the Japanese mathematician Mikio Sato [135] (see also [125]). To get a sense of what is involved without going far into details, we consider the Fourier series for the Dirac delta function.

The Fourier series for $\delta(x)$ is

$$\delta(x) = \sum_{n=-\infty}^{\infty} e^{inx}.$$

Replacing $e^{ix}$ with $z$, we get the Laurent series

$$-1 + \sum_{n=0}^{\infty} z^{-n} + \sum_{n=0}^{\infty} z^{n}.$$

The first sum converges for $|z| > 1$ and

$$\sum_{n=0}^{\infty} z^{-n} = \frac{1}{1 - z^{-n}} = \frac{z}{z - 1}.$$

The second sum converges for $|z| < 1$ and is

$$\sum_{n=0}^{\infty} z^n = \frac{1}{1 - z}.$$

The sum of these two functions is

$$\frac{z}{z - 1} + \frac{1}{1 - z} = \frac{z - 1}{z - 1},$$

so that

$$-1 + \frac{z}{z - 1} + \frac{1}{1 - z} = 0,$$

for $z \neq 1$. For $z = 1$, so that $x = 0$, the Laurent series sums to $+\infty$. We can see, therefore, that, in some sense, the Fourier series for $\delta(x)$ can be understood in terms of a Laurent series, but that the associated annulus of the Laurent series has zero width; there is no actual annulus within which both halves of the series converge. What we have, instead, are two functions, one analytic on the inside of the unit circle, and the other analytic on the outside. Sato's idea is to consider as a single object, a *hyperfunction*, the pair of functions so defined.

## 13.19 Fejér–Riesz Factorization

Sometimes we start with an analytic function and restrict it to the unit circle. Other times we start with a function $f(e^{i\theta})$ defined on the unit circle, or, equivalently, a function of the form $b(\theta)$ for $\theta$ in $[-\pi, \pi]$, and view this function as the restriction to the unit circle of a function that is analytic in a region containing the unit circle. One application of this idea is the Fejér–Riesz factorization theorem:

**Theorem 13.2** *Let $h(e^{i\theta})$ be a finite trigonometric polynomial*

$$h(e^{i\theta}) = \sum_{n=-N}^{N} h_n e^{in\theta},$$

such that $h(e^{i\theta}) \geq 0$ for all $\theta$ in the interval $[-\pi, \pi]$. Then there is

$$y(z) = \sum_{n=0}^{N} y_n z^n$$

with $h(e^{i\theta}) = |y(e^{i\theta})|^2$. The function $y(z)$ is unique if we require, in addition, that all its roots be outside $D$.

To prove this theorem we consider the function

$$h(z) = \sum_{n=-N}^{N} h_n z^n,$$

which is analytic in an annulus containing the unit circle. The rest of the proof is contained in the following exercise.

**Ex. 13.3** *Use the fact that $h_{-n} = \overline{h}_n$ to show that $z_j$ is a root of $h(z)$ if and only if $1/\overline{z}_j$ is also a root. From the nonnegativity of $h(e^{i\theta})$, conclude that if $h(z)$ has a root on the unit circle then it has even multiplicity. Take $y(z)$ to be proportional to the product of factors $z - z_j$ for all the $z_j$ outside $D$; for roots on $C$, include them with half their multiplicities.*

---

## 13.20   Burg Entropy

The Fejér–Riesz theorem is used in the derivation of Burg's maximum entropy method for spectrum estimation. The problem there is to estimate a function $R(\theta) > 0$ knowing only the values

$$r_n = \frac{1}{2\pi} \int_{-\pi}^{\pi} R(\theta) e^{-in\theta} d\theta,$$

for $|n| \leq N$. The approach is to estimate $R(\theta)$ by the function $S(\theta) > 0$ that maximizes the so-called Burg entropy, $\int_{-\pi}^{\pi} \log S(\theta) d\theta$, subject to the data constraints.

The Euler–Lagrange Equation from the calculus of variations allows us to conclude that $S(\theta)$ has the form

$$S(\theta) = 1 / \sum_{n=-N}^{N} h_n e^{in\theta}.$$

The function

$$h(\theta) = \sum_{n=-N}^{N} h_n e^{in\theta}$$

is nonnegative, so, by the Fejér–Riesz theorem, it factors as $h(\theta) = |y(\theta)|^2$. We then have $S(\theta)\overline{y(\theta)} = 1/y(\theta)$. Since all the roots of $y(z)$ lie outside $D$ and none are on $C$, the function $1/y(z)$ is analytic in a region containing $C$ and $D$ so it has a Taylor series expansion in that region. Restricting this Taylor series to $C$, we obtain a one-sided Fourier series having zero terms for the negative indices.

**Ex. 13.4** *Show that the coefficients $y_n$ in $y(z)$ satisfy a system of linear equations whose coefficients are the $r_n$. Hint: Compare the coefficients of the terms on both sides of the equation $S(\theta)\overline{y}(\theta) = 1/y(\theta)$ that correspond to negative indices.*

## 13.21  Some Eigenvector Methods

Prony's method shows that information about the signal can sometimes be obtained from the roots of certain polynomials formed from the data. Eigenvector methods are similar, as we shall see now.

Eigenvector methods assume the data are correlation values and involve polynomials formed from the eigenvectors of the correlation matrix. Schmidt's *multiple signal classification* (MUSIC) algorithm is one such method [136]. A related technique used in direction-of-arrival array processing is the *estimation of signal parameters by rotational invariance techniques* (ESPRIT) of Paulraj, Roy, and Kailath [123].

## 13.22  The Sinusoids-in-Noise Model

We suppose now that the function $f(t)$ being measured is signal plus noise, with the form

$$f(t) = \sum_{j=1}^{J} |A_j| e^{i\theta_j} e^{-i\omega_j t} + n(t) = s(t) + n(t),$$

where the phases $\theta_j$ are random variables, independent and uniformly distributed in the interval $[0, 2\pi)$, and $n(t)$ denotes the random complex stationary noise component. Assume that $E(n(t)) = 0$ for all $t$ and that the noise is independent of the signal components. We want to estimate $J$, the number of sinusoidal components, their magnitudes $|A_j|$ and their frequencies $\omega_j$.

---

## 13.23   Autocorrelation

The autocorrelation function associated with $s(t)$ is

$$r_s(\tau) = \sum_{j=1}^{J} |A_j|^2 e^{-i\omega_j \tau},$$

and the signal power spectrum is the Fourier transform of $r_s(\tau)$,

$$R_s(\omega) = \sum_{j=1}^{J} |A_j|^2 \delta(\omega - \omega_j).$$

The noise autocorrelation is denoted $r_n(\tau)$ and the noise power spectrum is denoted $R_n(\omega)$. For the remainder of this section we shall assume that the noise is *white noise*; that is, $R_n(\omega)$ is constant and $r_n(\tau) = 0$ for $\tau \neq 0$.

We collect samples of the function $f(t)$ and use them to estimate some of the values of $r_s(\tau)$. From these values of $r_s(\tau)$, we estimate $R_s(\omega)$, primarily looking for the locations $\omega_j$ at which there are delta functions.

We assume that the samples of $f(t)$ have been taken over an interval of time sufficiently long to take advantage of the independent nature of the phase angles $\theta_j$ and the noise. This means that when we estimate the $r_s(\tau)$ from products of the form $f(t + \tau)\overline{f(t)}$, the cross terms between one signal component and another, as well as between a signal component and the noise, are nearly zero, due to destructive interference coming from the random phases.

Suppose now that we have the values $r_f(m)$ for $m = -(M-1), ..., M-1$, where $M > J$, $r_f(m) = r_s(m)$ for $m \neq 0$, and $r_f(0) = r_s(0) + \sigma^2$, for $\sigma^2$ the variance (or *power*) of the noise. We form the $M$ by $M$ autocorrelation matrix $R$ with entries $R_{m,k} = r_f(m - k)$.

**Ex. 13.5** *Show that the matrix $R$ has the following form:*

$$R = \sum_{j=1}^{J} |A_j|^2 \mathbf{e}_j \mathbf{e}_j^\dagger + \sigma^2 I,$$

*where $\mathbf{e}_j$ is the column vector with entries $e^{-i\omega_j n}$, for $n = 0, 1, ..., M - 1$.*

Let $\mathbf{u}$ be an eigenvector of $R$ with $\|u\| = 1$ and associated eigenvalue $\lambda$. Then we have

$$\lambda = \mathbf{u}^\dagger R \mathbf{u} = \sum_{j=1}^{J} |A_j|^2 |\mathbf{e}_j^\dagger \mathbf{u}|^2 + \sigma^2 \geq \sigma^2.$$

Therefore, the smallest eigenvalue of $R$ is $\sigma^2$.

Because $M > J$, there must be non-zero $M$-dimensional vectors $\mathbf{v}$ that are orthogonal to all of the $\mathbf{e}_j$; in fact, we can say that there are $M - J$ linearly independent such $\mathbf{v}$. For each such vector $\mathbf{v}$ we have

$$R\mathbf{v} = \sum_{j=1}^{J} |A_j|^2 \mathbf{e}_j^\dagger \mathbf{v} \mathbf{e}_j + \sigma^2 \mathbf{v} = \sigma^2 \mathbf{v};$$

consequently, $\mathbf{v}$ is an eigenvector of $R$ with associated eigenvalue $\sigma^2$.

Let $\lambda_1 \geq \lambda_2 \geq ... \geq \lambda_M > 0$ be the eigenvalues of $R$ and let $\mathbf{u}^m$ be a norm-one eigenvector associated with $\lambda_m$. It follows from the previous paragraph that $\lambda_m = \sigma^2$, for $m = J + 1, ..., M$, while $\lambda_m > \sigma^2$ for $m = 1, ..., J$. This leads to the MUSIC method for determining the $\omega_j$.

---

## 13.24    Determining the Frequencies

By calculating the eigenvalues of $R$ and noting how many of them are greater than the smallest one, we find $J$. Now we seek the $\omega_j$.

For each $\omega$, we let $\mathbf{e}_\omega$ have the entries $e^{-i\omega n}$, for $n = 0, 1, ..., M - 1$ and form the function

$$T(\omega) = \sum_{m=J+1}^{M} |\mathbf{e}_\omega^\dagger \mathbf{u}^m|^2.$$

This function $T(\omega)$ will have zeros at precisely the values $\omega = \omega_j$, for $j = 1, ..., J$. Once we have determined $J$ and the $\omega_j$, we estimate the magnitudes $|A_j|$ using Fourier transform estimation techniques already discussed. This is basically Schmidt's MUSIC method.

We have made several assumptions here that may not hold in practice and we must modify this eigenvector approach somewhat. First, the time over which we are able to measure the function $f(t)$ may not be long enough to give good estimates of the $r_f(\tau)$. In that case we may work directly with the samples of $f(t)$. Second, the smallest eigenvalues will not be exactly equal to $\sigma^2$ and some will be larger than others. If the $\omega_j$ are not well separated, or if some of the $|A_j|$ are quite small, it may be hard to tell what the value of $J$ is. Third, we often have measurements of $f(t)$ that have errors other than those due to background noise; inexpensive sensors can introduce their own random phases that can complicate the estimation process. Finally, the noise may not be white, so that the estimated $r_f(\tau)$ will not equal $r_s(\tau)$ for $\tau \neq 0$, as before. If we know the noise power spectrum or have a decent idea what it is, we can perform a pre-whitening to $R$, which will then return us to the case considered above, although this can be a tricky procedure.

## 13.25   The Case of Non-White Noise

When the noise power spectrum has a component that is not white the eigenvalues and eigenvectors of $R$ behave somewhat differently from the white-noise case. The eigenvectors tend to separate into three groups. Those in the first group correspond to the smallest eigenvalues and are approximately orthogonal to both the signal components and the nonwhite noise component. Those in the second group, whose eigenvalues are somewhat larger than those in the previous group, tend to be orthogonal to the signal components but to have a sizable projection onto the nonwhite-noise component. Those in the third group, with the largest eigenvalues, have sizable projection onto both the signal and nonwhite noise components. Since the DFT estimate uses $R$, as opposed to $R^{-1}$, the DFT spectrum is determined largely by the eigenvectors in the third group. The MEM estimator, which uses $R^{-1}$, makes most use of the eigenvectors in the first group, but in the formation of the denominator. In the presence of a nonwhite-noise component, the orthogonality of those eigenvectors to both the signals and the nonwhite noise shows up as peaks throughout the region of interest, masking or distorting the signal peaks we wish to see.

There is a second problem exacerbated by the nonwhite component: sensitivity of nonlinear and eigenvector methods to phase errors. We have assumed up to now that the data we have obtained is accurate, but there isn't enough of it. In some cases the machinery used to obtain the measured data may not be of the highest quality; certain applications of sonar make use of relatively inexpensive hydrophones that will sink into the ocean after they have been used briefly. In such cases the complex numbers $r(n)$ will be distorted. Errors in the measurement of their phases are particularly damaging. Techniques for stabilizing high-resolution methods were presented in [28].

# Chapter 14

## Discrete Entropy Maximization

## 14.1  Chapter Summary

In Chapter 13 we considered the problem of estimating a nonnegative function of a continuous variable from finitely many of its Fourier coefficients. The estimate was again a function of a continuous variable. In such cases, we would convert the estimate to a finite vector just prior to graphing the estimate. In this chapter we discuss an alternative approach, in which the nonnegative function to be estimated is discretized at the outset. Discrete entropy maximization and related procedures are then used to reconstruct the nonnegative vector from finitely many linear-functional values. Unlike the MEM and the IPDFT methods, the algorithms we focus on here, primarily the *multiplicative algebraic reconstruction technique* (MART) and its simultaneous version, the SMART, are iterative.

We begin with the *algebraic reconstruction technique* (ART), which is not related to entropy maximization but which will help to motivate its multiplicative variant, the *multiplicative algebraic reconstruction technique* (MART). As we shall see, the MART is an iterative entropy-maximization method.

## 14.2 The Algebraic Reconstruction Technique

The ART, designed originally for the reconstruction of medical images in computerized tomography [84], is an iterative algorithm for finding a solution of a consistent system of linear equations, $Ax = b$, where $A$ is an arbitrary $I$ by $J$ complex matrix. In the tomography case the vector $x$ is a vectorization of a two- or three-dimensional discrete image, the vector $b$ is the vector of measured data, and the matrix $A$ describes the geometry of the sensing process.

In the ART we begin by choosing an arbitrary starting vector, denoted $x^0$. Having computed $x^k$, we calculate the next vector, $x^{k+1}$, using the formula

$$x_j^{k+1} = x_j^k + \alpha_i^{-1}\overline{A_{ij}}(b_i - (Ax^k)_i),$$

for $k = 0, 1, ...,$ $i = k(\bmod I) + 1$, and

$$\alpha_i = \sum_{j=1}^{J} |A_{ij}|^2.$$

When $Ax = b$ has a solution the sequence $\{x^k\}$ converges to the solution of the system closest to the starting vector, $x^0$; when $x^0 = 0$ the sequence converges to the minimum-two-norm solution.

## 14.3 The Multiplicative Algebraic Reconstruction Technique

The images to be reconstructed in transmission or emission tomography are necessarily nonnegative. The multiplicative algebraic reconstruction technique, MART, is a variant of the ART that incorporates the prior information that the image to be reconstructed is nonnegative [84]. Like the ART, the MART can be used to solve more general systems of linear equations, although, for the MART, the matrix and vectors involved must be nonnegative.

Let $P$ be an $I$ by $J$ matrix with nonnegative entries $P_{ij} \geq 0$, such that $s_j = \sum_{i=1}^{I} P_{ij} > 0$, for $j = 1, ..., J$. Let $y$ be the $I$-dimensional vector with entries $y_i > 0$, and suppose that the linear system of equations $y = Px$ has a nonnegative solution $x$.

For the MART we begin with a positive vector $x^0$. Having computed $x^k$, we calculate the next vector, $x^{k+1}$, using the formula

$$x_j^{k+1} = x_j^k \left( \frac{y_i}{(Px^k)_i} \right)^{m_i^{-1} P_{ij}}, \tag{14.1}$$

where $m_i = \max\{P_{ij} | j = 1, ..., J\}$, and $i = k(\mathrm{mod}\, I) + 1$. When there is a nonnegative solution for $y = Px$, the sequence $\{x^k\}$ converges to such a solution. When there are multiple nonnegative solutions, we would like to know which solution MART gives us; in particular, we want to know how the solution depends on the starting vector $x^0$. The answer involves the Kullback–Leibler, or cross-entropy, distance.

---

## 14.4　The Kullback–Leibler Distance

For real numbers $a > 0$ and $b > 0$ we define the Kullback–Leibler, or cross-entropy, distance from $a$ to $b$ to be

$$KL(a, b) = a \log \frac{a}{b} + b - a.$$

It follows from the inequality $\log t \leq t - 1$, with equality if and only if $t = 1$, that $KL(a, b) \geq 0$, and $KL(a, b) = 0$ if and only if $a = b$. We also let $KL(0, b) = b$ and $KL(a, 0) = +\infty$. We extend the KL distance to nonnegative vectors $x$ and $z$ by

$$KL(x, z) = \sum_{j=1}^{J} KL(x_j, z_j).$$

Since the function

$$f(t) = t - 1 - \log t$$

is convex, we have the following useful lemma:

**Lemma 14.1** *Let $a$ be a fixed positive number. If the set $\{KL(a, b) | b \in B\}$ is bounded, then so is $B$.*

Let $x$ and $z$ be nonnegative vectors, with $x_+ = \sum_{j=1}^{J} x_j$. Then a simple calculation shows that

$$KL(x_+, z_+) \leq KL(x, z). \tag{14.2}$$

When $x = u$, where $u_j = 1$ for all $j$, and $z$ is a probability vector, we have

$$KL(u, z) + J - 1 = -\sum_{j=1}^{J} \log z_j,$$

which is the negative of the Burg entropy of the probability vector $z$. Similarly,

$$KL(z, u) - J + 1 = \sum_{j=1}^{J} z_j \log z_j,$$

which is the negative of the Shannon entropy of the probability vector $z$.

When the system $y = Px$ has a nonnegative solution the MART sequence converges to the nonnegative solution $x$ for which the Kullback–Leibler distance $KL(x, x^0)$ is minimized. Therefore, when we select $x^0 = u$, the MART sequence converges to the solution of $y = Px$ maximizing the Shannon entropy.

## 14.5   The EMART

We see from Equation (14.1) that the MART is computationally more complicated than the ART. The EMART [37] is an iterative method that, like the MART, applies to nonnegative systems of linear equations, while, like the ART, requires no exponentiation.

Note that we can rewrite the right side of Equation (14.1) as a weighted geometric mean of two terms:

$$x_j^{k+1} = \left(x_j^k\right)^{1 - m_i^{-1} P_{ij}} \left(x_j^k \frac{y_i}{(Px^k)_i}\right)^{m_i^{-1} P_{ij}}.$$

In the EMART we exchange the weighted geometric means for weighted arithmetic means. The iterative step of the EMART is

$$x_j^{k+1} = (1 - m_i^{-1} P_{ij}) x_j^k + m_i^{-1} P_{ij} \left(x_j^k \frac{y_i}{(Px^k)_i}\right). \tag{14.3}$$

When the system $y = Px$ has nonnegative solutions, the sequence $\{x^k\}$ generated by Equation (14.3) converges to a nonnegative solution. Unlike the ART and MART, we do not know how the solution depends on the starting vector $x^0$.

## 14.6   Simultaneous Versions

All of the three methods discussed so far are *sequential*, in that only a single equation is used at each step of the iteration. There are simultaneous versions of these algorithms that use all of the equations at each step [41]. One reason for their use is that they also converge when the system of equations is not consistent.

### 14.6.1 The Landweber Algorithm

The simultaneous version of the ART is the Landweber algorithm [7]. The iterative step of the Landweber algorithm is the following:

$$x^{k+1} = x^k + \gamma A^\dagger (b - (Ax^k)), \tag{14.4}$$

with $0 < \gamma < \frac{2}{\rho(A^\dagger A)}$, and $\rho(A^\dagger A)$ is the *spectral radius* of $A^\dagger A$, which is its largest eigenvalue, in this case. When $Ax = b$ has solutions the sequence generated by Equation (14.4) converges to the solution closest to $x^0$ in the two-norm. When $Ax = b$ has no solutions, the sequence converges to the minimizer of $\|Ax - b\|_2$ for which $\|x - x^0\|_2$ is minimized.

### 14.6.2 The SMART

The simultaneous MART (SMART) is the simultaneous version of the MART. The iterative step of the SMART is the following:

$$x_j^{k+1} = x_j^k \exp\left(s_j^{-1} \sum_{i=1}^{I} P_{ij} \log \frac{y_i}{(Px)_i}\right). \tag{14.5}$$

When the system $y = Px$ has nonnegative solutions, the sequence generated by Equation (14.5) converges to the nonnegative solution that minimizes $KL(x, x^0)$, just as the MART does. In addition, when the system $y = Px$ is inconsistent, that is, has no nonnegative solution, the SMART sequence converges to the nonnegative minimizer of $KL(Px, y)$ for which $KL(x, x^0)$ is minimized.

### 14.6.3 The EMML Algorithm

Closely related to the SMART is the EMML algorithm, which is a special case of a more general method known in statistics as the EM algorithm. The iterative step of the EMML algorithm is the following:

$$x_j^{k+1} = x_j^k s_j^{-1} \sum_{i=1}^{I} P_{ij}\left(\frac{y_i}{(Px)_i}\right). \tag{14.6}$$

When the system $y = Px$ has nonnegative solutions, the sequence generated by Equation (14.6) converges to a nonnegative solution. When the system $y = Px$ has no nonnegative solutions, the sequence converges to a nonnegative minimizer of $KL(y, Px)$. In neither case can we say precisely how the limit of the sequence depends on the starting vector, $x^0$.

In the urn model for remote sensing, as discussed in Chapter 1, the entries of the vector $y$ are $y_i$, the proportion of the trials in which the $i$th color was drawn, so that $y_+ = 1$. In addition, we have $s_j = 1$. We want the solution $x$ to be a probability vector, as well. When $s_j = 1$ for each $j$ and there are exact nonnegative solutions of $y = Px$, the solutions provided by the SMART and the EMML method, although they may be different, will both be probability vectors. However, when $y = Px$ has no nonnegative solution, the limit of the SMART sequence will have $x_+ < 1$, while the limit of the EMML sequence will still be a probability vector. For details and references concerning the SMART and the EMML algorithm, consult [41].

### 14.6.4    Block-Iterative Versions

Simultaneous iterative algorithms tend to converge slowly. Sequential versions of these algorithms, which typically converge more rapidly, may make inefficient use of the machine architecture. Block-iterative versions of these algorithms permit the use of some, but perhaps not all, of the equations at each step. A block-iterative version of the EMML algorithm was used to obtain sub-pixel resolution from SAR image data [117].

### 14.6.5    Convergence of the SMART

We turn now to a proof of convergence of the SMART. Convergence of the MART and the block-iterative versions of SMART are proved similarly and we omit these proofs.

We shall assume, for notational convenience, that $s_j = 1$ for all $j$. As we have seen, this is sometimes the case in applications, and if it is not true, we can redefine both $P$ and $x$ to make it happen. Using Equation (14.2), we have

$$KL(x, z) \geq KL(Px, Pz),$$

for all nonnegative $x$ and $z$.

We use the *alternating minimization* formalism of [61]. For each nonnegative $x$ for which

$$(Px)_i = \sum_{j=1}^{J} P_{ij} x_j$$

is positive, for each $i$, we let $r(x)$ and $q(x)$ be the $I$ by $J$ matrices with entries

$$r(x)_{ij} = x_j P_{ij} \left( \frac{y_i}{(Px)_i} \right),$$

and
$$q(x)_{ij} = x_j P_{ij}.$$

The alternating minimization then involves the function

$$KL(q(x), r(z)) = \sum_{i=1}^{I} \sum_{j=1}^{J} KL(q(x)_{ij}, r(z)_{ij}).$$

The following *Pythagorean identities* are central to the proof:

$$KL(q(x), r(z)) = KL(q(x), r(x)) + KL(x, z) - KL(Px, Pz),$$

and
$$KL(q(x), r(z)) = KL(q(z'), r(z)) + KL(x, z'),$$

where
$$z'_j = z_j \exp\Big( \sum_{i=1}^{I} P_{ij} \log \frac{y_i}{(Pz)_i} \Big).$$

It follows, then, that, having calculated $x^k$, we get $x^{k+1}$ by minimizing $KL(q(x), r(x^k))$ over all nonnegative $x$. The remainder of the convergence proof is contained in the following sequence of exercises.

**Ex. 14.1** *Use the Pythagorean identities and the fact that*

$$KL(q(x), r(x)) = KL(Px, y)$$

*to show that the sequence $\{KL(Px^k, y)\}$ is decreasing, and the sequence $\{KL(x^k, x^{k+1})\}$ converges to zero.*

**Ex. 14.2** *Show that*

$$\sum_{j=1}^{J} x_j^k \le \sum_{i=1}^{I} y_i,$$

*so that the sequence $\{x^k\}$ is a bounded sequence.*

**Ex. 14.3** *Let $x^*$ be any cluster point of the sequence $\{x^k\}$. Show that $(x^*)' = x^*$.*

**Ex. 14.4** *Let $\hat{x}$ be a nonnegative minimizer of $KL(Px, y)$. Show that $(\hat{x})' = \hat{x}$.*

Note that, since $KL(x, z)$ is strictly convex in each variable, the vector $P\hat{x}$ is unique, even if $\hat{x}$ is not unique.

**Ex. 14.5** *Show that*

$$KL(\hat{x}, x^k) - KL(\hat{x}, x^{k+1}) = KL(Px^{k+1}, y) - KL(P\hat{x}, y)$$

$$+KL(P\hat{x}, Px^k) + KL(x^{k+1}, x^k) - KL(Px^{k+1}, Px^k), \qquad (14.7)$$

*so that* $KL(P\hat{x}, Px^*) = 0$, *the sequence* $\{KL(\hat{x}, x^k)\}$ *is decreasing, and* $KL(\hat{x}, x^*)$ *is finite.*

**Ex. 14.6** *Show that, for any cluster point* $x^*$, $KL(Px^*, y) = KL(P\hat{x}, y)$.

We know now that $x^*$ is a nonnegative minimizer of $KL(Px, y)$. Replacing $\hat{x}$ with $x^*$, we find that the sequence $\{KL(x^*, x^k)\}$ converges to zero. Since the right side of Equation (14.7) depends only on $P\hat{x}$ and not directly on $\hat{x}$, so does the left side. Consequently, summing the left side over $k$, we find that

$$KL(\hat{x}, x^0) - KL(\hat{x}, x^*)$$

is independent of the choice of $\hat{x}$. It follows that $x^*$ is the nonnegative minimizer of $KL(Px, y)$ for which $KL(x, x^0)$ is minimized.

# Chapter 15

## Analysis and Synthesis

## 15.1 Chapter Summary

Analysis and synthesis in signal processing refers to the effort to study complicated functions in terms of simpler ones. The basic building blocks are orthogonal bases and frames.

We begin with signal-processing problems arising in radar. Not only does radar provide an important illustration of the application of the theory of Fourier transforms and matched filters, but it also serves to motivate several of the mathematical concepts we shall encounter in our discussion of wavelets. The connection between radar signal processing and wavelets is discussed in some detail in Kaiser's book [97].

There are applications in which the frequency composition of the signal of interest will change over time. A good analogy is a piece of music, where notes at certain frequencies are heard for a while and then are replaced by notes at other frequencies. We do not usually care what the overall contribution of, say, middle C is to the song, but do want to know which notes are to be sounded when and for how long. Analyzing such non-stationary signals requires tools other than the Fourier transform: the short-time Fourier transform is one such tool; wavelet expansion is another.

## 15.2   The Basic Idea

An important theme that runs through most of mathematics, from the geometry of the early Greeks to modern signal processing, is *analysis and synthesis*, or, less formally, *breaking up and putting back together*. The Greeks estimated the area of a circle by breaking it up into sectors that approximated triangles. The Riemann approach to integration involves breaking up the area under a curve into pieces that approximate rectangles or other simple shapes. Viewed differently, the Riemann approach is first to approximate the function to be integrated by a step function and then to integrate the step function.

Along with geometry, Euclid includes a good deal of number theory, where, again, we find analysis and synthesis. His theorem that every positive integer is divisible by a prime is analysis; division does the breaking up and the simple pieces are the primes. The fundamental theorem of arithmetic, which asserts that every positive integer can be written in a unique way as the product of powers of primes, is synthesis, with the putting back together done by multiplication.

## 15.3   Polynomial Approximation

The individual power functions, $x^n$, are not particularly interesting by themselves, but when finitely many of them are scaled and added to form a polynomial, interesting functions can result, as the famous approximation theorem of Weierstrass confirms [101]:

**Theorem 15.1** *If $f : [a, b] \to \mathbb{R}$ is continuous and $\epsilon > 0$ is given, we can find a polynomial $P$ such that $|f(x) - P(x)| \le \epsilon$ for every $x$ in $[a, b]$.*

The idea of building complicated functions from powers is carried a step further with the use of infinite series, such as Taylor series. The sine function, for example, can be represented for all real $x$ by the infinite power series

$$\sin x = x - \frac{1}{3!}x^3 + \frac{1}{5!}x^5 - \frac{1}{7!}x^7 + \dots.$$

The most interesting thing to note about this is that the sine function has properties that none of the individual power functions possess; for example, it is bounded and periodic. So we see that an infinite sum of simple functions can be qualitatively different from the components in the sum. If we take the sum of only finitely many terms in the Taylor series for the sine function we get a polynomial, which cannot provide a good approximation of the sine function for all $x$; that is, the finite sum does not approximate the sine function uniformly over the real line. The approximation is better for $x$ near zero and poorer as we move away from zero. However, for any selected $x$ and for any $\epsilon > 0$, there is a positive integer $N$, depending on the $x$ and on the $\epsilon$, with the sum of the first $n$ terms of the series within $\epsilon$ of $\sin x$ for $n \geq N$; that is, the series converges pointwise to $\sin x$ for each real $x$. In Fourier analysis the trigonometric functions themselves are viewed as the simple functions, and we try to build more complicated functions as (possibly infinite) sums of trig functions. In wavelet analysis we have more freedom to design the simple functions to fit the problem at hand.

---

## 15.4 Signal Analysis

When we speak of *signal analysis*, we often mean that we believe the signal to be a superposition of simpler signals of a known type and we wish to know which of these simpler signals are involved and to what extent. For example, received sonar or radar data may be the superposition of individual components corresponding to spatially localized targets of interest. As we shall see in our discussion of the ambiguity function and of wavelets, we want to tailor the family of simpler signals to fit the physical problem being considered.

Sometimes it is not the individual components that are significant by themselves, but groupings of these components. For example, if our received signal is believed to consist of a lower frequency signal of interest plus a noise component employing both low and high frequencies, we can remove some of the noise by performing a low-pass filtering. This amounts to analyzing the received signal to determine what its low-pass and high-pass components are. We formulate this operation mathematically using the Fourier transform, which decomposes the received signal $f(t)$ into complex exponential function components corresponding to different frequencies.

More generally, we may analyze a signal $f(t)$ by calculating certain inner products $\langle f, g_n \rangle$, $n = 1, ..., N$. We may wish to encode the signal using these $N$ numbers, or to make a decision about the signal, such as recognizing a voice. If the signal is a two-dimensional image, say a fingerprint, we may want to construct a data-base of these $N$-dimensional vectors, for identification. In such a case we are not necessarily claiming that the signal $f(t)$ is a superposition of the $g_n(t)$ in any sense, nor do we necessarily expect to reconstruct $f(t)$ at some later date from the stored inner products. For example, one might identify a piece of music using only the upward or downward progression of the first few notes.

There are many cases, on the other hand, in which we do wish to reconstruct the signal $f(t)$ from measurements or stored compressed versions. In such cases we need to consider this when we design the measuring or compression procedures. For example, we may have values of the signal or its Fourier transform at some finite number of points and want to recapture $f(t)$ itself. Even in those cases mentioned previously in which reconstruction is not desired, such as the fingerprint case, we do wish to be reasonably sure that similar vectors of inner products correspond to similar signals and distinct vectors of inner products correspond to distinct signals, within the obvious limitations imposed by the finiteness of the stored inner products. The twin processes of analysis and synthesis are dealt with mathematically using the notions of *frames* and *bases*.

---

## 15.5　Practical Considerations in Signal Analysis

Perhaps the most basic problem in signal analysis is determining which sinusoidal components make up a given signal. Let the analog signal $f(t)$ be given for all real $t$ by

$$f(t) = \sum_{j=1}^{J} A_j e^{i\omega_j t},$$

where the $A_j$ are complex amplitudes and the $\omega_j$ are real numbers. If we view the variable $t$ as time, then the $\omega_j$ are frequencies. In theory, we can determine $J$, the $\omega_j$, and the $A_j$ simply by calculating the Fourier transform $F(\omega)$ of $f(t)$. The function $F(\omega)$ will have Dirac delta components at $\omega = \omega_j$ for each $j$, and will be zero elsewhere. Obviously, this is not a practical solution to the problem. The first step in developing a practical approach is to pass from analog signals, which are functions of the continuous variable $t$, to digital signals or sequences, which are functions of the integers.

In theoretical discussions of digital signal processing, analog signals are converted to discrete signals or sequences by sampling. We begin by choosing a positive sampling spacing $\Delta > 0$ and define the $n$th entry of the sequence $x = \{x(n)\}$ by

$$x(n) = f(n\Delta),$$

for all integers $n$.

## 15.5.1 The Discrete Model

Notice that, since

$$e^{i\omega_j n\Delta} = e^{i(\omega_j + \frac{2\pi}{\Delta})n\Delta}$$

for all $n$, we cannot distinguish frequency $\omega_j$ from $\omega_j + \frac{2\pi}{\Delta}$. We try to select $\Delta$ small enough so that each of the $\omega_j$ we seek lies in the interval $(-\frac{\pi}{\Delta}, \frac{\pi}{\Delta})$. If we fail to make $\Delta$ small enough we *under-sample*, with the result that some of the $\omega_j$ will be mistaken for lower frequencies; this is *aliasing*. Our goal now is to process the sequence $x$ to determine $J$, the $\omega_j$, and the $A_j$. We do this with matched filtering.

Every linear shift-invariant system operates through convolution; associated with the system is a sequence $h$, such that, when $x$ is the input sequence, the output sequence is $y$, with

$$y(n) = \sum_{k=-\infty}^{\infty} h(k)x(n-k),$$

for each integer $n$. In theoretical *matched filtering* we design a whole family of such systems or filters, one for each frequency $\omega$ in the interval $(-\frac{\pi}{\Delta}, \frac{\pi}{\Delta})$. We then use our sequence $x$ as input to each of these filters and use the outputs of each to solve our signal-analysis problem.

For each $\omega$ in the interval $(-\frac{\pi}{\Delta}, \frac{\pi}{\Delta})$ and each positive integer $K$, we consider the shift-invariant linear filter with $h = e_{K,\omega}$, where

$$e_\omega(k) = \frac{1}{2K+1} e^{i\omega k\Delta},$$

for $|k| \le K$ and $e_{K,\omega}(k) = 0$ otherwise. Using $x$ as input to this system, we find that the output value $y(0)$ is

$$y(0) = \sum_{j=1}^{J} A_j \left( \frac{1}{2K+1} \sum_{k=-K}^{K} e^{i(\omega-\omega_j)k\Delta} \right). \tag{15.1}$$

Recall the following identity for the Dirichlet kernel:

$$\sum_{k=-K}^{K} e^{ik\omega} = \frac{\sin((K+\frac{1}{2})\omega)}{\sin(\frac{\omega}{2})},$$

for $\sin(\frac{\omega}{2}) \ne 0$. As $K \to +\infty$, the inner sum in Equation (15.1) goes to zero for every $\omega$ except $\omega = \omega_j$. Therefore the limit, as $K \to +\infty$, of $y(0)$ is zero, if $\omega$ is not equal to any of the $\omega_j$, and equals $A_j$, if $\omega = \omega_j$. Therefore, in theory, at least, we can successfully decompose the digital signal into its constituent parts and distinguish one frequency component from another, no matter how close together the two frequencies may be.

It is important to note that, to achieve the perfect analysis described above, we require noise-free values $x(n)$ and we need to take $K$ to infinity; in practice, of course, neither of these conditions is realistic. We consider next the practical matter of having only finitely many values of $x(n)$; we leave the noisy case for another chapter.

## 15.5.2  The Finite-Data Problem

In reality we have only finitely many values of $x(n)$, say for $n = -N, ..., N$. In matched filtering we can only take $K \leq N$. For the choice of $K = N$, we get

$$y(0) = \sum_{j=1}^{J} A_j \left( \frac{1}{2N+1} \sum_{k=-N}^{N} e^{i(\omega - \omega_j)k\Delta} \right),$$

for each fixed $\omega$ different from the $\omega_j$, and $y(0) = A_j$ for $\omega = \omega_j$. We can then write

$$y(0) = \sum_{j=1}^{J} A_j \left( \frac{1}{2N+1} \frac{\sin((\omega - \omega_j)(N + \frac{1}{2})\Delta)}{\sin((\omega - \omega_j)(\frac{\Delta}{2}))} \right),$$

for $\omega$ not equal to $\omega_j$. The problem we face for finite data is that the $y(0)$ is not necessarily zero when $\omega$ is not one of the $\omega_j$.

In our earlier discussion of signal analysis it was shown that, if we are willing to make a simplifying assumption, we can continue as in the infinite-data case. The simplifying assumption is that the $\omega_j$ we seek are $J$ of the $2N + 1$ frequencies equally spaced in the interval $(-\frac{\pi}{\Delta}, \frac{\pi}{\Delta})$, beginning with $\alpha_1 = -\frac{\pi}{\Delta} + \frac{2\pi}{(2N+1)\Delta}$ and ending with $\alpha_{2N+1} = \frac{\pi}{\Delta}$. Therefore,

$$\alpha_m = -\frac{\pi}{\Delta} + \frac{2\pi m}{(2N+1)\Delta},$$

for $m = 1, ..., 2N + 1$.

Having made this simplifying assumption, we then design the matched filters corresponding to the frequencies $\alpha_n$, for $n = 1, ..., 2N + 1$. Because

$$\sum_{k=-N}^{N} e^{i(\alpha_m - \alpha_n)k\Delta} = \sum_{k=-N}^{N} e^{2\pi i \frac{m-n}{2N+1} k}$$

$$= \frac{\sin(2\pi \frac{m-n}{2N+1}(N + \frac{1}{2}))}{\sin(\pi \frac{m-n}{2N+1})},$$

it follows that

$$\sum_{k=-N}^{N} e^{i(\alpha_m - \alpha_n)k\Delta} = 0,$$

for $m \neq n$ and it is equal to $2N + 1$ when $m = n$. We conclude that, provided the frequencies we seek are among the $\alpha_m$, we can determine $J$ and the $\omega_j$. Once we have these pieces of information, we find the $A_j$ simply by solving a system of linear equations.

## 15.6 Frames

Although in practice we deal with finitely many measurements or inner product values, it is convenient, in theoretical discussions, to imagine that the signal $f(t)$ has been associated with an infinite sequence of inner products $\{\langle f, g_n \rangle, n = 1, 2, ...\}$. It is also convenient to assume that $||f||^2 = \int_{-\infty}^{\infty} |f(t)|^2 dt < +\infty$; that is, we assume that $f$ is in the Hilbert space $H = L^2$. The sequence $\{g_n | n = 1, 2, ...\}$ in any Hilbert space $H$ is called a *frame* for $H$ if there are positive constants $A \leq B$ such that, for all $f$ in $H$,

$$A||f||^2 \leq \sum_{n=1}^{\infty} |\langle f, g_n \rangle|^2 \leq B||f||^2. \tag{15.2}$$

The inequalities in (15.2) define the *frame property*. A frame is said to be *tight* if $A = B$.

To motivate this definition, suppose that $f = g - h$. If $g$ and $h$ are nearly equal, then $f$ is near zero, so that $||f||^2$ is near zero. Consequently, the numbers $|\langle f, g_n \rangle|^2$ are all small, meaning that $\langle g, g_n \rangle$ is nearly equal to $\langle h, g_n \rangle$ for each $n$. Conversely, if $\langle g, g_n \rangle$ is nearly equal to $\langle h, g_n \rangle$ for each $n$, then the numbers $|\langle f, g_n \rangle|^2$ are all small. Therefore, $||f||^2$ is small, from which we conclude that $g$ is close to $h$. The *analysis* operator is the one that takes us from $f$ to the sequence $\{\langle f, g_n \rangle\}$, while the *synthesis* operator takes us from the sequence $\{\langle f, g_n \rangle\}$ to $f$. This discussion of frames and related notions is based on the treatment in Christensen's book [54].

In the case of a finite-dimensional Hilbert space $H$, any finite set $\{g_n, n = 1, ..., N\}$ is a frame for the space $H$ of all $f$ that are linear combinations of the $g_n$.

**Ex. 15.1** *An interesting example of a frame in $H = \mathbb{R}^2$ is the so-called Mercedes frame: let $g_1 = (0, 1)$, $g_2 = (-\sqrt{3}/2, -1/2)$ and $g_3 = (\sqrt{3}/2, -1/2)$. Show that for this frame $A = B = 3/2$, so the Mercedes frame is tight.*

**Ex. 15.2** *Let $W = U \cup V$ be the union of two orthonormal bases for $\mathbb{C}^N$, $U$ and $V$. Show that $W$ is a tight frame with $A = B = 2$.*

For example, consider $U = \{u^1, u^2, ..., u^N\}$ the usual orthonormal basis for $\mathbb{C}^N$, where all the entries of $u^n$ are zero, except that $u^n_n = 1$, and $V = \{v^1, v^2, ..., v^N\}$ the Fourier basis, with the $m$th entry of $v^n$ given by

$$v^n_m = \frac{1}{\sqrt{N}} e^{2\pi i m n / N}.$$

This particular frame is used often in compressed sensing and compressed sampling, as discussed in Chapter 22.

The JPEG method for compressing images uses a similar frame that is the union of a discrete cosine basis and a discrete wavelet basis. The idea is that most images that we wish to compress can be represented as a linear combination of relatively few discrete cosine vectors and wavelet vectors.

The frame property in (15.2) provides a necessary condition for stable application of the decomposition and reconstruction operators. But it does more than that; it actually provides a reconstruction algorithm. The *frame operator* $S$ is given by

$$Sf = \sum_{n=1}^{\infty} \langle f, g_n \rangle g_n.$$

The frame property implies that the frame operator is invertible. The *dual frame* is the sequence $\{S^{-1}g_n, n = 1, 2, ...\}$.

**Ex. 15.3** *Use the definitions of the frame operator $S$ and the dual frame to obtain the following reconstruction formulas:*

$$f = \sum_{n=1}^{\infty} \langle f, g_n \rangle S^{-1} g_n;$$

*and*

$$f = \sum_{n=1}^{\infty} \langle f, S^{-1} g_n \rangle g_n.$$

If the frame is tight, then the dual frame is $\{\frac{1}{A} g_n, n = 1, 2, ...\}$; if the frame is not tight, inversion of the frame operator is done only approximately.

---

## 15.7 Bases, Riesz Bases, and Orthonormal Bases

A set of vectors $\{g_n, n = 1, 2, ...\}$ in $H$ is a *basis* for $H$ if, for every $f$ in $H$, there are unique constants $\{c_n, n = 1, 2, ...\}$ with

$$f = \sum_{n=1}^{\infty} c_n g_n.$$

A basis is called a *Riesz basis* if it is also a frame for $H$. It can be shown that a frame is a Riesz basis if the removal of any one element causes the loss of the frame property; since the second inequality in Inequality (15.2) is not lost, it follows that it is the first inequality that can now be violated for some $f$. A basis $\{g_n | n = 1, 2, ...\}$ is an *orthonormal basis* for $H$ if $\|g_n\| = 1$ for all $n$ and $\langle g_n, g_m \rangle = 0$ for distinct $m$ and $n$.

We know that the complex exponentials

$$\left\{ e_n(t) = \frac{1}{\sqrt{2\pi}} e^{int}, \ -\infty < n < \infty \right\}$$

form an orthonormal basis for the Hilbert space $L^2(-\pi, \pi)$ consisting of all $f$ supported on $(-\pi, \pi)$ with $\int_{-\pi}^{\pi} |f(t)|^2 dt < +\infty$. Every such $f$ can be written as

$$f(t) = \frac{1}{\sqrt{2\pi}} \sum_{n=-\infty}^{+\infty} a_n e^{int},$$

for

$$a_n = \langle f, e_n \rangle = \frac{1}{\sqrt{2\pi}} \int_{-\pi}^{\pi} f(t) e^{-int} dt.$$

Consequently, this is true for every $f$ in $L^2(-\pi/2, \pi/2)$, although the set of functions $\{g_n\}$ formed by restricting the $\{e_n\}$ to the interval $(-\pi/2, \pi/2)$ is no longer a basis for $H = L^2(-\pi/2, \pi/2)$. It is still a tight frame with $A = 1$, but is no longer normalized, since the norm of $g_n$ in $L^2(-\pi/2, \pi/2)$ is $1/\sqrt{2}$. An orthonormal basis can be characterized as any sequence with $\|g_n\| = 1$ for all $n$ that is a tight frame with $A = 1$. The sequence $\{\sqrt{2}g_{2k}, k = -\infty, ..., \infty\}$ is an orthonormal basis for $L^2(-\pi/2, \pi/2)$, as is the sequence $\{\sqrt{2}g_{2k+1}, k = -\infty, ..., \infty\}$. The sequence $\{\langle f, g_n \rangle, n = -\infty, ..., \infty\}$ is redundant; the half corresponding either to the odd $n$ or to the even $n$ suffices to recover $f$. Because of this redundancy we can tolerate more inaccuracy in measuring these values; indeed, this is one of the main attractions of frames in signal processing.

---

## 15.8  Radar Problems

In radar a real-valued function $\psi(t)$ representing a time-varying voltage is converted by an antenna in transmission mode into a propagating electromagnetic wave. When this wave encounters a reflecting target an echo is produced. The antenna, now in receiving mode, picks up the echo $f(t)$, which is related to the original signal by

$$f(t) = A\psi(t - d(t)),$$

where $d(t)$ is the time required for the original signal to make the round trip from the antenna to the target and return back at time $t$. The amplitude $A$ incorporates the reflectivity of the target as well as attenuation suffered by the signal. As we shall see shortly, the delay $d(t)$ depends on the distance from the antenna to the target and, if the target is moving, on its radial velocity. The main signal-processing problem here is to determine target range and radial velocity from knowledge of $f(t)$ and $\psi(t)$.

If the target is stationary, at a distance $r_0$ from the antenna, then $d(t) = 2r_0/c$, where $c$ is the speed of light. In this case the original signal and the received echo are related simply by

$$f(t) = A\psi(t - b),$$

for $b = 2r_0/c$. When the target is moving so that its distance to the antenna, $r(t)$, is time-dependent, the relationship between $f$ and $\psi$ is more complicated.

**Ex. 15.4** *Suppose the target is at a distance $r_0 > 0$ from the antenna at time $t = 0$, and has radial velocity $v$, with $v > 0$ indicating away from the antenna. Show that the delay function $d(t)$ is now*

$$d(t) = 2\frac{r_0 + vt}{c + v}$$

*and $f(t)$ is related to $\psi(t)$ according to*

$$f(t) = A\psi\left(\frac{t - b}{a}\right), \tag{15.3}$$

*for*

$$a = \frac{c + v}{c - v}$$

*and*

$$b = \frac{2r_0}{c - v}.$$

*Show also that if we select $A = (\frac{c-v}{c+v})^{1/2}$ then energy is preserved; that is, $||f|| = ||\psi||$.*

**Ex. 15.5** *Let $\Psi(\omega)$ be the Fourier transform of the signal $\psi(t)$. Show that the Fourier transform of the echo $f(t)$ in Equation (15.3) is then*

$$F(\omega) = Aae^{ib\omega}\Psi(a\omega).$$

The basic problem is to determine $a$ and $b$, and therefore the range and radial velocity of the target, from knowledge of $f(t)$ and $\psi(t)$. An obvious approach is to do a matched filter.

## 15.9 The Wideband Cross-Ambiguity Function

Note that the received echo $f(t)$ is related to the original signal by the operations of rescaling and shifting. We therefore match the received echo with all the shifted and rescaled versions of the original signal. For each $a > 0$ and real $b$, let

$$\psi_{a,b}(t) = \psi\left(\frac{t-b}{a}\right).$$

The *wideband cross-ambiguity function* (WCAF) is

$$W_\psi f)(b,a) = \frac{1}{\sqrt{a}} \int_{-\infty}^{\infty} f(t)\psi_{a,b}(t)dt.$$

In the ideal case the values of $a$ and $b$ for which the WCAF takes on its largest absolute value should be the true values of $a$ and $b$.

More generally, there will be many individual targets or sources of echoes, each having their own values of $a$, $b$, and $A$. The resulting received echo function $f(t)$ is a superposition of the individual functions $\psi_{a,b}(t)$, which, for technical reasons, we write as

$$f(t) = \int_{-\infty}^{\infty} \int_{0}^{\infty} D(b,a)\psi_{a,b}(t)\frac{dadb}{a^2}. \tag{15.4}$$

We then have the inverse problem of determining $D(b,a)$ from $f(t)$.

Equation (15.4) provides a representation of the echo $f(t)$ as a superposition of rescaled translates of a single function, namely the original signal $\psi(t)$. We shall encounter this representation again in our discussion of wavelets, where the signal $\psi(t)$ is called the *mother wavelet* and the WCAF is called the *integral wavelet transform*. One reason for discussing radar and ambiguity functions now is to motivate some of the wavelet theory. Our discussion here follows closely the treatment in [97], where Kaiser emphasizes the important connections between wavelets and radar ambiguity functions.

As we shall see when we study wavelets in Chapter 16, we can recover the signal $f(t)$ from the WCAF using the following inversion formula: at points $t$ where $f(t)$ is continuous we have

$$f(t) = \frac{1}{C_\psi} \int_{-\infty}^{\infty} \int_{-\infty}^{\infty} (W_\psi f)(b,a)\psi\left(\frac{t-b}{a}\right)\frac{dadb}{a^2},$$

with

$$C_\psi = \int_{-\infty}^{\infty} \frac{|\Psi(\omega)|^2}{|\omega|}d\omega$$

for $\Psi(\omega)$ the Fourier transform of $\psi(t)$. The obvious conjecture is then that the distribution function $D(b, a)$ is

$$D(b, a) = \frac{1}{C_\psi}(W_\psi f)(b, a).$$

However, this is not generally the case. Indeed, there is no particular reason why the physically meaningful function $D(b, a)$ must have the form $(W_\psi g)(b, a)$ for some function $g$. So the inverse problem of estimating $D(b, a)$ from $f(t)$ is more complicated. One approach mentioned in [97] involves transmitting more than one signal $\psi(t)$ and estimating $D(b, a)$ from the echoes corresponding to each of the several different transmitted signals.

## 15.10　The Narrowband Cross-Ambiguity Function

The real signal $\psi(t)$ with Fourier transform $\Psi(\omega)$ is said to be a *narrowband signal* if there are constants $\alpha$ and $\gamma$ such that the conjugate-symmetric function $\Psi(\omega)$ is concentrated on $\alpha \leq |\omega| \leq \gamma$ and $\frac{\gamma - \alpha}{\gamma + \alpha}$ is nearly equal to zero, which means that $\alpha$ is very much greater than $\beta = \frac{\gamma - \alpha}{2}$. The center frequency is $\omega_c = \frac{\gamma + \alpha}{2}$.

**Ex. 15.6** *Let $\phi = 2\omega_c v/c$. Show that $a\omega_c$ is approximately equal to $\omega_c + \phi$.*

It follows then that, for $\omega > 0$, $F(\omega)$, the Fourier transform of the echo $f(t)$, is approximately $Aae^{ib\omega}\Psi(\omega + \phi)$. Because the Doppler shift affects positive and negative frequencies differently, it is convenient to construct a related signal having only positive frequency components.

Let $G(\omega) = 2F(\omega)$ for $\omega > 0$ and $G(\omega) = 0$ otherwise. Let $g(t)$ be the inverse Fourier transform of $G(\omega)$. Then, the complex-valued function $g(t)$ is called the *analytic signal* associated with $f(t)$. The function $f(t)$ is the real part of $g(t)$; the imaginary part of $g(t)$ is the *Hilbert transform* of $f(t)$. Then, the *demodulated analytic signal* associated with $f(t)$ is $h(t)$ with Fourier transform $H(\omega) = G(\omega + \omega_c)$. Similarly, let $\gamma(t)$ be the demodulated analytic signal associated with $\psi(t)$.

**Ex. 15.7** *Show that the demodulated analytic signals $h(t)$ and $\gamma(t)$ are related by*

$$h(t) = Be^{i\phi t}\gamma(t - b) = B\gamma_{\phi, b}(t),$$

*for B a time-independent constant. Hint: Use the fact that* $\Psi(\omega) = 0$ *for* $0 \leq \omega < \alpha$ *and* $\phi < \alpha$.

To determine the range and radial velocity in the narrowband case we again use the matched filter, forming the *narrowband cross-ambiguity function* (NCAF)

$$N_h(\phi, b) = \langle h, \gamma_{\phi, b} \rangle = \int_{-\infty}^{\infty} h(t) e^{-i\phi t} \overline{\gamma(t - b)} dt.$$

Ideally, the values of $\phi$ and $b$ corresponding to the largest absolute value of $N_h(\phi, b)$ will be the true ones, from which the range and radial velocity can be determined. For each fixed value of $b$, the NCAF is the Fourier transform of the function $h(t) \overline{\gamma(t - b)}$, evaluated at $\omega = -\phi$; so the NCAF contains complete information about the function $h(t)$. In Chapter 16 on wavelets we shall consider the NCAF in a different light, with $\gamma$ playing the role of a window function and the NCAF the short-time Fourier transform of $h(t)$, describing the frequency content of $h(t)$ near the time $b$.

In the more general case in which the narrowband echo function $f(t)$ is a superposition of narrowband reflections,

$$f(t) = \int_{-\infty}^{\infty} \int_{0}^{\infty} D(b, a) \psi_{a,b}(t) \frac{dadb}{a^2},$$

we have

$$h(t) = \int_{-\infty}^{\infty} \int_{0}^{\infty} D_{NB}(b, \phi) e^{i\phi t} \gamma(t - b) d\phi db,$$

where $D_{NB}(b, \phi)$ is the narrowband distribution of reflecting target points, as a function of $b$ and $\phi = 2\omega_c v / c$. The inverse problem now is to estimate this distribution, given $h(t)$.

---

## 15.11   Range Estimation

If the transmitted signal is $\psi(t) = e^{i\omega t}$ and the target is stationary at range $r$, then the echo received is $f(t) = Ae^{i\omega(t-b)}$, where $b = 2r/c$. So our information about $r$ is that we know the value $e^{2i\omega r/c}$. Because of the periodicity of the complex exponential function, this is not enough information to determine $r$; we need $e^{2i\omega r/c}$ for a variety of values of $\omega$. To obtain these values we can transmit a signal whose frequency changes with time, such as a *chirp* of the form

$$\psi(t) = e^{i\omega t^2}$$

with the frequency $2\omega t$ at time $t$.

## 15.12   Time-Frequency Analysis

The inverse Fourier transform formula

$$f(t) = \frac{1}{2\pi} \int_{-\infty}^{\infty} F(\omega)e^{-i\omega t}d\omega$$

provides a representation of the function of time $f(t)$ as a superposition of sinusoids $e^{-i\omega t}$ with frequencies $\omega$. The value at $\omega$ of the Fourier transform

$$F(\omega) = \int_{-\infty}^{\infty} f(t)e^{i\omega t}dt$$

is the complex amplitude associated with the sinusoidal component $e^{-i\omega t}$. It quantifies the contribution to $f(t)$ made by that sinusoid, over all of $t$. To determine each individual number $F(\omega)$ we need $f(t)$ for all $t$. It is implicit that the frequency content has not changed over time.

## 15.13   The Short-Time Fourier Transform

To estimate the frequency content of the signal $f(t)$ around the time $t = b$, we could proceed as follows. Multiply $f(t)$ by the function that is equal to $\frac{1}{2\epsilon}$ on the interval $[b - \epsilon, b + \epsilon]$ and zero otherwise. Then take the Fourier transform. The multiplication step is called *windowing*.

To see how well this works, consider the case in which $f(t) = \exp(-i\omega_0 t)$ for all $t$. The Fourier transform of the windowed signal is then

$$\exp(i(\omega - \omega_0)b)\frac{\sin(\epsilon(\omega - \omega_0))}{\epsilon(\omega - \omega_0)}.$$

This function attains its maximum value of one at $\omega = \omega_0$. But, the first zeros of the function are at $|\omega - \omega_0| = \frac{\pi}{\epsilon}$, which says that as $\epsilon$ gets smaller the windowed Fourier transform spreads out more and more around $\omega = \omega_0$; that is, better time localization comes at the price of worse frequency localization. To achieve a somewhat better result we can change the window function.

The standard normal (or Gaussian) curve is

$$g(t) = \frac{1}{\sqrt{2\pi}} \exp\left(-\frac{1}{2}t^2\right),$$

which has its peak at $t = 0$ and falls off to zero symmetrically on either side. For $\sigma > 0$, let

$$g_\sigma(t) = \frac{1}{\sigma}g(t/\sigma).$$

Then the function $g_\sigma(t - b)$ is centered at $t = b$ and falls off on either side, more slowly for large $\sigma$, faster for smaller $\sigma$. Also we have

$$\int_{-\infty}^{\infty} g_\sigma(t - b)dt = 1$$

for each $b$ and $\sigma > 0$. Such functions were used by Gabor [79] for *windowing* signals and are called *Gabor windows*.

Gabor's idea was to multiply $f(t)$, the signal of interest, by the window $g_\sigma(t - b)$ and then to take the Fourier transform, obtaining the *short-time Fourier transform* (STFT)

$$G_b^\sigma(\omega) = \int_{-\infty}^{\infty} f(t)g_\sigma(t - b)e^{i\omega t}dt.$$

Since $g_\sigma(t - b)$ falls off to zero on either side of $t = b$, multiplying by this window essentially restricts the signal to a neighborhood of $t = b$. The STFT then measures the frequency content of the signal, near the time $t = b$. The STFT therefore performs a *time-frequency analysis* of the signal.

We focus more tightly around the time $t = b$ by choosing a small value for $\sigma$. Because of the uncertainty principle, the Fourier transform of the window $g_\sigma(t-b)$ grows wider as $\sigma$ gets smaller; the *time-frequency window* remains constant [55]. This causes the STFT to involve greater blurring in the frequency domain. In short, to get good resolution in frequency, we need to observe for a longer time; if we focus on a small time interval, we pay the price of reduced frequency resolution. This is unfortunate because when we focus on a short interval of time, it is to uncover a part of the signal that is changing within that short interval, which means it must have high frequency components within that interval. There is no reason to believe that the spacing is larger between those high frequencies we wish to resolve than between lower frequencies associated with longer time intervals. We would like to have the same resolving capability when focusing on a short time interval that we have when focusing on a longer one.

## 15.14 The Wigner–Ville Distribution

In [118] Meyer describes Ville's approach to determining the instantaneous power spectrum of the signal, that is, the energy in the signal $f(t)$

that corresponds to time $t$ and frequency $\omega$. The goal is to find a function $W_f(t, \omega)$ having the properties

$$\int W_f(t, \omega) d\omega / 2\pi = |f(t)|^2,$$

which is the total energy in the signal at time $t$, and

$$\int W_f(t, \omega) dt = |F(\omega)|^2,$$

which is the total energy in the Fourier transform at frequency $\omega$. Because these two properties do not specify a unique $W_f(t, \omega)$, two additional properties are usually required:

$$\int \int W_f(t, \omega) \overline{W_g(t, \omega)} dt d\omega / 2\pi = \left| \int f(t) \overline{g(t)} dt \right|^2$$

and, for $f(t) = g_\sigma(t - b) \exp(i\alpha t)$,

$$W_f(t, \omega) = 2 \exp(-\sigma^{-2}(t - b)^2) \exp(-\sigma^2(\omega - \alpha)^2).$$

The *Wigner–Ville distribution* of $f(t)$, given by

$$WV_f(t, \omega) = \int_{-\infty}^{\infty} f\left(t + \frac{\tau}{2}\right) \overline{f\left(t - \frac{\tau}{2}\right)} \exp(-i\omega\tau) d\tau,$$

has all four of the desired properties. The Wigner–Ville distribution is always real-valued, but its values need not be nonnegative.

In [65] De Bruijn defines the *score* of a signal $f(t)$ to be $H(x, y; f, f)$, where

$$H(x, y; f_1, f_2) = 2 \int_{-\infty}^{\infty} f_1(x + t) \overline{f_2(x - t)} e^{-4\pi i y t} dt.$$

**Ex. 15.8** *Relate the narrowband cross-ambiguity function to the De Bruijn's score and the Wigner–Ville distribution.*

# Chapter 16

# Wavelets

## 16.1 Chapter Summary

In this chapter we present an overview of the theory of wavelets, with particular emphasis on their use in signal processing..

## 16.2 Background

The fantastic increase in computer power over the last few decades has made possible, even routine, the use of digital procedures for solving problems that were believed earlier to be intractable, such as the modeling of large-scale systems. At the same time, it has created new applications unimagined previously, such as medical imaging. In some cases the mathematical formulation of the problem is known and progress has come with the introduction of efficient computational algorithms, as with the Fast Fourier Transform. In other cases, the mathematics is developed, or perhaps rediscovered, as needed by the people involved in the applications. Only later is it realized that the theory already existed, as with the development of computerized tomography without Radon's earlier work on reconstruction of functions from their line integrals.

It can happen that applications give a theoretical field of mathematics a rebirth; such seems to be the case with *wavelets* [95]. Sometime in the 1980s researchers working on various problems in electrical engineering, quantum mechanics, image processing, and other areas became aware that what the others were doing was related to their own work. As connections became established, similarities with the earlier mathematical theory of approximation in functional analysis were noticed. Meetings began to take place, and a common language began to emerge around this reborn area, now called wavelets. One of the most significant meetings took place in June of 1990, at the University of Massachusetts Lowell. The keynote speaker was Ingrid Daubechies; the lectures she gave that week were subsequently published in the book [64].

There are a number of good books on wavelets, such as [97], [11], and [159]. A recent issue of the IEEE Signal Processing Magazine has an interesting article on using wavelet analysis of paintings for artist identification [96].

Fourier analysis and synthesis concerns the decomposition, filtering, compressing, and reconstruction of signals using complex exponential functions as the building blocks; wavelet theory provides a framework in which other building blocks, better suited to the problem at hand, can be used. As always, efficient algorithms provide the bridge between theory and practice.

Since their development in the 1980s wavelets have been used for many purposes. In the discussion to follow, we focus on the problem of analyzing a signal whose frequency composition is changing over time. As we saw in our discussion of the narrowband cross-ambiguity function in radar, the need for such time-frequency analysis has been known for quite a while. Other methods, such as Gabor's short time Fourier transform and the Wigner-Ville distribution, have also been considered for this purpose.

## 16.3   A Simple Example

Imagine that $f(t)$ is defined for all real $t$ and we have sampled $f(t)$ every half-second. We focus on the time interval $[0, 2)$. Suppose that $f(0) = 1$,

$f(0.5) = -3$, $f(1) = 2$ and $f(1.5) = 4$. We approximate $f(t)$ within the interval $[0, 2)$ by replacing $f(t)$ with the step function that is 1 on $[0, 0.5)$, $-3$ on $[0.5, 1)$, 2 on $[1, 1.5)$, and 4 on $[1.5, 2)$; for notational convenience, we represent this step function by $(1, -3, 2, 4)$. We can decompose $(1, -3, 2, 4)$ into a sum of step functions

$$(1, -3, 2, 4) = 1(1, 1, 1, 1) - 2(1, 1, -1, -1) + 2(1, -1, 0, 0) - 1(0, 0, 1, -1).$$

The first basis element, $(1, 1, 1, 1)$, does not vary over a two-second interval. The second one, $(1, 1, -1, -1)$, is orthogonal to the first, and does not vary over a one-second interval. The other two, both orthogonal to the previous two and to each other, vary over half-second intervals. We can think of these basis functions as corresponding to different frequency components and time locations; that is, they are giving us a time-frequency decomposition.

Suppose we let $\phi_0(t)$ be the function that has the value 1 on the interval $[0, 1)$ and zero elsewhere, and $\psi_0(t)$ the function that has the value 1 on the interval $[0, 0.5)$, the value $-1$ on the interval $[0.5, 1)$, and zero elsewhere. Then we say that

$$\phi_0(t) = (1, 1, 0, 0),$$

and

$$\psi_0(t) = (1, -1, 0, 0).$$

We write

$$\phi_{-1}(t) = (1, 1, 1, 1) = \phi_0(0.5t) = \phi_0(2^{-1}t),$$

$$\psi_0(t - 1) = (0, 0, 1, -1),$$

and

$$\psi_{-1}(t) = (1, 1, -1, -1) = \psi_0(0.5t) = \psi_0(2^{-1}t).$$

So we have the decomposition of $(1, -3, 2, 4)$ as

$$(1, -3, 2, 4) = 1\phi_{-1}(t) - 2\psi_{-1}(t) + 2\psi_0(t) - 1\psi_0(t - 1).$$

In what follows we shall be interested in extending these ideas, to find other functions $\phi_0(t)$ and $\psi_0(t)$ that lead to bases consisting of functions of the form

$$\psi_{j,k}(t) = \psi_0(2^j t - k).$$

These will be our *wavelet bases*.

## 16.4    The Integral Wavelet Transform

For real numbers $b$ and $a \neq 0$, the *integral wavelet transform* (IWT) of the signal $f(t)$ relative to the *basic wavelet* (or *mother* wavelet) $\psi(t)$ is

$$(W_\psi f)(b, a) = |a|^{-\frac{1}{2}} \int_{-\infty}^{\infty} f(t) \psi \left( \frac{t - b}{a} \right) dt.$$

This function is also the wideband cross-ambiguity function in radar. The function $\psi(t)$ is also called a window function and, like Gaussian functions, it will be relatively localized in time. However, it must also have properties quite different from those of Gabor's Gaussian windows; in particular, we want

$$\int_{-\infty}^{\infty} \psi(t) dt = 0.$$

An example is the *Haar wavelet* $\psi_{Haar}(t)$ that has the value $+1$ for $0 \leq t < \frac{1}{2}$, $-1$ for $\frac{1}{2} \leq t < 1$, and $0$ otherwise.

As the scaling parameter $a$ grows larger the wavelet $\psi(t)$ grows wider, so choosing a small value of the scaling parameter permits us to focus on a neighborhood of the time $t = b$. The IWT then registers the contribution to $f(t)$ made by components with features on the scale determined by $a$, in the neighborhood of $t = b$. Calculations involving the uncertainty principle reveal that the IWT provides a flexible time-frequency window that narrows when we observe high frequency components and widens for lower frequencies [55].

Given the integral wavelet transform $(W_\psi f)(b, a)$, it is natural to ask how we might recover the signal $f(t)$. The following inversion formula answers that question: at points $t$ where $f(t)$ is continuous we have

$$f(t) = \frac{1}{C_\psi} \int_{-\infty}^{\infty} \int_{-\infty}^{\infty} (W_\psi f)(b, a) \psi \left( \frac{t - b}{a} \right) \frac{da}{a^2} db,$$

with

$$C_\psi = \int_{-\infty}^{\infty} \frac{|\Psi(\omega)|^2}{|\omega|} d\omega$$

for $\Psi(\omega)$, the Fourier transform of $\psi(t)$.

## 16.5    Wavelet Series Expansions

The Fourier series expansion of a function $f(t)$ on a finite interval is a representation of $f(t)$ as a sum of orthogonal complex exponentials. Localized alterations in $f(t)$ affect every one of the components of this sum. Wavelets, on the other hand, can be used to represent $f(t)$ so that localized alterations in $f(t)$ affect only a few of the components of the wavelet expansion. The simplest example of a wavelet expansion is with respect to the Haar wavelets.

**Ex. 16.1** *Let $w(t) = \psi_{Haar}(t)$. Show that the functions $w_{jk}(t) = w(2^j t - k)$ are mutually orthogonal on the interval $[0, 1]$, where $j = 0, 1, \ldots$ and $k = 0, 1, \ldots, 2^j - 1$.*

These functions $w_{jk}(t)$ are the *Haar wavelets*. Every continuous function $f(t)$ defined on $[0, 1]$ can be written as

$$f(t) = c_0 + \sum_{j=0}^{\infty} \sum_{k=0}^{2^j - 1} c_{jk} w_{jk}(t)$$

for some choice of $c_0$ and $c_{jk}$. Notice that the *support of the function $w_{jk}(t)$*, the interval on which it is nonzero, gets smaller as $j$ increases. Therefore, the components corresponding to higher values of $j$ in the Haar expansion of $f(t)$ come from features that are localized in the variable $t$; such features are transients that live for only a short time. Such transient components affect all of the Fourier coefficients but only those Haar wavelet coefficients corresponding to terms supported in the region of the disturbance. This ability to isolate localized features is the main reason for the popularity of wavelet expansions.

The orthogonal functions used in the Haar wavelet expansion are themselves discontinuous, which presents a bit of a problem when we represent continuous functions. Wavelets that are themselves continuous, or better still, differentiable, should do a better job representing smooth functions.

We can obtain other wavelet series expansions by selecting a basic wavelet $\psi(t)$ and defining $\psi_{jk}(t) = 2^{j/2} \psi(2^j t - k)$, for integers $j$ and $k$. We then say that the function $\psi(t)$ is an *orthogonal wavelet* if the family $\{\psi_{jk}\}$ is an orthonormal basis for the space of square-integrable functions on the real line, the Hilbert space $L^2(\mathbb{R})$. This implies that for every such $f(t)$ there are coefficients $c_{jk}$ so that

$$f(t) = \sum_{j=-\infty}^{\infty} \sum_{k=-\infty}^{\infty} c_{jk} \psi_{jk}(t),$$

with convergence in the mean-square sense. The coefficients $c_{jk}$ are found using the IWT:

$$c_{jk} = (W_\psi f)\left(\frac{k}{2^j}, \frac{1}{2^j}\right).$$

It is also of interest to consider wavelets $\psi$ for which $\{\psi_{jk}\}$ form a basis, but not an orthogonal one, or, more generally, form a *frame*, in which the series representations of $f(t)$ need not be unique.

As with Fourier series, wavelet series expansion permits the filtering of certain components, as well as signal compression. In the case of Fourier series, we might attribute high frequency components to noise and achieve a smoothing by setting to zero the coefficients associated with these high frequencies. In the case of wavelet series expansions, we might attribute to noise localized small-scale disturbances and remove them by setting to zero the coefficients corresponding to the appropriate $j$ and $k$. For both Fourier and wavelet series expansions we can achieve compression by ignoring those components whose coefficients are below some chosen level.

---

## 16.6 Multiresolution Analysis

One way to study wavelet series expansions is through *multiresolution analysis* (MRA) [115]. Let us begin with an example involving band-limited functions. This example is called the *Shannon* MRA.

### 16.6.1 The Shannon Multiresolution Analysis

Let $V_0$ be the collection of functions $f(t)$ whose Fourier transform $F(\omega)$ is zero for $|\omega| > \pi$; so $V_0$ is the collection of $\pi$-band-limited functions. Let $V_1$ be the collection of functions $f(t)$ whose Fourier transform $F(\omega)$ is zero for $|\omega| > 2\pi$; so $V_1$ is the collection of $2\pi$-band-limited functions. In general, for each integer $j$, let $V_j$ be the collection of functions $f(t)$ whose Fourier transform $F(\omega)$ is zero for $|\omega| > 2^j\pi$; so $V_j$ is the collection of $2^j\pi$-band-limited functions.

**Ex. 16.2** *Show that if the function $f(t)$ is in $V_j$ then the function $g(t) = f(2t)$ is in $V_{j+1}$.*

We then have a nested sequence of sets of functions $\{V_j\}$, with $V_j \subseteq V_{j+1}$ for each integer $j$. The intersection of all the $V_j$ is the set containing only the zero function. Every function in $L^2(\mathbb{R})$ is arbitrarily close to a function in at least one of the sets $V_j$; more mathematically, we say that the union of the $V_j$ is dense in $L^2(\mathbb{R})$. In addition, we have $f(t)$ in $V_j$ if and only if $g(t) = f(2t)$ is in $V_{j+1}$. In general, such a collection of sets of functions is called a *multiresolution analysis* for $L^2(\mathbb{R})$. Once we have a MRA for $L^2(\mathbb{R})$, how do we get a wavelet series expansion?

A function $\phi(t)$ is called a *scaling function* or sometimes the *father wavelet* for the MRA if the collection of integer translates $\{\phi(t-k)\}$ forms a basis for $V_0$ (more precisely, a Riesz basis). Then, for each fixed $j$, the functions $\phi_{jk}(t) = \phi(2^j t - k)$, for integer $k$, will form a basis for $V_j$. In the case of the Shannon MRA, the scaling function is $\phi(t) = \frac{\sin \pi t}{\pi t}$. But how do we get a basis for all of $L^2(\mathbb{R})$?

## 16.6.2 The Haar Multiresolution Analysis

To see how to proceed, it is helpful to return to the Haar wavelets. Let $\phi_{Haar}(t)$ be the function that has the value $+1$ for $0 \le t < 1$ and zero elsewhere. Let $V_0$ be the collection of all functions in $L^2(\mathbb{R})$ that are linear combinations of integer translates of $\phi(t)$; that is, all functions $f(t)$ that are constant on intervals of the form $[k, k+1)$, for all integers $k$. Now $V_1$ is the collection of all functions $g(t)$ of the form $g(t) = f(2t)$, for some $f(t)$ in $V_0$. Therefore, $V_1$ consists of all functions in $L^2(\mathbb{R})$ that are constant on intervals of the form $[k/2, (k+1)/2)$.

Every function in $V_0$ is also in $V_1$ and every function $g(t)$ in $V_1$ can be written uniquely as a sum of a function $f(t)$ in $V_0$ and a function $h(t)$ in $V_1$ that is orthogonal to every function in $V_0$. For example, the function $g(t)$ that takes the value $+3$ for $0 \le t < 1/2$, $-1$ for $1/2 \le t < 1$, and zero elsewhere can be written as $g(t) = f(t) + h(t)$, where $h(t)$ has the value $+2$ for $0 \le t < 1/2$, $-2$ for $1/2 \le t < 1$, and zero elsewhere, and $f(t)$ takes the value $+1$ for $0 \le t < 1$ and zero elsewhere. Clearly, $h(t)$, which is twice the Haar wavelet function, is orthogonal to all functions in $V_0$.

**Ex. 16.3** *Show that the function $f(t)$ can be written uniquely as $f(t) = d(t) + e(t)$, where $d(t)$ is in $V_{-1}$ and $e(t)$ is in $V_0$ and is orthogonal to every function in $V_{-1}$. Relate the function $e(t)$ to the Haar wavelet function.*

## 16.6.3 Wavelets and Multiresolution Analysis

To get an orthogonal wavelet expansion from a general MRA, we write the set $V_1$ as the direct sum $V_1 = V_0 \oplus W_0$, so every function $g(t)$ in $V_1$ can be uniquely written as $g(t) = f(t) + h(t)$, where $f(t)$ is a function in $V_0$ and $h(t)$ is a function in $W_0$, with $f(t)$ and $h(t)$ orthogonal. Since the scaling function or father wavelet $\phi(t)$ is in $V_1$, it can be written as

$$\phi(t) = \sum_{k=-\infty}^{\infty} p_k \phi(2t - k), \qquad (16.1)$$

for some sequence $\{p_k\}$ called the *two-scale sequence* for $\phi(t)$. This most important identity is the *scaling relation* for the father wavelet. The mother wavelet is defined using a similar expression

$$\psi(t) = \sum_k (-1)^k \overline{p_{1-k}} \phi(2t - k). \tag{16.2}$$

We define

$$\phi_{jk}(t) = 2^{j/2} \phi(2^j t - k) \tag{16.3}$$

and

$$\psi_{jk}(t) = 2^{j/2} \psi(2^j t - k). \tag{16.4}$$

The collection $\{\psi_{jk}(t), -\infty < j, k < \infty\}$ then forms an orthogonal wavelet basis for $L^2(\mathbb{R})$. For the Haar MRA, the two-scale sequence is $p_0 = p_1 = 1$ and $p_k = 0$ for the rest.

**Ex. 16.4** *Show that the two-scale sequence $\{p_k\}$ has the properties*

$$p_k = 2 \int \phi(t) \overline{\phi(2t - k)} dt;$$

$$\sum_{k=-\infty}^{\infty} p_{k-2m} \overline{p_k} = 0,$$

*for $m \neq 0$ and equals 2 when $m = 0$.*

---

## 16.7    Signal Processing Using Wavelets

Once we have an orthogonal wavelet basis for $L^2(\mathbb{R})$, we can use the basis to represent and process a signal $f(t)$. Suppose, for example, that $f(t)$ is band-limited but essentially zero for $t$ not in $[0, 1]$ and we have samples $f(\frac{k}{M})$, $k = 0, ..., M$. We assume that the sampling rate $\Delta = \frac{1}{M}$ is faster than the Nyquist rate so that the Fourier transform of $f(t)$ is zero outside, say, the interval $[0, 2\pi M]$. Roughly speaking, the $W_j$ component of $f(t)$, given by

$$g_j(t) = \sum_{k=0}^{2^j - 1} \beta_k^j \psi_{jk}(t),$$

with $\beta_k^j = \langle f(t), \psi_{jk}(t) \rangle$, corresponds to the components of $f(t)$ with frequencies $\omega$ between $2^{j-1}$ and $2^j$. For $2^j > 2\pi M$ we have $\beta_k^j = 0$, so

$g_j(t) = 0$. Let $J$ be the smallest integer greater than $\log_2(2\pi) + \log_2(M)$. Then, $f(t)$ is in the space $V_J$ and has the expansion

$$f(t) = \sum_{k=0}^{2^J-1} \alpha_k^J \phi_{Jk}(t),$$

for $\alpha_k^J = \langle f(t), \phi_{Jk}(t) \rangle$. It is common practice, but not universally approved, to take $M = 2^J$ and to estimate the $\alpha_k^J$ by the samples $f(\frac{k}{M})$. Once we have the sequence $\{\alpha_k^J\}$, we can begin the decomposition of $f(t)$ into components in $V_j$ and $W_j$ for $j < J$. As we shall see, the algorithms for the decomposition and subsequent reconstruction of the signal are quite similar to the FFT.

## 16.7.1 Decomposition and Reconstruction

The decomposition and reconstruction algorithms both involve the equation

$$\sum_k a_k^j \phi_{jk} = \sum_m a_m^{j-1} \phi_{(j-1),m} + b_m^{j-1} \psi_{(j-1),m} ; \qquad (16.5)$$

in the decomposition step we know the $\{a_k^j\}$ and want the $\{a_m^{j-1}\}$ and $\{b_m^{j-1}\}$, while in the reconstruction step we know the $\{a_m^{j-1}\}$ and $\{b_m^{j-1}\}$ and want the $\{a_k^j\}$.

Using Equations (16.1) and (16.3), we obtain

$$\phi_{(j-1),l} = 2^{-1/2} \sum_k p_k \phi_{j,(k+2l)} = 2^{-1/2} \sum_k p_{k-2l} \phi_{jk}; \qquad (16.6)$$

using Equations (16.2), (16.3) and (16.4), we get

$$\psi_{(j-1),l} = 2^{-1/2} \sum_k (-1)^k \overline{p_{1-k+2l}} \phi_{jk}. \qquad (16.7)$$

Therefore,

$$\langle \phi_{jk}, \phi_{(j-1),l} \rangle = 2^{-1/2} \overline{p_{k-2l}}; \qquad (16.8)$$

this comes from substituting $\phi_{(j-1),l}$ as in Equation (16.6) into the second term in the inner product. Similarly, we have

$$\langle \phi_{jk}, \psi_{(j-1),l} \rangle = 2^{-1/2} (-1)^k p_{1-k+2l}. \qquad (16.9)$$

These relationships are then used to derive the decomposition and reconstruction algorithms.

#### 16.7.1.1  The Decomposition Step

To find $a_l^{j-1}$ we take the inner product of both sides of Equation (16.5) with the function $\phi_{(j-1),l}$. Using Equation (16.8) and the fact that $\phi_{(j-1),l}$ is orthogonal to all the $\phi_{(j-1),m}$ except for $m = l$ and is orthogonal to all the $\psi_{(j-1),m}$, we obtain

$$2^{-1/2} \sum_k a_k^j \overline{p_{k-2l}} = a_l^{j-1};$$

similarly, using Equation (16.9), we get

$$2^{-1/2} \sum_k a_k^j (-1)^k p_{1-k+2l} = b_l^{j-1}.$$

The decomposition step is to apply these two equations to get the $\{a_l^{j-1}\}$ and $\{b_l^{j-1}\}$ from the $\{a_k^j\}$.

#### 16.7.1.2  The Reconstruction Step

Now we use Equations (16.6) and (16.7) to substitute into the right hand side of Equation (16.5). Combining terms, we get

$$a_k^j = 2^{-1/2} \sum_l a_l^{j-1} p_{k-2l} + b_l^{j-1} (-1)^k \overline{p_{1-k+2l}}.$$

This takes us from the $\{a_l^{j-1}\}$ and $\{b_l^{j-1}\}$ to the $\{a_k^j\}$.

We have assumed that we have already obtained the scaling function $\phi(t)$ with the property that $\{\phi(t - k)\}$ is an orthogonal basis for $V_0$. But how do we actually obtain such functions?

---

## 16.8  Generating the Scaling Function

The scaling function $\phi(t)$ is generated from the two-scale sequence $\{p_k\}$ using the following iterative procedure. Start with $\phi_0(t) = \phi_{Haar}(t)$, the Haar scaling function that is 1 on $[0, 1]$ and 0 elsewhere. Now, for each $n = 1, 2, ...$, define

$$\phi_n(t) = \sum_{k=-\infty}^{\infty} p_k \phi_{n-1}(2t - k).$$

Provided that the sequence $\{p_k\}$ has certain properties to be discussed below, this sequence of functions converges and the limit is the desired scaling function.

The properties of $\{p_k\}$ that are needed can be expressed in terms of properties of the function

$$P(z) = \frac{1}{2} \sum_{k=-\infty}^{\infty} p_k z^k.$$

For the Haar MRA, this function is $P(z) = \frac{1}{2}(1 + z)$. We require that

1. $P(1) = 1$,

2. $|P(e^{i\theta})|^2 + |P(e^{i(\theta+\pi)})|^2 = 1$, for $0 \le \theta \le \pi$, and

3. $|P(e^{i\theta})| > 0$ for $-\frac{\pi}{2} \le \theta \le \frac{\pi}{2}$.

---

## 16.9   Generating the Two-Scale Sequence

The final piece of the puzzle is the generation of the sequence $\{p_k\}$ itself, or, equivalently, finding a function $P(z)$ with the properties listed above. The following example, also used in [11], illustrates Ingrid Daubechies' method [63].

We begin with the identity

$$\cos^2 \frac{\theta}{2} + \sin^2 \frac{\theta}{2} = 1$$

and then raise both sides to an odd power $n = 2N - 1$. Here we use $N = 2$, obtaining

$$1 = \cos^6 \frac{\theta}{2} + 3 \cos^4 \frac{\theta}{2} \sin^2 \frac{\theta}{2} + \cos^6 \frac{(\theta + \pi)}{2} + 3 \cos^4 \frac{(\theta + \pi)}{2} \sin^2 \frac{(\theta + \pi)}{2}.$$

We then let

$$|P(e^{i\theta})|^2 = \cos^6 \frac{\theta}{2} + 3 \cos^4 \frac{\theta}{2} \sin^2 \frac{\theta}{2},$$

so that

$$|P(e^{i\theta})|^2 + |P(e^{i(\theta+\pi)})|^2 = 1$$

for $0 \le \theta \le \pi$. Now we have to find $P(e^{i\theta})$. Writing

$$|P(e^{i\theta})|^2 = \cos^4 \frac{\theta}{2} \left( \cos^2 \frac{\theta}{2} + 3 \sin^2 \frac{\theta}{2} \right),$$

we have

$$P(e^{i\theta}) = \cos^2 \frac{\theta}{2} \left( \cos \frac{\theta}{2} + \sqrt{3}i \sin \frac{\theta}{2} \right) e^{i\alpha(\theta)},$$

where the real function $\alpha(\theta)$ is arbitrary. Selecting $\alpha(\theta) = 3\frac{\theta}{2}$, we get

$$P(e^{i\theta}) = p_0 + p_1 e^{i\theta} + p_2 e^{2i\theta} + p_3 e^{3i\theta},$$

for

$$p_0 = \frac{1 + \sqrt{3}}{4},$$

$$p_1 = \frac{3 + \sqrt{3}}{4},$$

$$p_2 = \frac{3 - \sqrt{3}}{4},$$

$$p_3 = \frac{1 - \sqrt{3}}{4},$$

and all the other coefficients are zero. The resulting Daubechies' wavelet is compactly supported and continuous, but not differentiable [11, 63]. Figure 16.1 shows the scaling function and mother wavelet for $N = 2$. When larger values of $N$ are used, the resulting wavelet, often denoted $\psi_N(t)$, which is again compactly supported, has approximately $N/5$ continuous derivatives.

These notions extend to nonorthogonal wavelet bases and to frames. Algorithms similar to the fast Fourier transform provide the wavelet decomposition and reconstruction of signals. The recent text by Boggess and Narcowich [11] is a nice introduction to this fast-growing area; the more advanced book by Chui [55] is also a good source. Wavelets in the context of Riesz bases and frames are discussed in Christensen's book [54]. Applications of wavelets to medical imaging are found in [127], as well as in the other papers in that special issue.

---

## 16.10   Wavelets and Filter Banks

In [152] Strang and Nguyen take a somewhat different approach to wavelets, emphasizing the role of filters and matrices. To illustrate one of their main points, we consider the two-point moving average filter.

The two-point moving average filter transforms an input sequence $x = \{x(n)\}$ to output $y = \{y(n)\}$, with $y(n) = \frac{1}{2}x(n) + \frac{1}{2}x(n-1)$. The filter $h = \{h(k)\}$ has $h(0) = h(1) = \frac{1}{2}$ and all the remaining $h(n)$ are zero. This filter is a *finite impulse response* (FIR) low-pass filter and is not invertible; the input sequence with $x(n) = (-1)^n$ has output zero. Similarly, the two-point moving difference filter $g = \{g(k)\}$, with $g(0) = \frac{1}{2}$, $g(1) = -\frac{1}{2}$, and the rest zero, is a FIR high-pass filter, also not invertible. However, if we

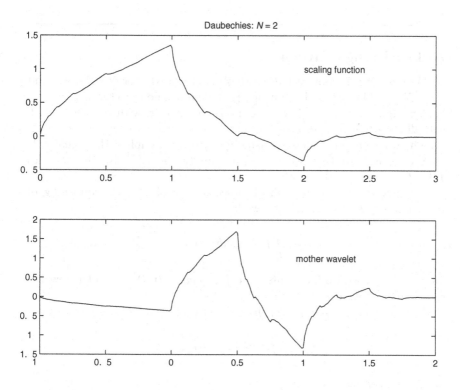

**FIGURE 16.1**: Daubechies' scaling function and mother wavelet for $N = 2$.

perform these filters in parallel, as a filter bank, no information is lost and the input can be completely reconstructed, with a unit delay. In addition, the outputs of the two filters contain redundancy that can be removed by *decimation*, which is taken here to mean *downsampling*, that is, throwing away every other term of a sequence.

The authors treat the more general problem of obtaining perfect reconstruction of the input from the output of a filter bank of low- and high-pass filters followed by downsampling. The properties that must be required of the filters are those we encountered earlier with regard to the two-scale sequences for the father and mother wavelets. When the filter operations are construed as matrix multiplications, the decomposition and reconstruction algorithms become matrix factorizations.

## 16.11   Using Wavelets

We consider the Daubechies mother wavelet $\psi_N(t)$, for $N = 1, 2, ...,$ and $n = 2N - 1$. The two-scale sequence $\{p_k\}$ then has nonzero terms $p_0, ..., p_n$. For example, when $N = 1$, we get the Haar wavelet, with $p_0 = p_1 = 1/2$, and all the other $p_k = 0$.

The wavelet signal analysis usually begins by sampling the signal $f(t)$ closely enough so that we can approximate the $a_k^{j+1}$ by the samples $f(k/2^{j+1})$.

An important aspect of the Daubechies wavelets is the vanishing of moments. For $k = 0, 1, ..., N - 1$ we have

$$\int t^k \psi_N(t)dt = 0;$$

for the Haar case we have only that $\int \psi_1(t)dt = 0$. We consider now the significance of vanishing moments for detection.

For an arbitrary signal $f(t)$ the wavelet coefficients $b_k^j$ are given by

$$b_k^j = \int f(t)2^{j/2}\psi_N(2^j t - k)dt.$$

We focus on $N = 2$.

The function $\psi_2(2^j t - k)$ is supported on the interval $[k/2^j, (k+3)/2^j]$ so we have

$$b_k^j = \int_0^{3/2^j} f(t + k/2^j)\psi_2(2^j t)dt.$$

If $f(t)$ is smooth near $t = k/2^j$, and $j$ is large enough, then

$$f(t + k/2^j) = f(k/2^j) + f'(k/2^j)t + \frac{1}{2!}f''(k/2^j)t^2 + \cdots,$$

and so

$$b_k^j \simeq 2^{j/2}\left( f(k/2^j)\int_0^{3/2^j} \psi_2(2^j t)dt \right.$$
$$\left. + f'(k/2^j)\int_0^{3/2^j} t\psi_2(2^j t)dt + f''(k/2^j)\int_0^{3/2^j} t^2\psi_2(2^j t)dt \right).$$

Since

$$\int \psi_2(t)dt = \int t\psi_2(t)dt = 0$$

and

$$\int t^2 \psi_2(t) dt \simeq -\frac{1}{8}\sqrt{\frac{3}{2\pi}},$$

we have

$$b_k^j \simeq -\frac{1}{16}\sqrt{\frac{3}{2\pi}} 2^{-5j/2} f''(k/2^j).$$

On the other hand, if $f(t)$ is not smooth near $t = k/2^j$, we expect the $b_k^j$ to have a larger magnitude.

**Example 1** Suppose that $f(t)$ is piecewise linear. Then $f''(t) = 0$, except at the places where the lines meet. So we expect the $b_k^j$ to be zero, except at the nodes.

**Example 2** Let $f(t) = t(1-t)$, for $t \in [0,1]$, and zero elsewhere. We might begin with the sample values $f(k/2^7)$ and then consider $b_k^6$. Again using $N = 2$, we find that $b_k^6 \simeq f''(k/2^6) = 2$, independent of $k$, except near the endpoints $t = 0$ and $t = 1$. The discontinuity of $f'(t)$ at the ends will make the $b_k^6$ there larger.

**Example 3** Now let $g(t) = t^2(1-t)^2$, for $t \in [0,1]$, and zero elsewhere. The first derivative is continuous at the endpoints $t = 0$ and $t = 1$, but the second derivative is discontinuous there. Using $N = 2$, we won't be able to detect this discontinuity, but using $N = 3$ we will.

**Example 4** Suppose that $f(t) = e^{i\omega t}$. Then we have

$$b_k^j = 2^{-j/2} e^{i\omega k/2^j} \Psi_N(\omega/2^j),$$

independent of $k$, where $\Psi_N$ denotes the Fourier transform of $\psi_N$ . If we plot these values for various $j$, the maximum is reached when

$$\omega/2^j = \operatorname{argmax} \Psi_N,$$

from which we can find $\omega$.

# Chapter 17

# The BLUE and the Kalman Filter

## 17.1 Chapter Summary

In most signal- and image-processing applications the measured data includes (or may include) a signal component we want and unwanted components called *noise*. Estimation involves determining the precise nature and strength of the signal component; deciding if that strength is zero or not is detection.

Noise often appears as an additive term, which we then try to remove. If we knew precisely the noisy part added to each data value we would simply subtract it; of course, we never have such information. How then do we remove something when we don't know what it is? Statistics provides a way out.

The basic idea in statistics is to use procedures that perform well on average, when applied to a class of problems. The procedures are built using properties of that class, usually involving probabilistic notions, and are evaluated by examining how they would have performed had they been applied to every problem in the class. To use such methods to remove additive noise, we need a description of the class of noises we expect to encounter, not specific values of the noise component in any one particular instance. We also need some idea about what signal components look like. In this chapter we discuss solving this noise removal problem using the *best linear unbiased estimation* (BLUE). We begin with the simplest case and then proceed to discuss increasingly complex scenarios.

An important application of the BLUE is in Kalman filtering. The connection between the BLUE and Kalman filtering is best understood by considering the case of the BLUE with a prior estimate of the signal component, and mastering the various matrix manipulations that are involved in this problem. These calculations then carry over, almost unchanged, to the Kalman filtering.

Kalman filtering is usually presented in the context of estimating a sequence of vectors evolving in time. Kalman filtering for image processing is derived by analogy with the temporal case, with certain parts of the image considered to be in the "past" of a fixed pixel.

## 17.2 The Simplest Case

Suppose our data is $z_j = c + v_j$, for $j = 1, ..., J$, where $c$ is an unknown constant to be estimated and the $v_j$ are additive noise. We assume that $E(v_j) = 0, E(v_j \overline{v_k}) = 0$ for $j \neq k$, and $E(\|v_j\|^2) = \sigma_j^2$. So, the additive noises are assumed to have mean zero and to be independent (or at least uncorrelated). In order to estimate $c$, we adopt the following rules:

1. The estimate $\hat{c}$ is *linear* in the data $\mathbf{z} = (z_1, ..., z_J)^T$; that is, $\hat{c} = \mathbf{k}^\dagger \mathbf{z}$, for some vector $\mathbf{k} = (k_1, ..., k_J)^T$.

2. The estimate is *unbiased*; $E(\hat{c}) = c$. This means $\sum_{j=1}^J k_j = 1$.

3. The estimate is best in the sense that it minimizes the expected error squared; that is, $E(|\hat{c} - c|^2)$ is minimized.

**Ex. 17.1** *Show that the resulting vector* $\mathbf{k}$ *is*

$$k_i = \sigma_i^{-2} / \left( \sum_{j=1}^J \sigma_j^{-2} \right),$$

*and the BLUE estimator of* $c$ *is then*

$$\hat{c} = \sum_{i=1}^J z_i \sigma_i^{-2} / \left( \sum_{j=1}^J \sigma_j^{-2} \right).$$

**Ex. 17.2** *Suppose we have data $z_1 = c + v_1$ and $z_2 = c + v_2$ and we want to estimate the constant $c$. Assume that $E(v_1) = E(v_2) = 0$ and $E(v_1 v_2) = \rho$, with $0 < |\rho| < 1$. Find the BLUE estimate of $c$.*

**Ex. 17.3** *The concentration of a substance in solution decreases exponentially during an experiment. Noisy measurements of the concentration are made at times $t_1$ and $t_2$, giving the data*

$$z_i = x_0 e^{-t_i} + v_i, \ i = 1, 2,$$

*where the $v_i$ have mean zero, and are uncorrelated. Find the BLUE for the initial concentration $x_0$.*

---

## 17.3 A More General Case

Suppose now that our data vector is $\mathbf{z} = H\mathbf{x} + \mathbf{v}$. Here, $\mathbf{x}$ is an unknown vector whose value is to be estimated. The random vector $\mathbf{v}$ is additive noise whose mean is $E(\mathbf{v}) = 0$ and whose known covariance matrix $Q = E(\mathbf{v}\mathbf{v}^\dagger)$ is invertible and not necessarily diagonal. The known matrix $H$ is $J$ by $N$, with $J > N$. We seek an estimate of the vector $\mathbf{x}$, using the following rules:

1. The estimate $\hat{\mathbf{x}}$ must have the form $\hat{\mathbf{x}} = K^\dagger \mathbf{z}$, where the matrix $K$ is to be determined.

2. The estimate is unbiased; that is, $E(\hat{\mathbf{x}}) = \mathbf{x}$.

3. The $K$ is determined as the minimizer of the expected squared error; that is, once again we minimize $E(\|\hat{\mathbf{x}} - \mathbf{x}\|^2)$.

**Ex. 17.4** *Show that for the estimator to be unbiased we need $K^\dagger H = I$, the identity matrix.*

**Ex. 17.5** *Show that*

$$E(\|\hat{\mathbf{x}} - \mathbf{x}\|^2) = \text{trace } K^\dagger Q K.$$

*Hints: Write the left side as*

$$E(\text{trace }((\hat{\mathbf{x}} - \mathbf{x})(\hat{\mathbf{x}} - \mathbf{x})^\dagger)).$$

*Also use the fact that the trace and expected-value operations commute.*

The problem then is to minimize trace $K^\dagger Q K$ subject to the constraint equation $K^\dagger H = I$. We solve this problem using a technique known as *prewhitening*.

Since the noise covariance matrix $Q$ is Hermitian and nonnegative definite, we have $Q = UDU^\dagger$, where the columns of $U$ are the (mutually orthogonal) eigenvectors of $Q$ and $D$ is a diagonal matrix whose diagonal entries are the (necessarily nonnegative) eigenvalues of $Q$; therefore, $U^\dagger U = I$. We call $C = UD^{1/2}U^\dagger$ the Hermitian square root of $Q$, since $C^\dagger = C$ and $C^2 = Q$. We assume that $Q$ is invertible, so that $C$ is also. Given the system of equations

$$\mathbf{z} = H\mathbf{x} + \mathbf{v},$$

as before, we obtain a new system

$$\mathbf{y} = G\mathbf{x} + \mathbf{w}$$

by multiplying both sides by $C^{-1} = Q^{-1/2}$; here, $G = C^{-1}H$ and $\mathbf{w} = C^{-1}\mathbf{v}$. The new noise correlation matrix is

$$E(\mathbf{w}\mathbf{w}^\dagger) = C^{-1}QC^{-1} = I,$$

so the new noise is white. For this reason the step of multiplying by $C^{-1}$ is called *prewhitening*.

With $J = CK$ and $M = C^{-1}H$, we have

$$K^\dagger Q K = J^\dagger J$$

and

$$K^\dagger H = J^\dagger M.$$

Our problem then is to minimize trace $J^\dagger J$, subject to $J^\dagger M = I$. Recall that the trace of the matrix $A^\dagger A$ is simply the square of the 2-norm of the vectorization of $A$.

Our solution method is to transform the original problem into a simpler problem, where the answer is obvious.

First, for any given matrices $L$ and $M$ such that $J$ and $ML$ have the same dimensions, the minimum value of

$$f(J) = \text{trace}[(J^\dagger - L^\dagger M^\dagger)(J - ML)]$$

is zero and occurs when $J = ML$.

Now let $L = L^\dagger = (M^\dagger M)^{-1}$. The solution is again $J = ML$, but now this choice for $J$ has the additional property that $J^\dagger M = I$. So, minimizing $f(J)$ is equivalent to minimizing $f(J)$ subject to the constraint $J^\dagger M = I$ and both problems have the solution $J = ML$.

Now using $J^\dagger M = I$, we expand $f(J)$ to get

$$
\begin{aligned}
f(J) &= \text{trace}[J^\dagger J - J^\dagger ML - L^\dagger M^\dagger J + L^\dagger M^\dagger ML] \\
&= \text{trace}[J^\dagger J - L - L^\dagger + L^\dagger M^\dagger ML].
\end{aligned}
$$

The only term here that involves the unknown matrix $J$ is the first one. Therefore, minimizing $f(J)$ subject to $J^\dagger M = I$ is equivalent to minimizing trace $J^\dagger J$ subject to $J^\dagger M = I$, which is our original problem. Therefore, the optimal choice for $J$ is $J = ML$. Consequently, the optimal choice for $K$ is

$$
K = Q^{-1}HL = Q^{-1}H(H^\dagger Q^{-1}H)^{-1},
$$

and the BLUE estimate of $\mathbf{x}$ is

$$
\mathbf{x}_{BLUE} = \hat{\mathbf{x}} = K^\dagger \mathbf{z} = (H^\dagger Q^{-1}H)^{-1}H^\dagger Q^{-1}\mathbf{z}.
$$

The simplest case can be obtained from this more general formula by taking $N = 1$, $H = (1, 1, ..., 1)^T$ and $\mathbf{x} = c$.

Note that if the noise is *white*, that is, $Q = \sigma^2 I$, then $\hat{\mathbf{x}} = (H^\dagger H)^{-1}H^\dagger \mathbf{z}$, which is the least-squares solution of the equation $\mathbf{z} = H\mathbf{x}$. The effect of requiring that the estimate be unbiased is that, in this case, we simply ignore the presence of the noise and calculate the least-squares solution of the noise-free equation $\mathbf{z} = H\mathbf{x}$.

The BLUE estimator involves nested inversion, making it difficult to calculate, especially for large matrices. In the exercise that follows, we discover an approximation of the BLUE that is easier to calculate.

**Ex. 17.6** *Show that for $\epsilon > 0$ we have*

$$
(H^\dagger Q^{-1}H + \epsilon I)^{-1}H^\dagger Q^{-1} = H^\dagger (HH^\dagger + \epsilon Q)^{-1}. \tag{17.1}
$$

*Hint: Use the identity*

$$
H^\dagger Q^{-1}(HH^\dagger + \epsilon Q) = (H^\dagger Q^{-1}H + \epsilon I)H^\dagger.
$$

It follows from Equation (17.1) that

$$
\mathbf{x}_{BLUE} = \lim_{\epsilon \to 0} H^\dagger (HH^\dagger + \epsilon Q)^{-1}\mathbf{z}.
$$

Therefore, we can get an approximation of the BLUE estimate by selecting $\epsilon > 0$ near zero, solving the system of linear equations

$$
(HH^\dagger + \epsilon Q)\mathbf{a} = \mathbf{z}
$$

for $\mathbf{a}$ and taking $\mathbf{x} = H^\dagger \mathbf{a}$.

## 17.4   Some Useful Matrix Identities

In the exercise that follows we consider several matrix identities that are useful in developing the Kalman filter.

**Ex. 17.7** *Establish the following identities, assuming that all the products and inverses involved are defined:*

$$CDA^{-1}B(C^{-1} - DA^{-1}B)^{-1} = (C^{-1} - DA^{-1}B)^{-1} - C; \qquad (17.2)$$

$$(A - BCD)^{-1} = A^{-1} + A^{-1}B(C^{-1} - DA^{-1}B)^{-1}DA^{-1}; \qquad (17.3)$$

$$A^{-1}B(C^{-1} - DA^{-1}B)^{-1} = (A - BCD)^{-1}BC; \qquad (17.4)$$

$$(A - BCD)^{-1} = (I + GD)A^{-1}, \qquad (17.5)$$

*for*

$$G = A^{-1}B(C^{-1} - DA^{-1}B)^{-1}.$$

*Hints: To get Equation (17.2) use*

$$C(C^{-1} - DA^{-1}B) = I - CDA^{-1}B.$$

*For the second identity, multiply both sides of Equation (17.3) on the left by $A - BCD$ and at the appropriate step use Equation (17.2). For Equation (17.4) show that*

$$BC(C^{-1} - DA^{-1}B) = B - BCDA^{-1}B = (A - BCD)A^{-1}B.$$

*For Equation (17.5), substitute what G is and use Equation (17.3).*

## 17.5   The BLUE with a Prior Estimate

In Kalman filtering we want to estimate an unknown vector $\mathbf{x}$ given measurements $\mathbf{z} = H\mathbf{x} + \mathbf{v}$, but also given a prior estimate $\mathbf{y}$ of $\mathbf{x}$. It is the case there that $E(\mathbf{y}) = E(\mathbf{x})$, so we write $\mathbf{y} = \mathbf{x} + \mathbf{w}$, with $\mathbf{w}$ independent of both $\mathbf{x}$ and $\mathbf{v}$ and $E(\mathbf{w}) = \mathbf{0}$. The covariance matrix for $\mathbf{w}$ we denote by

$E(\mathbf{ww}^\dagger) = R$. We now require that the estimate $\hat{\mathbf{x}}$ be linear in both $\mathbf{z}$ and $\mathbf{y}$; that is, the estimate has the form

$$\hat{\mathbf{x}} = C^\dagger \mathbf{z} + D^\dagger \mathbf{y},$$

for matrices $C$ and $D$ to be determined.

Our approach is to apply the BLUE to the combined system of linear equations

$$\mathbf{z} = H\mathbf{x} + \mathbf{v} \quad \text{and}$$

$$\mathbf{y} = \mathbf{x} + \mathbf{w}.$$

In matrix language this combined system becomes $\mathbf{u} = J\mathbf{x} + \mathbf{n}$, with $\mathbf{u}^T = [\mathbf{z}^T \ \mathbf{y}^T]$, $J^T = [H^T \ I^T]$, and $\mathbf{n}^T = [\mathbf{v}^T \ \mathbf{w}^T]$. The noise covariance matrix becomes

$$P = \begin{bmatrix} Q & 0 \\ 0 & R \end{bmatrix}.$$

The BLUE estimate is $K^\dagger \mathbf{u}$, with $K^\dagger J = I$. Minimizing the variance, we find that the optimal $K^\dagger$ is

$$K^\dagger = (J^\dagger P^{-1} J)^{-1} J^\dagger P^{-1}.$$

The optimal estimate is then

$$\hat{\mathbf{x}} = (H^\dagger Q^{-1} H + R^{-1})^{-1}(H^\dagger Q^{-1}\mathbf{z} + R^{-1}\mathbf{y}).$$

Therefore,

$$C^\dagger = (H^\dagger Q^{-1} H + R^{-1})^{-1} H^\dagger Q^{-1}$$

and

$$D^\dagger = (H^\dagger Q^{-1} H + R^{-1})^{-1} R^{-1}.$$

Using the matrix identities in Equations (17.3) and (17.4) we can rewrite this estimate in the more useful form

$$\hat{\mathbf{x}} = \mathbf{y} + G(\mathbf{z} - H\mathbf{y}),$$

for

$$G = RH^\dagger (Q + HRH^\dagger)^{-1}. \tag{17.6}$$

The covariance matrix of the optimal estimator is $K^\dagger P K$, which can be written as

$$K^\dagger P K = (R^{-1} + H^\dagger Q^{-1} H)^{-1} = (I - GH)R.$$

In the context of the Kalman filter, $R$ is the covariance of the prior estimate of the current state, $G$ is the Kalman gain matrix, and $K^\dagger P K$ is the posterior covariance of the current state. The algorithm proceeds recursively from one state to the next in time.

## 17.6    Adaptive BLUE

We have assumed so far that we know the covariance matrix $Q$ corresponding to the measurement noise. If we do not, then we may attempt to estimate $Q$ from the measurements themselves; such methods are called *noise-adaptive*. To illustrate, let the *innovations* vector be $\mathbf{e} = \mathbf{z} - H\mathbf{y}$. Then the covariance matrix of $\mathbf{e}$ is $S = HRH^\dagger + Q$. Having obtained an estimate $\hat{S}$ of $S$ from the data, we use $\hat{S} - HRH^\dagger$ in place of $Q$ in Equation (17.6).

## 17.7    The Kalman Filter

So far in this chapter we have focused on the filtering problem: Given the data vector $\mathbf{z}$, estimate $\mathbf{x}$, assuming that $\mathbf{z}$ consists of noisy measurements of $H\mathbf{x}$; that is, $\mathbf{z} = H\mathbf{x} + \mathbf{v}$. An important extension of this problem is that of stochastic prediction. Shortly, we discuss the Kalman-filter method for solving this more general problem. One area in which prediction plays an important role is the tracking of moving targets, such as ballistic missiles, using radar. The range to the target, its angle of elevation, and its azimuthal angle are all functions of time governed by linear differential equations. The *state vector* of the system at time $t$ might then be a vector with nine components, the three functions just mentioned, along with their first and second derivatives. In theory, if we knew the initial state perfectly and our differential equations model of the physics was perfect, that would be enough to determine the future states. In practice neither of these is true, and we need to assist the differential equation by taking radar measurements of the state at various times. The problem then is to estimate the state at time $t$ using both the measurements taken prior to time $t$ and the estimate based on the physics.

When such tracking is performed digitally, the functions of time are replaced by discrete sequences. Let the state vector at time $k\Delta t$ be denoted by $\mathbf{x}_k$, for $k$ an integer and $\Delta t > 0$. Then, with the derivatives in the differential equation approximated by divided differences, the physical model for the evolution of the system in time becomes

$$\mathbf{x}_k = A_{k-1}\mathbf{x}_{k-1} + \mathbf{m}_{k-1}.$$

The matrix $A_{k-1}$, which we assume is known, is obtained from the differential equation, which may have nonconstant coefficients, as well as from the divided difference approximations to the derivatives. The random vector sequence $\mathbf{m}_{k-1}$ represents the error in the physical model due to the discretization and necessary simplification inherent in the original differential equation itself. We assume that the expected value of $\mathbf{m_k}$ is zero for each $k$. The covariance matrix is $E(\mathbf{m}_k\mathbf{m}_k^\dagger) = M_k$.

At time $k\Delta t$ we have the measurements

$$\mathbf{z}_k = H_k\mathbf{x}_k + \mathbf{v}_k,$$

where $H_k$ is a known matrix describing the nature of the linear measurements of the state vector and the random vector $\mathbf{v}_k$ is the noise in these measurements. We assume that the mean value of $\mathbf{v}_k$ is zero for each $k$. The covariance matrix is $E(\mathbf{v}_k\mathbf{v}_k^\dagger) = Q_k$. We assume that the initial state vector $\mathbf{x}_0$ is arbitrary.

Given an unbiased estimate $\hat{\mathbf{x}}_{k-1}$ of the state vector $\mathbf{x}_{k-1}$, our prior estimate of $\mathbf{x}_k$ based solely on the physics is

$$\mathbf{y}_k = A_{k-1}\hat{\mathbf{x}}_{k-1}.$$

**Ex. 17.8** *Show that $E(\mathbf{y}_k - \mathbf{x}_k) = 0$, so the prior estimate of $\mathbf{x}_k$ is unbiased. We can then write $\mathbf{y}_k = \mathbf{x}_k + \mathbf{w}_k$, with $E(\mathbf{w}_k) = \mathbf{0}$.*

---

## 17.8   Kalman Filtering and the BLUE

The *Kalman filter* [98, 81, 56] is a recursive algorithm to estimate the state vector $\mathbf{x}_k$ at time $k\Delta t$ as a linear combination of the vectors $\mathbf{z}_k$ and $\mathbf{y}_k$. The estimate $\hat{\mathbf{x}}_k$ will have the form

$$\hat{\mathbf{x}}_k = C_k^\dagger\mathbf{z}_k + D_k^\dagger\mathbf{y}_k,$$

for matrices $C_k$ and $D_k$ to be determined. As we shall see, this estimate can also be written as

$$\hat{\mathbf{x}}_k = \mathbf{y}_k + G_k(\mathbf{z}_k - H_k\mathbf{y}_k),$$

which shows that the estimate involves a prior prediction step, the $\mathbf{y}_k$, followed by a correction step, in which $H_k\mathbf{y}_k$ is compared to the measured data vector $\mathbf{z}_k$; such estimation methods are sometimes called *predictor-corrector methods*.

In our discussion of the BLUE, we saw how to incorporate a prior estimate of the vector to be estimated. The trick was to form a larger matrix equation and then to apply the BLUE to that system. The Kalman filter does just that.

The correction step in the Kalman filter uses the BLUE to solve the combined linear system

$$\mathbf{z}_k = H_k \mathbf{x}_k + \mathbf{v}_k$$

and

$$\mathbf{y}_k = \mathbf{x}_k + \mathbf{w}_k.$$

The covariance matrix of $\hat{\mathbf{x}}_{k-1} - \mathbf{x}_{k-1}$ is denoted by $P_{k-1}$, and we let $Q_k = E(\mathbf{w}_k \mathbf{w}_k^\dagger)$. The covariance matrix of $\mathbf{y}_k - \mathbf{x}_k$ is

$$\text{cov}(\mathbf{y}_k - \mathbf{x}_k) = R_k = M_{k-1} + A_{k-1} P_{k-1} A_{k-1}^\dagger.$$

It follows from our earlier discussion of the BLUE that the estimate of $\mathbf{x}_k$ is

$$\hat{\mathbf{x}}_k = \mathbf{y}_k + G_k(\mathbf{z}_k - H\mathbf{y}_k),$$

with

$$G_k = R_k H_k^\dagger (Q_k + H_k R_k H_k^\dagger)^{-1}.$$

Then, the covariance matrix of $\hat{\mathbf{x}}_k - \mathbf{x}_k$ is

$$P_k = (I - G_k H_k) R_k.$$

The recursive procedure is to go from $P_{k-1}$ and $M_{k-1}$ to $R_k$, then to $G_k$, from which $\hat{\mathbf{x}}_k$ is formed, and finally to $P_k$, which, along with the known matrix $M_k$, provides the input to the next step. The time-consuming part of this recursive algorithm is the matrix inversion in the calculation of $G_k$. Simpler versions of the algorithm are based on the assumption that the matrices $Q_k$ are diagonal, or on the convergence of the matrices $G_k$ to a limiting matrix $G$ [56].

There are many variants of the Kalman filter, corresponding to variations in the physical model, as well as in the statistical assumptions. The differential equation may be nonlinear, so that the matrices $A_k$ depend on $\mathbf{x}_k$. The system noise sequence $\{\mathbf{w}_k\}$ and the measurement noise sequence $\{\mathbf{v}_k\}$ may be correlated. For computational convenience the various functions that describe the state may be treated separately. The model may include known external inputs to drive the differential system, as in the tracking of spacecraft capable of firing booster rockets. Finally, the noise covariance matrices may not be known *a priori* and adaptive filtering may be needed. We discuss this last issue briefly in the next section.

## 17.9 Adaptive Kalman Filtering

As in [56] we consider only the case in which the covariance matrix $Q_k$ of the measurement noise $\mathbf{v}_k$ is unknown. As we saw in the discussion of adaptive BLUE, the covariance matrix of the innovations vector $\mathbf{e}_k = \mathbf{z}_k - H_k\mathbf{y}_k$ is

$$S_k = H_k R_k H_k^\dagger + Q_k.$$

Once we have an estimate for $S_k$, we estimate $Q_k$ using

$$\hat{Q}_k = \hat{S}_k - H_k R_k H_k^\dagger.$$

We might assume that $S_k$ is independent of $k$ and estimate $S_k = S$ using past and present innovations; for example, we could use

$$\hat{S} = \frac{1}{k-1} \sum_{j=1}^{k} (\mathbf{z}_j - H_j\mathbf{y}_j)(\mathbf{z}_j - H_j\mathbf{y}_j)^\dagger.$$

## 17.10 Difficulties with the BLUE

As we just saw, the best linear unbiased estimate of $\mathbf{x}$, given the observed vector $\mathbf{z} = H\mathbf{x} + \mathbf{v}$, is

$$\mathbf{x}_{BLUE} = (H^\dagger Q^{-1} H)^{-1} H^\dagger Q^{-1} \mathbf{z}, \tag{17.7}$$

where $Q$ is the invertible covariance matrix of the mean zero noise vector $\mathbf{v}$ and $H$ is a $J$ by $N$ matrix with $J \geq N$ and $H^\dagger H$ invertible. Even if we know $Q$ exactly, the double inversion in Equation (17.7) makes it difficult to calculate the BLUE estimate, especially for large vectors $\mathbf{z}$. It is often the case in practice that we do not know $Q$ precisely and must estimate or model it. Because good approximations of $Q$ do not necessarily lead to good approximations of $Q^{-1}$, the calculation of the BLUE is further complicated. For these reasons one may decide to use the least-squares estimate

$$\mathbf{x}_{LS} = (H^\dagger H)^{-1} H^\dagger \mathbf{z}$$

instead. We are therefore led to consider when the two estimation methods produce the same answers; that is, when we have

$$(H^\dagger H)^{-1} H^\dagger = (H^\dagger Q^{-1} H)^{-1} H^\dagger Q^{-1}. \tag{17.8}$$

We turn now to a theorem that answers this question. The proof of this theorem relies on the results of several exercises, useful in themselves, that involve basic facts of linear algebra.

## 17.11  Preliminaries from Linear Algebra

We begin with some definitions. Let $S$ be a subspace of finite-dimensional Euclidean space $\mathbb{C}^J$ and $Q$ a $J$ by $J$ Hermitian matrix. We denote by $Q(S)$ the set

$$Q(S) = \{\mathbf{t}|\text{there exists } \mathbf{s} \in S \text{ with } \mathbf{t} = Q\mathbf{s}\}$$

and by $Q^{-1}(S)$ the set

$$Q^{-1}(S) = \{\mathbf{u}|Q\mathbf{u} \in S\}.$$

Note that the set $Q^{-1}(S)$ is defined whether or not $Q$ is invertible.

We denote by $S^\perp$ the set of vectors $\mathbf{u}$ that are orthogonal to every member of $S$; that is,

$$S^\perp = \{\mathbf{u}|\mathbf{u}^\dagger\mathbf{s} = 0, \text{for every } \mathbf{s} \in S\}.$$

Let $H$ be a $J$ by $N$ matrix. Then $CS(H)$, the column space of $H$, is the subspace of $\mathbb{C}^J$ consisting of all the linear combinations of the columns of $H$. The null space of $H^\dagger$, denoted $NS(H^\dagger)$, is the subspace of $\mathbb{C}^J$ containing all the vectors $\mathbf{w}$ for which $H^\dagger\mathbf{w} = 0$.

**Ex. 17.9** *Show that $CS(H)^\perp = NS(H^\dagger)$. Hint: If $\mathbf{v} \in CS(H)^\perp$, then $\mathbf{v}^\dagger H\mathbf{x} = 0$ for all $\mathbf{x}$, including $\mathbf{x} = H^\dagger\mathbf{v}$.*

**Ex. 17.10** *Show that $CS(H) \cap NS(H^\dagger) = \{\mathbf{0}\}$. Hint: If $\mathbf{y} = H\mathbf{x} \in NS(H^\dagger)$ consider $\|\mathbf{y}\|^2 = \mathbf{y}^\dagger\mathbf{y}$.*

**Ex. 17.11** *Let $S$ be any subspace of $\mathbb{C}^J$. Show that if $Q$ is invertible and $Q(S) = S$ then $Q^{-1}(S) = S$. Hint: If $Q\mathbf{t} = Q\mathbf{s}$ then $\mathbf{t} = \mathbf{s}$.*

**Ex. 17.12** *Let $Q$ be Hermitian. Show that $Q(S)^\perp = Q^{-1}(S^\perp)$ for every subspace $S$. If $Q$ is also invertible then $Q^{-1}(S)^\perp = Q(S^\perp)$. Find an example of a non-invertible $Q$ for which $Q^{-1}(S)^\perp$ and $Q(S^\perp)$ are different.*

We assume that $Q$ is Hermitian and invertible and that the matrix $H^\dagger H$ is invertible. Note that the matrix $H^\dagger Q^{-1} H$ need not be invertible under these assumptions. We shall denote by $S$ an arbitrary subspace of $\mathbb{C}^J$.

**Ex. 17.13** *Show that $Q(S) = S$ if and only if $Q(S^\perp) = S^\perp$. Hint: Use Exercise 17.12.*

**Ex. 17.14** *Show that if $Q(CS(H)) = CS(H)$ then $H^\dagger Q^{-1} H$ is invertible. Hint: Show that $H^\dagger Q^{-1} H\mathbf{x} = \mathbf{0}$ if and only if $\mathbf{x} = \mathbf{0}$. Recall that $Q^{-1} H\mathbf{x} \in CS(H)$, by Exercise 17.12. Then use Exercise 17.10.*

## 17.12 When Are the BLUE and the LS Estimator the Same?

We are looking for conditions on $Q$ and $H$ that imply Equation (17.8), which we rewrite as

$$H^\dagger = (H^\dagger Q^{-1} H)(H^\dagger H)^{-1} H^\dagger Q \qquad (17.9)$$

or

$$H^\dagger T \mathbf{x} = \mathbf{0}$$

for all $\mathbf{x}$, where

$$T = I - Q^{-1} H (H^\dagger H)^{-1} H^\dagger Q.$$

In other words, we want $T\mathbf{x} \in NS(H^\dagger)$ for all $\mathbf{x}$. The theorem is the following:

**Theorem 17.1** *We have $T\mathbf{x} \in NS(H^\dagger)$ for all $\mathbf{x}$ if and only if we have $Q(CS(H)) = CS(H)$.*

An equivalent form of this theorem was proven by Anderson in [1]; he attributes a portion of the proof to Magness and McQuire [114]. The proof we give here is due to Kheifets [100] and is much simpler than Anderson's proof. The proof of the theorem is simplified somewhat by first establishing the result in the next exercise.

**Ex. 17.15** *Show that if Equation (17.9) holds, then the matrix $H^\dagger Q^{-1} H$ is invertible. Hint: Recall that we have assumed that $CS(H^\dagger) = \mathbb{C}^J$ when we assumed that $H^\dagger H$ is invertible. From Equation (17.9) it follows that $CS(H^\dagger Q^{-1} H) = \mathbb{C}^J$.*

**A Proof of Theorem 17.1:** Assume first that $Q(CS(H)) = CS(H)$, which, as we now know, also implies $Q(NS(H^\dagger)) = NS(H^\dagger)$, as well as $Q^{-1}(CS(H)) = CS(H)$, $Q^{-1}(NS(H^\dagger)) = NS(H^\dagger)$, and the invertibility of the matrix $H^\dagger Q^{-1} H$. Every $\mathbf{x} \in \mathbb{C}^J$ has the form $\mathbf{x} = H\mathbf{a} + \mathbf{w}$, for some $\mathbf{a}$ and $\mathbf{w} \in NS(H^\dagger)$. We show that $T\mathbf{x} = \mathbf{w}$, so that $T\mathbf{x} \in NS(H^\dagger)$ for all $\mathbf{x}$. We have

$$T\mathbf{x} = TH\mathbf{a} + T\mathbf{w} =$$

$$\mathbf{x} - Q^{-1} H (H^\dagger H)^{-1} H^\dagger Q H\mathbf{a} - Q^{-1} H (H^\dagger H)^{-1} H^\dagger Q\mathbf{w}.$$

We know that $QH\mathbf{a} = H\mathbf{b}$ for some $\mathbf{b}$, so that $H\mathbf{a} = Q^{-1} H\mathbf{b}$. We also know that $Q\mathbf{w} = \mathbf{v} \in NS(H^\dagger)$, so that $\mathbf{w} = Q^{-1}\mathbf{v}$. Then, continuing our calculations, we have

$$T\mathbf{x} = \mathbf{x} - Q^{-1} H\mathbf{b} - \mathbf{0} = \mathbf{x} - H\mathbf{a} = \mathbf{w},$$

so $T\mathbf{x} \in NS(H^\dagger)$.

Conversely, suppose now that $T\mathbf{x} \in NS(H^{\dagger})$ for all $\mathbf{x}$, which, as we have seen, is equivalent to Equation (17.9). We show that $Q^{-1}(NS(H^{\dagger})) = NS(H^{\dagger})$. First, let $\mathbf{v} \in Q^{-1}(NS(H^{\dagger}))$; we show $\mathbf{v} \in NS(H^{\dagger})$. We have

$$H^{\dagger}\mathbf{v} = (H^{\dagger}Q^{-1}H)(H^{\dagger}H)^{-1}H^{\dagger}Q\mathbf{v},$$

which is zero, since $H^{\dagger}Q\mathbf{v} = \mathbf{0}$. So, we have shown that $Q^{-1}(NS(H^{\dagger})) \subseteq NS(H^{\dagger})$. To complete the proof, we take an arbitrary member $\mathbf{v}$ of $NS(H^{\dagger})$ and show that $\mathbf{v}$ is in $Q^{-1}(NS(H^{\dagger}))$; that is, $Q\mathbf{v} \in NS(H^{\dagger})$. We know that $Q\mathbf{v} = H\mathbf{a} + \mathbf{w}$, for $\mathbf{w} \in NS(H^{\dagger})$, and

$$\mathbf{a} = (H^{\dagger}H)^{-1}H^{\dagger}Q\mathbf{v},$$

so that

$$H\mathbf{a} = H(H^{\dagger}H)^{-1}H^{\dagger}Q\mathbf{v}.$$

Then, using Exercise 17.15, we have

$$\begin{aligned} Q\mathbf{v} &= H(H^{\dagger}H)^{-1}H^{\dagger}Q\mathbf{v} + \mathbf{w} \\ &= H(H^{\dagger}Q^{-1}H)^{-1}H^{\dagger}Q^{-1}Q\mathbf{v} + \mathbf{w} \\ &= H(H^{\dagger}Q^{-1}H)^{-1}H^{\dagger}\mathbf{v} + \mathbf{w} = \mathbf{w}. \end{aligned}$$

So $Q\mathbf{v} = \mathbf{w}$, which is in $NS(H^{\dagger})$. This completes the proof. ∎

## 17.13 A Recursive Approach

In array processing and elsewhere, it sometimes happens that the matrix $Q$ is estimated from several measurements $\{\mathbf{v}^n, n = 1, ..., N\}$ of the noise vector $\mathbf{v}$ as

$$Q = \frac{1}{N}\sum_{n=1}^{N}\mathbf{v}^n(\mathbf{v}^n)^{\dagger}.$$

Then, the inverses of $Q$ and of $H^{\dagger}Q^{-1}H$ can be obtained recursively, using the Sherman–Morrison–Woodbury matrix-inversion identity.

**Ex. 17.16 The Sherman–Morrison–Woodbury Identity** *Let $B$ be an invertible matrix. Show that*

$$(B - \mathbf{u}\mathbf{v}^{\dagger})^{-1} = B^{-1} + \alpha^{-1}(B^{-1}\mathbf{u})(\mathbf{v}^{\dagger}B^{-1}), \qquad (17.10)$$

*whenever*

$$\alpha = 1 - \mathbf{v}^{\dagger}B^{-1}\mathbf{u} \neq 0.$$

*Show that, if $\alpha = 0$, then the matrix $B - \mathbf{u}\mathbf{v}^{\dagger}$ has no inverse.*

Since the matrices involved here are nonnegative definite this denominator will always be at least one. The idea is to define $Q_0 = \epsilon I$, for some $\epsilon > 0$, and, for $n = 1, ..., N$,

$$Q_n = Q_{n-1} + \mathbf{v}^n (\mathbf{v}^n)^\dagger.$$

Then, $Q_n^{-1}$ can be obtained from $Q_{n-1}^{-1}$ and $(H^\dagger Q_n^{-1} H)^{-1}$ from $(H^\dagger Q_{n-1}^{-1} H)^{-1}$ using the identity in Equation (17.10).

# Chapter 18

# Signal Detection and Estimation

## 18.1    Chapter Summary

In this chapter we consider the problem of deciding whether or not a particular signal is present in the measured data; this is the *detection* problem. The underlying framework for the detection problem is optimal estimation and statistical hypothesis testing [81].

## 18.2    The Model of Signal in Additive Noise

The basic model used in detection is that of a signal in additive noise. The complex data vector is $\mathbf{x} = (x_1, x_2, ..., x_N)^T$. We assume that there are two possibilities:

**Case 1: Noise only**

$$x_n = z_n, \; n = 1, ..., N,$$

or

**Case 2: Signal in noise**

$$x_n = \gamma s_n + z_n,$$

where $\mathbf{z} = (z_1, z_2, ..., z_N)^T$ is a complex vector whose entries $z_n$ are values of random variables that we call *noise*, about which we have only statistical information (that is to say, information about the average behavior), $\mathbf{s} = (s_1, s_2, ..., s_N)^T$ is a complex signal vector that we may know exactly, or at least for which we have a specific parametric model, and $\gamma$ is a scalar that may be viewed either as deterministic or random (but unknown, in either case). Unless otherwise stated, we shall assume that $\gamma$ is deterministic.

The *detection problem* is to decide which case we are in, based on some calculation performed on the data $\mathbf{x}$. Since Case 1 can be viewed as a special case of Case 2 in which the value of $\gamma$ is zero, the detection problem is closely related to the problem of estimating $\gamma$, which we discussed in the chapter dealing with the best linear unbiased estimator, the BLUE.

We shall assume throughout that the entries of $\mathbf{z}$ correspond to random variables with means equal to zero. What the variances are and whether or not these random variables are mutually correlated will be discussed next. In all cases we shall assume that this information has been determined previously and is available to us in the form of the covariance matrix $Q = E(\mathbf{z}\mathbf{z}^\dagger)$ of the vector $\mathbf{z}$; the symbol $E$ denotes expected value, so the entries of $Q$ are the quantities $Q_{mn} = E(z_m \bar{z}_n)$. The diagonal entries of $Q$ are $Q_{nn} = \sigma_n^2$, the variance of $z_n$. As in Chapter 17, we assume here that $Q$ is invertible, which is the typical case.

Note that we have adopted the common practice of using the same symbols, $z_n$, when speaking about the random variables and about the specific values of these random variables that are present in our data. The context should make it clear to which we are referring.

In Case 2 we say that the *signal power* is equal to $|\gamma|^2 \frac{1}{N} \sum_{n=1}^{N} |s_n|^2 = \frac{1}{N} |\gamma|^2 \mathbf{s}^\dagger \mathbf{s}$ and the *noise power* is $\frac{1}{N} \sum_{n=1}^{N} \sigma_n^2 = \frac{1}{N} tr(Q)$, where $tr(Q)$ is the trace of the matrix $Q$, that is, the sum of its diagonal terms; therefore, the noise power is the average of the variances $\sigma_n^2$. The *input signal-to-noise ratio* (SNR$_{\text{in}}$) is the ratio of the signal power to that of the noise, prior to processing the data; that is,

$$\text{SNR}_{\text{in}} = \frac{1}{N} |\gamma|^2 \mathbf{s}^\dagger \mathbf{s} / \frac{1}{N} tr(Q) = |\gamma|^2 \mathbf{s}^\dagger \mathbf{s} / tr(Q).$$

## 18.3    Optimal Linear Filtering for Detection

In each case to be considered next, our detector will take the form of a linear estimate of $\gamma$; that is, we shall compute the estimate $\hat{\gamma}$ given by

$$\hat{\gamma} = \sum_{n=1}^{N} \overline{b}_n x_n = \mathbf{b}^\dagger \mathbf{x},$$

where $\mathbf{b} = (b_1, b_2, ..., b_N)^T$ is a vector to be determined. The objective is to use what we know about the situation to select the optimal $\mathbf{b}$, which will depend on $\mathbf{s}$ and $Q$.

For any given vector $\mathbf{b}$, the quantity

$$\hat{\gamma} = \mathbf{b}^\dagger \mathbf{x} = \gamma \mathbf{b}^\dagger \mathbf{s} + \mathbf{b}^\dagger \mathbf{z}$$

is a random variable whose mean value is equal to $\gamma \mathbf{b}^\dagger \mathbf{s}$ and whose variance is

$$var(\hat{\gamma}) = E(|\mathbf{b}^\dagger \mathbf{z}|^2) = E(\mathbf{b}^\dagger \mathbf{z} \mathbf{z}^\dagger \mathbf{b}) = \mathbf{b}^\dagger E(\mathbf{z}\mathbf{z}^\dagger)\mathbf{b} = \mathbf{b}^\dagger Q \mathbf{b}.$$

Therefore, the *output signal-to-noise ratio* (SNR$_{\text{out}}$) is defined as

$$\text{SNR}_{\text{out}} = |\gamma \mathbf{b}^\dagger \mathbf{s}|^2 / \mathbf{b}^\dagger Q \mathbf{b}.$$

The advantage we obtain from processing the data is called the *gain* associated with $\mathbf{b}$ and is defined to be the ratio of the SNR$_{\text{out}}$ to SNR$_{\text{in}}$; that is,

$$\text{gain}(\mathbf{b}) = \frac{|\gamma \mathbf{b}^\dagger \mathbf{s}|^2/(\mathbf{b}^\dagger Q \mathbf{b})}{|\gamma|^2 (\mathbf{s}^\dagger \mathbf{s})/tr(Q)} = \frac{|\mathbf{b}^\dagger \mathbf{s}|^2 \, tr(Q)}{(\mathbf{b}^\dagger Q \mathbf{b})(\mathbf{s}^\dagger \mathbf{s})}.$$

The best $\mathbf{b}$ to use will be the one for which gain($\mathbf{b}$) is the largest. So, ignoring the terms in the gain formula that do not involve $\mathbf{b}$, we see that the problem becomes *maximize* $\frac{|\mathbf{b}^\dagger \mathbf{s}|^2}{\mathbf{b}^\dagger Q \mathbf{b}}$, *for fixed signal vector* $\mathbf{s}$ *and fixed noise covariance matrix* $Q$.

The Cauchy Inequality plays a major role in optimal filtering and detection.

**Cauchy's Inequality:** For any vectors $\mathbf{a}$ and $\mathbf{b}$ we have

$$|\mathbf{a}^\dagger \mathbf{b}|^2 \leq (\mathbf{a}^\dagger \mathbf{a})(\mathbf{b}^\dagger \mathbf{b}),$$

with equality if and only if $\mathbf{a}$ is proportional to $\mathbf{b}$; that is, there is a scalar $\beta$ such that $\mathbf{b} = \beta \mathbf{a}$.

**Ex. 18.1** *Use Cauchy's Inequality to show that, for any fixed vector* $\mathbf{a}$, *the choice* $\mathbf{b} = \beta \mathbf{a}$ *maximizes the quantity* $|\mathbf{b}^\dagger \mathbf{a}|^2 / \mathbf{b}^\dagger \mathbf{b}$, *for any constant* $\beta$.

**Ex. 18.2** *Use the definition of the covariance matrix* $Q$ *to show that* $Q$ *is Hermitian and that, for any vector* $\mathbf{y}$, $\mathbf{y}^\dagger Q \mathbf{y} \geq 0$. *Therefore,* $Q$ *is a nonnegative definite matrix and, using its eigenvector decomposition, can be written as* $Q = CC^\dagger$, *for some invertible square matrix* $C$.

**Ex. 18.3** *Consider now the problem of maximizing* $|\mathbf{b}^\dagger \mathbf{s}|^2 / \mathbf{b}^\dagger Q \mathbf{b}$. *Using the two previous exercises, show that the solution is* $\mathbf{b} = \beta Q^{-1} \mathbf{s}$, *for some arbitrary constant* $\beta$.

We can now use the results of these exercises to continue our discussion. We choose the constant $\beta = 1/(\mathbf{s}^\dagger Q^{-1} \mathbf{s})$ so that the optimal $\mathbf{b}$ has $\mathbf{b}^\dagger \mathbf{s} = 1$; that is, the *optimal filter* $\mathbf{b}$ is

$$\mathbf{b} = (1/(\mathbf{s}^\dagger Q^{-1} \mathbf{s}))Q^{-1}\mathbf{s},$$

and the *optimal estimate* of $\gamma$ is

$$\hat{\gamma} = \mathbf{b}^\dagger \mathbf{x} = (1/(\mathbf{s}^\dagger Q^{-1} \mathbf{s}))(\mathbf{s}^\dagger Q^{-1} \mathbf{x}).$$

The mean of the random variable $\hat{\gamma}$ is equal to $\gamma \mathbf{b}^\dagger \mathbf{s} = \gamma$, and the variance is equal to $1/(\mathbf{s}^\dagger Q^{-1} \mathbf{s})$. Therefore, the output signal power is $|\gamma|^2$, the output noise power is $1/(\mathbf{s}^\dagger Q^{-1} \mathbf{s})$, and so the *output signal-to-noise ratio* ($\text{SNR}_{\text{out}}$) is

$$\text{SNR}_{\text{out}} = |\gamma|^2 (\mathbf{s}^\dagger Q^{-1} \mathbf{s}).$$

The gain associated with the optimal vector $\mathbf{b}$ is then

$$\text{maximum gain} = \frac{(\mathbf{s}^\dagger Q^{-1} \mathbf{s})\, tr(Q)}{(\mathbf{s}^\dagger \mathbf{s})}.$$

The calculation of the vector $C^{-1}\mathbf{x}$ is sometimes called *prewhitening* since $C^{-1}\mathbf{x} = \gamma C^{-1}\mathbf{s} + C^{-1}\mathbf{z}$ and the new noise vector, $C^{-1}\mathbf{z}$, has the identity matrix for its covariance matrix. The new signal vector is $C^{-1}\mathbf{s}$. The filtering operation that gives $\hat{\gamma} = \mathbf{b}^\dagger \mathbf{x}$ can be written as

$$\hat{\gamma} = (1/(\mathbf{s}^\dagger Q^{-1} \mathbf{s}))(C^{-1}\mathbf{s})^\dagger C^{-1}\mathbf{x};$$

the term $(C^{-1}\mathbf{s})^\dagger C^{-1}\mathbf{x}$ is described by saying that we *prewhiten, then do a matched filter*. Now we consider some special cases of noise.

## 18.4   The Case of White Noise

We say that the noise is *white noise* if the covariance matrix is $Q = \sigma^2 I$, where $I$ denotes the identity matrix, whose entries on the main diagonal have the value 1 and the other entries have the value 0, and $\sigma > 0$ is the common standard deviation of the $z_n$. This means that the $z_n$ are mutually uncorrelated (independent, in the Gaussian case) and share a common variance.

In this case the optimal vector **b** is $\mathbf{b} = \frac{1}{(\mathbf{s}^\dagger \mathbf{s})}\mathbf{s}$ and the gain is $N$. Notice that $\hat{\gamma}$ now involves only a matched filter. We consider now some special cases of the signal vectors **s**.

### 18.4.1   Constant Signal

Suppose that the vector **s** is constant; that is, $\mathbf{s} = \mathbf{1} = (1, 1, ..., 1)^T$. Then, we have

$$\hat{\gamma} = \frac{1}{N}\sum_{n=1}^{N} x_n.$$

This is the same result we found in our discussion of the BLUE, when we estimated the mean value and the noise was white.

### 18.4.2   Sinusoidal Signal, Frequency Known

Suppose that

$$\mathbf{s} = \mathbf{e}(\omega_0) = (\exp(-i\omega_0), \exp(-2i\omega_0), ..., \exp(-Ni\omega_0))^T,$$

where $\omega_0$ denotes a known frequency in $[-\pi, \pi)$. Then, $\mathbf{b} = \frac{1}{N}\mathbf{e}(\omega_0)$ and

$$\hat{\gamma} = \frac{1}{N}\sum_{n=1}^{N} x_n \exp(in\omega_0);$$

so, we see yet another occurrence of the DFT.

### 18.4.3   Sinusoidal Signal, Frequency Unknown

If we do not know the value of the signal frequency $\omega_0$, a reasonable thing to do is to calculate the $\hat{\gamma}$ for each (actually, finitely many) of the possible frequencies within $[-\pi, \pi)$ and base the detection decision on the largest value; that is, we calculate the DFT as a function of the variable $\omega$. If there is only a single $\omega_0$ for which there is a sinusoidal signal present in the data, the values of $\hat{\gamma}$ obtained at frequencies other than $\omega_0$ provide estimates of the noise power $\sigma^2$, against which the value of $\hat{\gamma}$ for $\omega_0$ can be compared.

## 18.5   The Case of Correlated Noise

We say that the noise is *correlated* if the covariance matrix $Q$ is not a multiple of the identity matrix. This means either that the $z_n$ are mutually correlated (dependent, in the Gaussian case) or that they are uncorrelated, but have different variances.

In this case, as we saw previously, the optimal vector **b** is

$$\mathbf{b} = \frac{1}{(\mathbf{s}^\dagger Q^{-1}\mathbf{s})} Q^{-1}\mathbf{s}$$

and the gain is

$$\text{maximum gain} = \frac{(\mathbf{s}^\dagger Q^{-1}\mathbf{s})\, tr(Q)}{(\mathbf{s}^\dagger \mathbf{s})}.$$

How large or small the gain is depends on how the signal vector **s** relates to the matrix $Q$.

For sinusoidal signals, the quantity $\mathbf{s}^\dagger \mathbf{s}$ is the same, for all values of the parameter $\omega$; this is not always the case, however. In passive detection of sources in acoustic array processing, for example, the signal vectors arise from models of the acoustic medium involved. For far-field sources in an (acoustically) isotropic deep ocean, planewave models for **s** will have the property that $\mathbf{s}^\dagger \mathbf{s}$ does not change with source location. However, for near-field or shallow-water environments, this is usually no longer the case.

It follows from Exercise 18.3 that the quantity $\frac{\mathbf{s}^\dagger Q^{-1}\mathbf{s}}{\mathbf{s}^\dagger \mathbf{s}}$ achieves its maximum value when **s** is an eigenvector of $Q$ associated with its smallest eigenvalue, $\lambda_N$; in this case, we are saying that the signal vector does not look very much like a typical noise vector. The maximum gain is then $\lambda_N^{-1} tr(Q)$. Since $tr(Q)$ equals the sum of its eigenvalues, multiplying by $tr(Q)$ serves to normalize the gain, so that we cannot get larger gain simply by having all the eigenvalues of $Q$ small.

On the other hand, if **s** should be an eigenvector of $Q$ associated with its largest eigenvalue, say $\lambda_1$, then the maximum gain is $\lambda_1^{-1} tr(Q)$. If the noise is signal-like, that is, has one dominant eigenvalue, then $tr(Q)$ is approximately $\lambda_1$ and the maximum gain is around one, so we have lost the maximum gain of $N$ we were able to get in the white-noise case. This makes sense, in that it says that we cannot significantly improve our ability to discriminate between signal and noise by taking more samples, if the signal and noise are very similar.

## 18.5.1 Constant Signal with Unequal-Variance Uncorrelated Noise

Suppose that the vector $\mathbf{s}$ is constant; that is, $\mathbf{s} = \mathbf{1} = (1, 1, ..., 1)^T$. Suppose also that the noise covariance matrix is $Q = \text{diag}\{\sigma_1, ..., \sigma_N\}$.

In this case the optimal vector $\mathbf{b}$ has entries

$$b_m = \frac{1}{\left(\sum_{n=1}^{N} \sigma_n^{-1}\right)} \sigma_m^{-1},$$

for $m = 1, ..., N$, and we have

$$\hat{\gamma} = \frac{1}{\left(\sum_{n=1}^{N} \sigma_n^{-1}\right)} \sum_{m=1}^{N} \sigma_m^{-1} x_m.$$

This is the BLUE estimate of $\gamma$ in this case.

## 18.5.2 Sinusoidal Signal, Frequency Known, in Correlated Noise

Suppose that

$$\mathbf{s} = \mathbf{e}(\omega_0) = (\exp(-i\omega_0), \exp(-2i\omega_0), ..., \exp(-Ni\omega_0))^T,$$

where $\omega_0$ denotes a known frequency in $[-\pi, \pi)$. In this case the optimal vector $\mathbf{b}$ is

$$\mathbf{b} = \frac{1}{\mathbf{e}(\omega_0)^\dagger Q^{-1} \mathbf{e}(\omega_0)} Q^{-1} \mathbf{e}(\omega_0)$$

and the gain is

$$\text{maximum gain} = \frac{1}{N} [\mathbf{e}(\omega_0)^\dagger Q^{-1} \mathbf{e}(\omega_0)] tr(Q).$$

How large or small the gain is depends on the quantity $q(\omega_0)$, where

$$q(\omega) = \mathbf{e}(\omega)^\dagger Q^{-1} \mathbf{e}(\omega).$$

The function $1/q(\omega)$ can be viewed as a sort of noise power spectrum, describing how the noise power appears when decomposed over the various frequencies in $[-\pi, \pi)$. The maximum gain will be large if this *noise power spectrum* is relatively small near $\omega = \omega_0$; however, when the noise is similar to the signal, that is, when the noise power spectrum is relatively large near $\omega = \omega_0$, the maximum gain can be small. In this case the noise power spectrum plays a role analogous to that played by the eigenvalues of $Q$ earlier.

To see more clearly why it is that the function $1/q(\omega)$ can be viewed as a sort of noise power spectrum, consider what we get when we apply the optimal filter associated with $\omega$ to data containing only noise. The average output should tell us how much power there is in the component of the noise that resembles $\mathbf{e}(\omega)$; this is essentially what is meant by a noise power spectrum. The result is $\mathbf{b}^{\dagger}\mathbf{z} = (1/q(\omega))\mathbf{e}(\omega)^{\dagger}Q^{-1}\mathbf{z}$. The expected value of $|\mathbf{b}^{\dagger}\mathbf{z}|^2$ is then $1/q(\omega)$.

### 18.5.3   Sinusoidal Signal, Frequency Unknown, in Correlated Noise

Again, if we do not know the value of the signal frequency $\omega_0$, a reasonable thing to do is to calculate the $\hat{\gamma}$ for each (actually, finitely many) of the possible frequencies within $[-\pi, \pi)$ and base the detection decision on the largest value. For each $\omega$ the corresponding value of $\hat{\gamma}$ is

$$\hat{\gamma}(\omega) = [1/(\mathbf{e}(\omega)^{\dagger}Q^{-1}\mathbf{e}(\omega))] \sum_{n=1}^{N} a_n \exp(in\omega),$$

where $\mathbf{a} = (a_1, a_2, ..., a_N)^T$ satisfies the linear system $Q\mathbf{a} = \mathbf{x}$ or $\mathbf{a} = Q^{-1}\mathbf{x}$. It is interesting to note the similarity between this estimation procedure and the PDFT discussed earlier; to see the connection, view $[1/(\mathbf{e}(\omega)^{\dagger}Q^{-1}\mathbf{e}(\omega))]$ in the role of $P(\omega)$ and $Q$ its corresponding matrix of Fourier-transform values. The analogy breaks down when we notice that $Q$ need not be Toeplitz, as in the PDFT case; however, the similarity is intriguing.

---

## 18.6   Capon's Data-Adaptive Method

When the noise covariance matrix $Q$ is not available, perhaps because we cannot observe the background noise in the absence of any signals that may also be present, we may use the signal-plus-noise covariance matrix $R$ in place of $Q$.

**Ex. 18.4** *Show that for*

$$R = |\gamma|^2 \mathbf{s}\mathbf{s}^{\dagger} + Q$$

*maximizing the ratio*

$$|bfb^{\dagger}\mathbf{s}|^2/\mathbf{b}^{\dagger}R\mathbf{b}$$

*is equivalent to maximizing the ratio*

$$|\mathbf{b}^\dagger \mathbf{s}|^2 / \mathbf{b}^\dagger Q \mathbf{b}.$$

In [49] Capon offered a high-resolution method for detecting and resolving sinusoidal signals with unknown frequencies in noise. His estimator has the form

$$1/\mathbf{e}(\omega)^\dagger R^{-1} \mathbf{e}(\omega).$$

The idea here is to fix an arbitrary $\omega$, and then to find the vector $\mathbf{b}(\omega)$ that minimizes $\mathbf{b}(\omega)^\dagger R \mathbf{b}(\omega)$, subject to $\mathbf{b}(\omega)^\dagger \mathbf{e}(\omega) = 1$. The vector $\mathbf{b}(\omega)$ turns out to be

$$\mathbf{b}(\omega) = \frac{1}{\mathbf{e}(\omega)^\dagger R^{-1} \mathbf{e}(\omega)} R^{-1} \mathbf{e}(\omega).$$

Now we allow $\omega$ to vary and compute the expected output of the filter $\mathbf{b}(\omega)$, operating on the signal plus noise input. This expected output is then

$$1/\mathbf{e}(\omega)^\dagger R^{-1} \mathbf{e}(\omega).$$

The reason that this estimator resolves closely spaced delta functions better than linear methods such as the DFT is that, when $\omega$ is fixed, we obtain an optimal filter using $R$ as the noise covariance matrix, which then includes all sinusoids not at the frequency $\omega$ in the "noise" component. This is actually a good thing, since, when we are looking at a frequency $\omega$ that does not correspond to a frequency actually present in the data, we want the sinusoidal components present at nearby frequencies to be filtered out, to improve resolution. We lose resolution of two nearby peaks in estimators like the DFT when the estimator gives a larger value between two actual peaks than it does at the peaks themselves. Methods such as Capon's reduce the estimator's value between the two peaks.

# Chapter 19

# Inner Products

## 19.1   Chapter Summary

Many methods for analyzing measured signals are based on the idea of matching the data against various potential signals to see which ones match best. The role of *inner products* in this matching approach is the topic of this chapter.

## 19.2   Cauchy's Inequality

The matching is done using the complex dot product, $\mathbf{e}_\omega^\dagger \mathbf{d}$. In the ideal case this dot product is large, for those values of $\omega$ that correspond to an actual component of the signal; otherwise it is small. Why this should be the case is the Cauchy-Schwarz Inequality (or sometimes, depending on the context, just Cauchy's Inequality, just Schwarz's Inequality, or, in the Russian literature, Bunyakovsky's Inequality). The proof of Cauchy's

Inequality rests on four basic properties of the complex dot product. These properties can then be used to obtain the more general notion of an inner product.

---

## 19.3 The Complex Vector Dot Product

Let $\mathbf{u} = (a, b)$ and $\mathbf{v} = (c, d)$ be two vectors in two-dimensional space. Let $\mathbf{u}$ make the angle $\alpha > 0$ with the positive $x$-axis and $\mathbf{v}$ the angle $\beta > 0$. Let $||\mathbf{u}|| = \sqrt{a^2 + b^2}$ denote the length of the vector $\mathbf{u}$. Then $a = ||\mathbf{u}|| \cos \alpha$, $b = ||\mathbf{u}|| \sin \alpha$, $c = ||\mathbf{v}|| \cos \beta$ and $d = ||\mathbf{v}|| \sin \beta$. So $\mathbf{u} \cdot \mathbf{v} = ac + bd = ||\mathbf{u}||\,||\mathbf{v}||(\cos \alpha \cos \beta + \sin \alpha \sin \beta = ||\mathbf{u}||\,||\mathbf{v}|| \cos(\alpha - \beta)$. Therefore, we have

$$\mathbf{u} \cdot \mathbf{v} = ||\mathbf{u}||\,||\mathbf{v}|| \cos \theta, \tag{19.1}$$

where $\theta = \alpha - \beta$ is the angle between $\mathbf{u}$ and $\mathbf{v}$. Cauchy's Inequality is

$$|\mathbf{u} \cdot \mathbf{v}| \leq ||\mathbf{u}||\,||\mathbf{v}||,$$

with equality if and only if $\mathbf{u}$ and $\mathbf{v}$ are parallel.

Cauchy's Inequality extends to vectors of any size with complex entries. For example, the complex $M$-dimensional vectors $\mathbf{e}_\omega$ and $\mathbf{e}_\theta$ defined earlier both have length equal to $\sqrt{M}$ and

$$|\mathbf{e}_\omega^\dagger \mathbf{e}_\theta| \leq M,$$

with equality if and only if $\omega$ and $\theta$ differ by an integer multiple of $\pi$.

From Equation (19.1) we know that the dot product $\mathbf{u} \cdot \mathbf{v}$ is zero if and only if the angle between these two vectors is a right angle; we say then that $\mathbf{u}$ and $\mathbf{v}$ are mutually *orthogonal*. The idea of using the dot product to measure how similar two vectors are is called *matched filtering*; it is a popular method in signal detection and estimation of parameters.

**Proof of Cauchy's Inequality:** To prove Cauchy's Inequality for the complex vector dot product, we write $\mathbf{u} \cdot \mathbf{v} = |\mathbf{u} \cdot \mathbf{v}|e^{i\theta}$. Let $t$ be a real variable and consider

$$
\begin{aligned}
0 \leq ||e^{-i\theta}\mathbf{u} - t\mathbf{v}||^2 &= (e^{-i\theta}\mathbf{u} - t\mathbf{v}) \cdot (e^{-i\theta}\mathbf{u} - t\mathbf{v}) \\
&= ||\mathbf{u}||^2 - t[(e^{-i\theta}\mathbf{u}) \cdot \mathbf{v} + \mathbf{v} \cdot (e^{-i\theta}\mathbf{u})] + t^2||\mathbf{v}||^2 \\
&= ||\mathbf{u}||^2 - t[(e^{-i\theta}\mathbf{u}) \cdot \mathbf{v} + \overline{(e^{-i\theta}\mathbf{u}) \cdot \mathbf{v}}] + t^2||\mathbf{v}||^2 \\
&= ||\mathbf{u}||^2 - 2Re(te^{-i\theta}(\mathbf{u} \cdot \mathbf{v})) + t^2||\mathbf{v}||^2 \\
&= ||\mathbf{u}||^2 - 2Re(t|\mathbf{u} \cdot \mathbf{v}|) + t^2||\mathbf{v}||^2 \\
&= ||\mathbf{u}||^2 - 2t|\mathbf{u} \cdot \mathbf{v}| + t^2||\mathbf{v}||^2.
\end{aligned}
$$

This is a nonnegative quadratic polynomial in the variable $t$, so it cannot have two distinct real roots. Therefore, the discriminant $4|\mathbf{u} \cdot \mathbf{v}|^2 - 4||\mathbf{v}||^2||\mathbf{u}||^2$ must be non-positive; that is, $|\mathbf{u} \cdot \mathbf{v}|^2 \leq ||\mathbf{u}||^2||\mathbf{v}||^2$. This is Cauchy's Inequality. ∎

**Ex. 19.1** *Use Cauchy's Inequality to show that*

$$||\mathbf{u} + \mathbf{v}|| \leq ||\mathbf{u}|| + ||\mathbf{v}||;$$

*this is called the* triangle inequality.

A careful examination of the proof just presented shows that we did not explicitly use the definition of the complex vector dot product, but only some of its properties. This suggested to mathematicians the possibility of abstracting these properties and using them to define a more general concept, an *inner product*, between objects more general than complex vectors, such as infinite sequences, random variables, and matrices. Such an inner product can then be used to define the *norm* of these objects and thereby a distance between such objects. Once we have an inner product defined, we also have available the notions of orthogonality and best approximation. We shall address all of these topics shortly.

---

## 19.4 Orthogonality

Consider the problem of writing the two-dimensional real vector $(3, -2)$ as a linear combination of the vectors $(1, 1)$ and $(1, -1)$; that is, we want to find constants $a$ and $b$ so that $(3, -2) = a(1, 1) + b(1, -1)$. One way to do this, of course, is to compare the components: $3 = a + b$ and $-2 = a - b$; we can then solve this simple system for the $a$ and $b$. In higher dimensions this way of doing it becomes harder, however. A second way is to make use of the dot product and orthogonality.

The dot product of two vectors $(x, y)$ and $(w, z)$ in $\mathbb{R}^2$ is $(x, y) \cdot (w, z) = xw + yz$. If the dot product is zero then the vectors are said to be *orthogonal*; the two vectors $(1, 1)$ and $(1, -1)$ are orthogonal. We take the dot product of both sides of $(3, -2) = a(1, 1) + b(1, -1)$ with $(1, 1)$ to get

$$1 = (3, -2) \cdot (1, 1) = a(1, 1) \cdot (1, 1) + b(1, -1) \cdot (1, 1) = a(1, 1) \cdot (1, 1) + 0 = 2a,$$

so we see that $a = \frac{1}{2}$. Similarly, taking the dot product of both sides with $(1, -1)$ gives

$$5 = (3, -2) \cdot (1, -1) = a(1, 1) \cdot (1, -1) + b(1, -1) \cdot (1, -1) = 2b,$$

so $b = \frac{5}{2}$. Therefore, $(3, -2) = \frac{1}{2}(1, 1) + \frac{5}{2}(1, -1)$. The beauty of this approach is that it does not get much harder as we go to higher dimensions.

Since the cosine of the angle $\theta$ between vectors $\mathbf{u}$ and $\mathbf{v}$ is

$$\cos \theta = \mathbf{u} \cdot \mathbf{v} / \|\mathbf{u}\| \, \|\mathbf{v}\|,$$

where $\|\mathbf{u}\|^2 = \mathbf{u} \cdot \mathbf{u}$, the projection of vector $\mathbf{v}$ on to the line through the origin parallel to $\mathbf{u}$ is

$$\text{Proj}_{\mathbf{u}}(\mathbf{v}) = \frac{\mathbf{u} \cdot \mathbf{v}}{\mathbf{u} \cdot \mathbf{u}} \mathbf{u}.$$

Therefore, the vector $\mathbf{v}$ can be written as

$$\mathbf{v} = \text{Proj}_{\mathbf{u}}(\mathbf{v}) + (\mathbf{v} - \text{Proj}_{\mathbf{u}}(\mathbf{v})),$$

where the first term on the right is parallel to $\mathbf{u}$ and the second one is orthogonal to $\mathbf{u}$.

How do we find vectors that are mutually orthogonal? Suppose we begin with $(1, 1)$. Take a second vector, say $(1, 2)$, that is not parallel to $(1, 1)$ and write it as we did $\mathbf{v}$ earlier, that is, as a sum of two vectors, one parallel to $(1, 1)$ and the second orthogonal to $(1, 1)$. The projection of $(1, 2)$ onto the line parallel to $(1, 1)$ passing through the origin is

$$\frac{(1, 1) \cdot (1, 2)}{(1, 1) \cdot (1, 1)}(1, 1) = \frac{3}{2}(1, 1) = \left(\frac{3}{2}, \frac{3}{2}\right)$$

so

$$(1, 2) = \left(\frac{3}{2}, \frac{3}{2}\right) + \left((1, 2) - \left(\frac{3}{2}, \frac{3}{2}\right)\right) = \left(\frac{3}{2}, \frac{3}{2}\right) + \left(-\frac{1}{2}, \frac{1}{2}\right).$$

The vectors $(-\frac{1}{2}, \frac{1}{2}) = -\frac{1}{2}(1, -1)$ and, therefore, $(1, -1)$ are then orthogonal to $(1, 1)$. This approach is the basis for the *Gram-Schmidt* method for constructing a set of mutually orthogonal vectors.

**Ex. 19.2** *Use the Gram-Schmidt approach to find a third vector in $\mathbb{R}^3$ orthogonal to both $(1, 1, 1)$ and $(1, 0, -1)$.*

Orthogonality is a convenient tool that can be exploited whenever we have an inner product defined.

## 19.5    Generalizing the Dot Product: Inner Products

The proof of Cauchy's Inequality rests not on the actual definition of the complex vector dot product, but rather on four of its most basic properties. We use these properties to extend the concept of the complex vector dot product to that of *inner product*. Later in this chapter we shall give several examples of inner products, applied to a variety of mathematical objects, including infinite sequences, functions, random variables, and matrices. For now, let us denote our mathematical objects by $\mathbf{u}$ and $\mathbf{v}$ and the inner product between them as $\langle \mathbf{u}, \mathbf{v} \rangle$. The objects will then be said to be members of an *inner-product space*. We are interested in inner products because they provide a notion of orthogonality, which is fundamental to best approximation and optimal estimation.

**Defining an inner product:** The four basic properties that will serve to define an inner product are:

1. $\langle \mathbf{u}, \mathbf{u} \rangle \geq 0$, with equality if and only if $\mathbf{u} = \mathbf{0}$;

2. $\langle \mathbf{v}, \mathbf{u} \rangle = \overline{\langle \mathbf{u}, \mathbf{v} \rangle}$;

3. $\langle \mathbf{u}, \mathbf{v} + \mathbf{w} \rangle = \langle \mathbf{u}, \mathbf{v} \rangle + \langle \mathbf{u}, \mathbf{w} \rangle$;

4. $\langle c\mathbf{u}, \mathbf{v} \rangle = c\langle \mathbf{u}, \mathbf{v} \rangle$ for any complex number $c$.

The inner product is the basic ingredient in Hilbert space theory. Using the inner product, we define the *norm* of $\mathbf{u}$ to be

$$||\mathbf{u}|| = \sqrt{\langle \mathbf{u}, \mathbf{u} \rangle}$$

and the distance between $\mathbf{u}$ and $\mathbf{v}$ to be $||\mathbf{u} - \mathbf{v}||$.

**The Cauchy–Schwarz Inequality:** Because these four properties were all we needed to prove the Cauchy Inequality for the complex vector dot product, we obtain the same inequality whenever we have an inner product. This more general inequality is the Cauchy-Schwarz Inequality:

$$|\langle \mathbf{u}, \mathbf{v} \rangle| \leq \sqrt{\langle \mathbf{u}, \mathbf{u} \rangle}\sqrt{\langle \mathbf{v}, \mathbf{v} \rangle}$$

or

$$|\langle \mathbf{u}, \mathbf{v} \rangle| \leq ||\mathbf{u}|| \, ||\mathbf{v}||,$$

with equality if and only if there is a scalar $c$ such that $\mathbf{v} = c\mathbf{u}$. We say that the vectors $\mathbf{u}$ and $\mathbf{v}$ are *orthogonal* if $\langle \mathbf{u}, \mathbf{v} \rangle = 0$.

---

## 19.6   Another View of Orthogonality

We can develop orthogonality and the Cauchy-Schwarz Inequality in another way. For simplicity, we assume that the inner product is defined on a real vector space. From the definition of the norm we have

$$\|x + y\|^2 = \langle x + y, x + y \rangle = \|x\|^2 + \|y\|^2 + 2\langle x, y \rangle.$$

We say that Pythagoras' Theorem holds for $x \neq 0$ and $y \neq 0$ if

$$\|x + y\|^2 = \langle x + y, x + y \rangle = \|x\|^2 + \|y\|^2.$$

Clearly, Pythagoras' Theorem holds if and only if $\langle x, y \rangle = 0$.

Now, we say that nonzero vectors $x$ and $y$ are *orthogonal* if

$$\|x + y\| = \|x - y\|.$$

It is an easy exercise to show that $x \neq 0$ and $y \neq 0$ are orthogonal if and only if $\langle x, y \rangle = 0$ and if and only if Pythagoras' Theorem holds.

For nonzero $x$ and $y$, let $p = \gamma y$ be the vector in the span of $y$ for which

$$\|x - p\| \leq \|x - \beta y\|,$$

for all real $\beta$. Minimizing the function

$$f(\beta) = \|x - \beta y\|^2$$

with respect to the variable $\beta$, we find that the optimal $\gamma$ is

$$\gamma = \frac{\langle x, y \rangle}{\|y\|^2}.$$

A simple calculation shows that the vectors $x - p$ and $p$ are orthogonal, so that, by Pythagoras' Theorem,

$$\|x\|^2 = \|x - p\|^2 + \|p\|^2.$$

It follows, therefore, that

$$\|x\| \geq \|p\|,$$

and so

$$|\langle x, y \rangle| \leq \|x\| \|y\|,$$

with equality if and only if $x = p$. This is the Cauchy-Schwarz Inequality once again.

For nonzero vectors in $\mathbb{R}^2$ or $\mathbb{R}^3$ we know that

$$x \cdot y = \|x\| \|y\| \cos(\theta),$$

where $\theta$ is the angle between the two vectors when they are viewed as directed line segments placed so that they have a common starting point. Using the Cauchy-Schwarz Inequality, we can mimic what happens in $\mathbb{R}^2$ and $\mathbb{R}^3$ by defining the *angle between nonzero vectors* in an arbitrary inner product space to be

$$\theta(x, y) = \arccos\left(\frac{\langle x, y \rangle}{\|x\| \|y\|}\right).$$

We turn now to some examples of inner products.

## 19.7   Examples of Inner Products

In this section we illustrate the notion of inner product with several examples.

### 19.7.1   An Inner Product for Infinite Sequences

Let $\mathbf{u} = \{u_n\}$ and $\mathbf{v} = \{v_n\}$ be infinite sequences of complex numbers. The inner product is then

$$\langle \mathbf{u}, \mathbf{v} \rangle = \sum u_n \overline{v_n},$$

and

$$\|\mathbf{u}\| = \sqrt{\sum |u_n|^2}.$$

The sums are assumed to be finite; the index of summation $n$ is singly or doubly infinite, depending on the context. The Cauchy-Schwarz Inequality says that

$$\left| \sum u_n \overline{v_n} \right| \leq \sqrt{\sum |u_n|^2} \sqrt{\sum |v_n|^2}.$$

### 19.7.2   An Inner Product for Functions

Now suppose that $\mathbf{u} = f(x)$ and $\mathbf{v} = g(x)$. Then the $L^2$ inner product is

$$\langle \mathbf{u}, \mathbf{v} \rangle = \int f(x)\overline{g(x)}dx$$

and the $L^2$ norm of $u$ is

$$\|\mathbf{u}\| = \sqrt{\int |f(x)|^2 dx}.$$

The integrals are assumed to be finite; the limits of integration depend on the support of the functions involved. The Cauchy-Schwarz Inequality now says that

$$\left| \int f(x)\overline{g(x)}dx \right| \leq \sqrt{\int |f(x)|^2 dx} \sqrt{\int |g(x)|^2 dx}.$$

### 19.7.3 An Inner Product for Random Variables

Now suppose that $\mathbf{u} = X$ and $\mathbf{v} = Y$ are random variables. Then,

$$\langle \mathbf{u}, \mathbf{v} \rangle = E(X\overline{Y})$$

and

$$||\mathbf{u}|| = \sqrt{E(|X|^2)},$$

which is the standard deviation of $X$ if the mean of $X$ is zero. The expected values are assumed to be finite. The Cauchy-Schwarz Inequality now says that

$$|E(X\overline{Y})| \leq \sqrt{E(|X|^2)}\sqrt{E(|Y|^2)}.$$

If $E(X) = 0$ and $E(Y) = 0$, the random variables $X$ and $Y$ are orthogonal if and only if they are *uncorrelated*.

### 19.7.4 An Inner Product for Complex Matrices

Now suppose that $\mathbf{u} = A$ and $\mathbf{v} = B$ are complex matrices. Then,

$$\langle \mathbf{u}, \mathbf{v} \rangle = \mathrm{trace}(B^{\dagger}A)$$

and

$$||\mathbf{u}|| = \sqrt{\mathrm{trace}(A^{\dagger}A)},$$

where the trace of a square matrix is the sum of the entries on the main diagonal. This inner product is simply the complex vector dot product of the vectorized versions of the matrices involved. The Cauchy-Schwarz Inequality now says that

$$|\mathrm{trace}(B^{\dagger}A)| \leq \sqrt{\mathrm{trace}(A^{\dagger}A)}\sqrt{\mathrm{trace}(B^{\dagger}B)}.$$

### 19.7.5 A Weighted Inner Product for Complex Vectors

Let $\mathbf{u}$ and $\mathbf{v}$ be complex vectors and let $Q$ be a Hermitian positive-definite matrix; that is, $Q^{\dagger} = Q$ and $\mathbf{u}^{\dagger}Q\mathbf{u} > 0$ for all nonzero vectors $\mathbf{u}$. The $Q$-inner product is then

$$\langle \mathbf{u}, \mathbf{v} \rangle = \mathbf{v}^{\dagger}Q\mathbf{u}$$

and the $Q$-norm of $u$ is

$$||\mathbf{u}|| = \sqrt{\mathbf{u}^{\dagger}Q\mathbf{u}}.$$

We know from the eigenvector decomposition of $Q$ that $Q = C^{\dagger}C$ for some matrix $C$. Therefore, the inner product is simply the complex vector dot product of the vectors $C\mathbf{u}$ and $C\mathbf{v}$. The Cauchy-Schwarz Inequality says that

$$|\mathbf{v}^{\dagger}Q\mathbf{u}| \leq \sqrt{\mathbf{u}^{\dagger}Q\mathbf{u}}\sqrt{\mathbf{v}^{\dagger}Q\mathbf{v}}.$$

### 19.7.6 A Weighted Inner Product for Functions

Now suppose that $\mathbf{u} = f(x)$, $\mathbf{v} = g(x)$, and $w(x) > 0$. Then define

$$\langle \mathbf{u}, \mathbf{v} \rangle = \int f(x)\overline{g(x)}w(x)dx$$

and

$$\|\mathbf{u}\| = \sqrt{\int |f(x)|^2 w(x)dx}.$$

The integrals are assumed to be finite; the limits of integration depend on the support of the functions involved. This inner product is simply the $L^2$ inner product of the functions $f(x)\sqrt{w(x)}$ and $g(x)\sqrt{w(x)}$. The Cauchy-Schwarz Inequality now says that

$$\left| \int f(x)\overline{g(x)}w(x)dx \right| \leq \sqrt{\int |f(x)|^2 w(x)dx} \sqrt{\int |g(x)|^2 w(x)dx}.$$

Once we have an inner product defined, we can speak about orthogonality and best approximation. Important in that regard is the orthogonality principle.

## 19.8   The Orthogonality Principle

Imagine that you are standing and looking down at the floor. The point $B$ on the floor that is closest to $N$, the tip of your nose, is the unique point on the floor such that the vector from $B$ to any other point $A$ on the floor is perpendicular to the vector from $N$ to $B$; that is, $\langle BN, BA \rangle = 0$. This is a simple illustration of the *orthogonality principle*. Whenever we have an inner product defined we can speak of orthogonality and apply the orthogonality principle to find best approximations.

**The orthogonality principle:** Let $\mathbf{u}$ and $\mathbf{v}^1, ..., \mathbf{v}^N$ be members of an inner-product space. For all choices of scalars $a_1, ..., a_N$, we can compute the distance from $\mathbf{u}$ to the member $a_1\mathbf{v}^1 + ...a_N\mathbf{v}^N$. Then, we minimize this distance over all choices of the scalars; let $b_1, ..., b_N$ be this best choice. The *orthogonality principle* tells us that the member $\mathbf{u} - (b_1\mathbf{v}^1 + ...b_N\mathbf{v}^N)$ is orthogonal to the member $(a_1\mathbf{v}^1 + ... + a_N\mathbf{v}^N) - (b_1\mathbf{v}^1 + ...b_N\mathbf{v}^N)$, that is,

$$\langle \mathbf{u} - (b_1\mathbf{v}^1 + ...b_N\mathbf{v}^N), (a_1\mathbf{v}^1 + ... + a_N\mathbf{v}^N) - (b_1\mathbf{v}^1 + ...b_N\mathbf{v}^N) \rangle = 0,$$

for every choice of scalars $a_n$. We can then use the orthogonality principle to find the best choice $b_1., , , .b_N$.

For each fixed index value $j$ in the set $\{1, ..., N\}$, let $a_n = b_n$ if $j$ is not equal to $n$ and $a_j = b_j + 1$. Then we have

$$0 = \langle \mathbf{u} - (b_1 \mathbf{v}^1 + ...b_N \mathbf{v}^N), \mathbf{v}^j \rangle,$$

or

$$\langle \mathbf{u}, \mathbf{v}^j \rangle = \sum_{n=1}^{N} b_n \langle \mathbf{v}^n, \mathbf{v}^j \rangle,$$

for each $j$. The $\mathbf{v}^n$ are known, so we can calculate the inner products $\langle \mathbf{v}^n, \mathbf{v}^j \rangle$ and solve this system of equations for the best $b_n$.

We shall encounter a number of particular cases of the orthogonality principle in subsequent chapters. The example of the *least-squares* solution of a system of linear equations provides a good example of the use of this principle.

**The least-squares solution:** Let $V\mathbf{a} = \mathbf{u}$ be a system of $M$ linear equations in $N$ unknowns. For $n = 1, ..., N$ let $\mathbf{v}^n$ be the $n$th column of the matrix $V$. For any choice of the vector $\mathbf{a}$ with entries $a_n$, $n = 1, ..., N$, the vector $V\mathbf{a}$ is

$$V\mathbf{a} = \sum_{n=1}^{N} a_n \mathbf{v}^n.$$

Solving $V\mathbf{a} = \mathbf{u}$ amounts to representing the vector $\mathbf{u}$ as a linear combination of the columns of $V$.

If there is no solution of $V\mathbf{a} = \mathbf{u}$ then we can look for the best choice of coefficients so as to minimize the distance $||\mathbf{u} - (a_1 \mathbf{v}^1 + ... + a_N \mathbf{v}^N)||$. The matrix with entries $\langle \mathbf{v}^n, \mathbf{v}^j \rangle$ is $V^\dagger V$, and the vector with entries $\langle \mathbf{u}, \mathbf{v}^j \rangle$ is $V^\dagger \mathbf{u}$. According to the orthogonality principle, we must solve the system of equations $V^\dagger \mathbf{u} = V^\dagger V \mathbf{a}$, which leads to the least-squares solution.

**Ex. 19.3** *Find polynomial functions $f(x)$, $g(x)$ and $h(x)$ that are orthogonal in the sense of the $L^2$ inner product on the interval $[0, 1]$ and have the property that every polynomial of degree two or less can be written as a linear combination of these three functions.*

**Ex. 19.4** *Show that the functions $e^{inx}$, $n$ an integer, are orthogonal in the sense of the $L^2$ inner product on the interval $[-\pi, \pi]$. Let $f(x)$ have the Fourier expansion*

$$f(x) = \sum_{n=-\infty}^{\infty} a_n e^{inx}, \ |x| \le \pi.$$

*Use orthogonality to find the coefficients $a_n$.*

We have seen that orthogonality can be used to determine the coefficients in the Fourier series representation of a function. There are other useful representations in which orthogonality also plays a role; wavelets is one example. Let $f(x)$ be defined on some closed interval $[a, b]$. Suppose that we change the function $f(x)$ to a new function $g(x)$ by altering the values for $x$ within a small interval, keeping the remaining values the same: then all of the Fourier coefficients change. Looked at another way, a localized disturbance in the function $f(x)$ affects all of its Fourier coefficients. It would be helpful to be able to represent $f(x)$ as a sum of orthogonal functions in such a way that localized changes in $f(x)$ affect only a small number of the components in the sum. One way to do this is with wavelets, as we saw in Chapter 18.

# Chapter 20

# Wiener Filtering

## 20.1   Chapter Summary

The vector Wiener filter (VWF) is similar to the BLUE and provides another method for estimating the vector $\mathbf{x}$ given noisy measurements $\mathbf{z}$ in $\mathbb{C}^J$, where

$$\mathbf{z} = H\mathbf{x} + \mathbf{v},$$

with $\mathbf{x}$ and $\mathbf{v}$ independent random vectors and $H$ a known matrix. We shall assume throughout this chapter that $E(\mathbf{v}) = \mathbf{0}$ and let $Q = E(\mathbf{v}\mathbf{v}^\dagger)$.

When the data is a finite vector composed of signal plus noise the vector Wiener filter can be used to estimate the signal component, provided we know something about the possible signals and possible noises. In theoretical discussion of filtering signal from signal plus noise, it is traditional to assume that both components are doubly infinite sequences of random variables. In this case the Wiener filter is a convolution filter that operates on the input signal plus noise sequence to produce the output estimate of the signal-only sequence. The derivation of the Wiener filter is in terms of the autocorrelation sequences of the two components, as well as their respective power spectra.

## 20.2　The Vector Wiener Filter in Estimation

It is common to formulate the VWF in the context of filtering a signal vector **s** from signal plus noise. The data is the vector

$$\mathbf{z} = \mathbf{s} + \mathbf{v},$$

and we want to estimate **s**. Each entry of our estimate of the vector **s** will be a linear combination of the data values; that is, our estimate is $\hat{\mathbf{s}} = B^{\dagger}\mathbf{z}$ for some matrix $B$ to be determined. This $B$ will be called the *vector Wiener filter*. To extract the signal from the noise, we must know something about possible signals and possible noises. We consider several stages of increasing complexity and correspondence with reality.

## 20.3　The Simplest Case

Suppose, initially, and unrealistically, that all signals must have the form $\mathbf{s} = a\mathbf{u}$, where $a$ is an unknown scalar and **u** is a known vector. Suppose that all noises must have the form $\mathbf{v} = b\mathbf{w}$, where $b$ is an unknown scalar and **w** is a known vector. Then, to estimate **s**, we must find $a$. So long as $J \geq 2$, we should be able to solve for $a$ and $b$. We form the two equations

$$\mathbf{u}^{\dagger}\mathbf{z} = a\mathbf{u}^{\dagger}\mathbf{u} + b\mathbf{u}^{\dagger}\mathbf{w}$$

and

$$\mathbf{w}^{\dagger}\mathbf{z} = a\mathbf{w}^{\dagger}\mathbf{u} + b\mathbf{w}^{\dagger}\mathbf{w}.$$

This system of two equations in two unknowns will have a unique solution unless **u** and **w** are proportional, in which case we cannot expect to distinguish signal from noise.

## 20.4　A More General Case

We move now to a somewhat more complicated, but still unrealistic, model. Suppose that all signals must have the form

$$\mathbf{s} = \sum_{n=1}^{N} a_n \mathbf{u}^n,$$

where the $a_n$ are unknown scalars and the $\mathbf{u}^n$ are known linearly independent vectors. Suppose that all noises must have the form

$$\mathbf{v} = \sum_{m=1}^{M} b_m \mathbf{w}^m,$$

where the $b_m$ are unknown scalars and $\mathbf{w}^m$ are known linearly independent vectors. Then, to estimate $\mathbf{s}$, we must find the $a_n$. So long as $J \geq N + M$, we should be able to solve for the unique $a_n$ and $b_m$. However, we usually do not know a great deal about the signal and the noise, so it is better to assume that we are in the situation in which the $N$ and $M$ are large and $J < N + M$.

Let $U$ be the $J$ by $N$ matrix whose $n$th column is $\mathbf{u}^n$ and $W$ the $J$ by $M$ matrix whose $m$th column is $\mathbf{w}^m$. Let $V$ be the $J$ by $N + M$ matrix whose first $N$ columns contain $U$ and whose last $M$ columns contain $W$; so, $V = [U \ \ W]$. Let $\mathbf{c}$ be the $N + M$ by $1$ column vector whose first $N$ entries are the $a_n$ and whose last $M$ entries are the $b_m$. We want to solve $\mathbf{z} = V\mathbf{c}$.

The system of linear equations $\mathbf{z} = V\mathbf{c}$ has too many unknowns when $N + M > J$, so we seek the minimum-norm solution. In closed form this solution is

$$\hat{\mathbf{c}} = V^\dagger (VV^\dagger)^{-1} \mathbf{z}.$$

The first $N$ entries of $\hat{\mathbf{c}}$ are our estimates of the $a_n$. Once we have these, we estimate the signal itself by multiplying by the matrix $U$; that is, our estimate of $\mathbf{s}$ is

$$\hat{\mathbf{s}} = UU^\dagger (UU^\dagger + WW^\dagger)^{-1} \mathbf{z}.$$

The matrix $VV^\dagger = (UU^\dagger + WW^\dagger)$ involves what we shall call the *signal correlation matrix* $UU^\dagger$ and the *noise correlation matrix* $WW^\dagger$, by analogy with the statistical terminology.

Consider $UU^\dagger$. The matrix $UU^\dagger$ is $J$ by $J$ and the $(i, j)$ entry of $UU^\dagger$ is given by

$$UU_{ij}^\dagger = \sum_{n=1}^{N} u_i^n \overline{u_j^n}.$$

The matrix $\frac{1}{N}UU^\dagger$ has for its entries the average, over all the $n = 1, ..., N$, of the product of the $i$th and $j$th entries of the vectors $\mathbf{u}^n$. Therefore, $\frac{1}{N}UU^\dagger$ is statistical information about the signal; it tells us how these products look, on average, over all members of the family $\{\mathbf{u}^n\}$, the *ensemble*, to use the statistical word.

## 20.5    The Stochastic Case

To pass to a more formal statistical framework, we let the coefficient vectors $\mathbf{a} = (a_1, a_2, ..., a_N)^T$ and $\mathbf{b} = (b_1, b_2, ..., b_M)^T$ be independent random white-noise vectors, both with mean zero and covariance matrices $E(\mathbf{aa}^\dagger) = I$ and $E(\mathbf{bb}^\dagger) = I$. Now the matrices $UU^\dagger$ and $WW^\dagger$ are defined statistically;

$$UU^\dagger = E(\mathbf{ss}^\dagger) = R_s$$

and

$$WW^\dagger = E(\mathbf{vv}^\dagger) = Q = R_v.$$

The estimate of $\mathbf{s}$ is the result of applying the vector Wiener filter to the vector $\mathbf{z}$ and is once again given by

$$\hat{\mathbf{s}} = UU^\dagger (UU^\dagger + WW^\dagger)^{-1} \mathbf{z}.$$

**Ex. 20.1** *Apply the vector Wiener filter to the simplest problem discussed earlier in the chapter on the BLUE; let $N = 1$ and assume that $c$ is a random variable with mean zero and variance one. It will help to use the matrix-inversion identity*

$$(Q + \mathbf{uu}^\dagger)^{-1} = Q^{-1} - (1 + \mathbf{u}^\dagger Q^{-1} \mathbf{u})^{-1} Q^{-1} \mathbf{uu}^\dagger Q^{-1}; \qquad (20.1)$$

*see also Equation (17.10).*

## 20.6    The VWF and the BLUE

To apply the VWF to the problem considered in the discussion of the BLUE, let the vector $\mathbf{s}$ be $H\mathbf{x}$. We assume, in addition, that the vector $\mathbf{x}$ is a white-noise vector; that is, $E(\mathbf{xx}^\dagger) = \sigma^2 I$. Then, $R_s = \sigma^2 HH^\dagger$.

In the VWF approach we estimate $\mathbf{s}$ using

$$\hat{\mathbf{s}} = B^\dagger \mathbf{z},$$

where the matrix $B$ is chosen so as to minimize the mean squared error, $E||\hat{\mathbf{s}} - \mathbf{s}||^2$. This is equivalent to minimizing

$$\text{trace } E((B^\dagger \mathbf{z} - \mathbf{s})(B^\dagger \mathbf{z} - \mathbf{s})^\dagger).$$

Expanding the matrix products and using the previous definitions, we see that we must minimize

$$\text{trace}\,(B^\dagger(R_s + R_v)B - R_sB - B^\dagger R_s + R_s).$$

Differentiating with respect to the matrix $B$ using Equations (21.15) and (21.16), we find

$$(R_s + R_v)B - R_s = 0,$$

so that

$$B = (R_s + R_v)^{-1}R_s.$$

Our estimate of the signal component is then

$$\hat{\mathbf{s}} = R_s(R_s + R_v)^{-1}\mathbf{z}.$$

With $\mathbf{s} = H\mathbf{x}$, our estimate of $\mathbf{s}$ is

$$\hat{\mathbf{s}} = \sigma^2 HH^\dagger(\sigma^2 HH^\dagger + Q)^{-1}\mathbf{z},$$

and the VWF estimate of $\mathbf{x}$ is

$$\hat{\mathbf{x}} = \sigma^2 H^\dagger(\sigma^2 HH^\dagger + Q)^{-1}\mathbf{z}.$$

How does this estimate relate to the one we got from the BLUE?
The BLUE estimate of $\mathbf{x}$ is

$$\hat{\mathbf{x}} = (H^\dagger Q^{-1}H)^{-1}H^\dagger Q^{-1}\mathbf{z}.$$

From the matrix identity in Equation (17.4), we know that

$$(H^\dagger Q^{-1}H + \sigma^{-2}I)^{-1}H^\dagger Q^{-1} = \sigma^2 H^\dagger(\sigma^2 HH^\dagger + Q)^{-1}.$$

Therefore, the VWF estimate of $\mathbf{x}$ is

$$\hat{\mathbf{x}} = (H^\dagger Q^{-1}H + \sigma^{-2}I)^{-1}H^\dagger Q^{-1}\mathbf{z}.$$

Note that the BLUE estimate is unbiased and unaffected by changes in the signal strength or the noise strength. In contrast, the VWF is not unbiased and does depend on the signal-to-noise ratio; that is, it depends on the ratio $\sigma^2/\text{trace}\,(Q)$. The BLUE estimate is the limiting case of the VWF estimate, as the signal-to-noise ratio goes to infinity.

The BLUE estimates $\mathbf{s} = H\mathbf{x}$ by first finding the BLUE estimate of $\mathbf{x}$ and then multiplying it by $H$ to get the estimate of the signal $\mathbf{s}$.

**Ex. 20.2** *Show that the mean-squared error in the estimation of* $\mathbf{s}$ *is*

$$E(||\hat{\mathbf{s}} - \mathbf{s}||^2) = \text{trace}\,(H(H^\dagger Q^{-1}H)^{-1}H^\dagger).$$

The VWF finds the linear estimate of $\mathbf{s} = H\mathbf{x}$ that minimizes the mean-squared error $E(||\hat{\mathbf{s}} - \mathbf{s}||^2)$. Consequently, the mean squared error in the VWF is less than that in the BLUE.

**Ex. 20.3** *Assume that* $E(\mathbf{x}\mathbf{x}^\dagger) = \sigma^2 I$. *Show that the mean squared error for the VWF estimate is*

$$E(||\hat{\mathbf{s}} - \mathbf{s}||^2) = \text{trace}\,(H(H^\dagger Q^{-1} H + \sigma^{-2} I)^{-1} H^\dagger).$$

## 20.7   Wiener Filtering of Functions

The Wiener filter is often presented in the context of random functions of, say, time. In this model the signal is $s(t)$ and the noise is $q(t)$, where these functions of time are viewed as random functions (stochastic processes). The data is taken to be $z(t)$, a function of $t$, so that the matrices $UU^\dagger$ and $WW^\dagger$ are now *infinite matrices*; the discrete index $j = 1, ..., J$ is now replaced by the continuous index variable $t$. Instead of the finite family $\{\mathbf{u}^n, n = 1..., N\}$, we now have an infinite family of functions $u(t)$ in $\mathcal{U}$. The entries of $UU^\dagger$ are essentially the average values of the products $u(t_1)\overline{u(t_2)}$ over all the members of $\mathcal{U}$. It is often assumed that this average of products is a function not of $t_1$ and $t_2$ separately, but only of their difference $t_1 - t_2$; this is called *stationarity*. So, $aver\{u(t_1)\overline{u(t_2)}\} = r_s(t_1 - t_2)$ comes from a function $r_s(\tau)$ of a single variable. The Fourier transform of $r_s(\tau)$ is $R_s(\omega)$, the signal power spectrum. The matrix $UU^\dagger$ is then an infinite Toeplitz matrix, constant on each diagonal. The Wiener filtering can actually be achieved by taking Fourier transforms and multiplying and dividing by power spectra, instead of inverting infinite matrices. It is also common to discretize the time variable and to consider the Wiener filter operating on infinite sequences, as we see in the next section.

## 20.8   Wiener Filter Approximation: The Discrete Stationary Case

Suppose now that the discrete stationary random process to be filtered is the doubly infinite sequence $\{z_n = s_n + q_n\}_{n=-\infty}^\infty$, where $\{s_n\}$ is the signal component with autocorrelation function $r_s(k) = E(s_{n+k}\overline{s_n})$ and power spectrum $R_s(\omega)$ defined for $\omega$ in the interval $[-\pi, \pi]$, and $\{q_n\}$ is the noise component with autocorrelation function $r_q(k)$ and power spectrum $R_q(\omega)$ defined for $\omega$ in $[-\pi, \pi]$. We assume that for each $n$ the random variables $s_n$ and $q_n$ have mean zero and that the signal and noise are independent of one another. Then the autocorrelation function for the signal-plus-noise sequence $\{z_n\}$ is

$$r_z(n) = r_s(n) + r_q(n)$$

for all $n$ and

$$R_z(\omega) = R_s(\omega) + R_q(\omega)$$

is the signal-plus-noise power spectrum.

Let $h = \{h_k\}_{k=-\infty}^{\infty}$ be a linear filter with *transfer function*

$$H(\omega) = \sum_{k=-\infty}^{\infty} h_k e^{ik\omega},$$

for $\omega$ in $[-\pi, \pi]$. Given the sequence $\{z_n\}$ as input to this filter, the output is the sequence

$$y_n = \sum_{k=-\infty}^{\infty} h_k z_{n-k}. \tag{20.2}$$

The goal of Wiener filtering is to select the filter $h$ so that the output sequence $y_n$ approximates the signal $s_n$ sequence as well as possible. Specifically, we seek $h$ so as to minimize the expected squared error, $E(|y_n - s_n|^2)$, which, because of stationarity, is independent of $n$. We have

$$
\begin{aligned}
E(|y_n|^2) &= \sum_{k=-\infty}^{\infty} h_k \left( \sum_{j=-\infty}^{\infty} \overline{h_j}(r_s(j-k) + r_q(j-k)) \right) \\
&= \sum_{k=-\infty}^{\infty} h_k \overline{(r_z * \overline{h})_k},
\end{aligned}
$$

which, by the Parseval Equation (2.17), equals

$$\frac{1}{2\pi} \int H(\omega) R_z(\omega) \overline{H(\omega)} d\omega = \frac{1}{2\pi} \int |H(\omega)|^2 R_z(\omega) d\omega.$$

Similarly,

$$E(s_n \overline{y_n}) = \sum_{j=-\infty}^{\infty} \overline{h_j} r_s(j),$$

which equals

$$\frac{1}{2\pi} \int R_s(\omega) \overline{H(\omega)} d\omega,$$

and

$$E(|s_n|^2) = \frac{1}{2\pi} \int R_s(\omega)d\omega.$$

Therefore,

$$E(|y_n - s_n|^2) = \frac{1}{2\pi} \int |H(\omega)|^2 R_z(\omega)d\omega - \frac{1}{2\pi} \int R_s(\omega)\overline{H(\omega)}d\omega$$
$$- \frac{1}{2\pi} \int R_s(\omega)H(\omega)d\omega + \frac{1}{2\pi} \int R_s(\omega)d\omega.$$

As we shall see shortly, minimizing $E(|y_n - s_n|^2)$ with respect to the function $H(\omega)$ leads to the equation

$$R_z(\omega)H(\omega) = R_s(\omega),$$

so that the transfer function of the optimal filter is

$$H(\omega) = R_s(\omega)/R_z(\omega).$$

The *Wiener filter* is then the sequence $\{h_k\}$ of the Fourier coefficients of this function $H(\omega)$.

To prove that this choice of $H(\omega)$ minimizes $E(|y_n - s_n|^2)$, we note that

$$|H(\omega)|^2 R_z(\omega) - R_s(\omega)\overline{H(\omega)} - R_s(\omega)H(\omega) + R_s(\omega)$$
$$= R_z|H(\omega) - R_s(\omega)/R_z(\omega)|^2 + R_s(\omega) - R_s(\omega)^2/R_z(\omega).$$

Only the first term involves the function $H(\omega)$.

---

## 20.9   Approximating the Wiener Filter

Since $H(\omega)$ is a nonnegative function of $\omega$, therefore real-valued, its Fourier coefficients $h_k$ will be *conjugate symmetric*; that is, $h_{-k} = \overline{h_k}$. This poses a problem when the random process $z_n$ is a discrete time series, with $z_n$ denoting the measurement recorded at time $n$. From Equation (20.2) we see that to produce the output $y_n$ corresponding to time $n$ we need the input for every time, past and future. To remedy this we can obtain the best causal approximation of the Wiener filter $h$.

A filter $g = \{g_k\}_{k=-\infty}^{\infty}$ is said to be *causal* if $g_k = 0$ for $k < 0$; this means that given the input sequence $\{z_n\}$, the output

$$w_n = \sum_{k=-\infty}^{\infty} g_k z_{n-k} = \sum_{k=0}^{\infty} g_k z_{n-k}$$

requires only values of $z_m$ up to $m = n$. To obtain the causal filter $g$ that best approximates the Wiener filter, we find the coefficients $g_k$ that minimize the quantity $E(|y_n - w_n|^2)$, or, equivalently, we minimize

$$\int_{-\pi}^{\pi} \left| H(\omega) - \sum_{k=0}^{+\infty} g_k e^{ik\omega} \right|^2 R_z(\omega) d\omega.$$

The orthogonality principle tells us that the optimal coefficients must satisfy the equations

$$r_s(m) = \sum_{k=0}^{+\infty} g_k r_z(m - k),$$

for all $m$. These are the *Wiener–Hopf equations* [122].

Even having a causal filter does not completely solve the problem, since we would have to record and store the infinite past. Instead, we can decide to use a filter $f = \{f_k\}_{k=-\infty}^{\infty}$ for which $f_k = 0$ unless $-K \leq k \leq L$ for some positive integers $K$ and $L$. This means we must store $L$ values and wait until time $n + K$ to obtain the output for time $n$. Such a linear filter is a *finite memory, finite delay* filter, also called a *finite impulse response* (FIR) filter. Given the input sequence $\{z_n\}$ the output of the FIR filter is

$$v_n = \sum_{k=-K}^{L} f_k z_{n-k}.$$

To obtain such an FIR filter $f$ that best approximates the Wiener filter, we find the coefficients $f_k$ that minimize the quantity $E(|y_n - v_n|^2)$, or, equivalently, we minimize

$$\int_{-\pi}^{\pi} \left| H(\omega) - \sum_{k=-K}^{L} f_k e^{ik\omega} \right|^2 R_z(\omega) d\omega. \tag{20.3}$$

The orthogonality principle tells us that the optimal coefficients must satisfy the equations

$$r_s(m) = \sum_{k=-K}^{L} f_k r_z(m - k), \tag{20.4}$$

for $-K \leq m \leq L$.

In [31] it was pointed out that the linear equations that arise in Wiener-filter approximation also occur in image reconstruction from projections, with the image to be reconstructed playing the role of the power spectrum to be approximated. The methods of Wiener-filter approximation were then used to derive linear and nonlinear image-reconstruction procedures.

## 20.10    Adaptive Wiener Filters

Once again, we consider a stationary random process $z_n = s_n + v_n$ with autocorrelation function $E(z_n\overline{z_{n-m}}) = r_z(m) = r_s(m) + r_v(m)$. The finite causal Wiener filter (FCWF) $\mathbf{f} = (f_0, f_1, ..., f_L)^T$ is convolved with $\{z_n\}$ to produce an estimate of $s_n$ given by

$$\hat{s}_n = \sum_{k=0}^{L} f_k z_{n-k}.$$

With $\mathbf{y}_n^\dagger = (z_n, z_{n-1}, ..., z_{n-L})$ we can write $\hat{s}_n = \mathbf{y}_n^\dagger \mathbf{f}$. The FCWF $\mathbf{f}$ minimizes the expected squared error

$$J(\mathbf{f}) = E(|s_n - \hat{s}_n|^2)$$

and is obtained as the solution of the equations

$$r_s(m) = \sum_{k=0}^{L} f_k r_z(m - k),$$

for $0 \leq m \leq L$. Therefore, to use the FCWF we need the values $r_s(m)$ and $r_z(m - k)$ for $m$ and $k$ in the set $\{0, 1, ..., L\}$. When these autocorrelation values are not known, we can use adaptive methods to approximate the FCWF.

### 20.10.1    An Adaptive Least-Mean-Square Approach

We assume now that we have $z_0, z_1, ..., z_N$ and $p_0, p_1, ..., p_N$, where $p_n$ is a prior estimate of $s_n$, but that we do not know the correlation functions $r_z$ and $r_s$.

The gradient of the function $J(\mathbf{f})$ is

$$\nabla J(\mathbf{f}) = R_{zz}\mathbf{f} - \mathbf{r}_s,$$

where $R_{zz}$ is the square matrix with entries $r_z(m - n)$ and $\mathbf{r}_s$ is the vector with entries $r_s(m)$. An iterative gradient descent method for solving the system of equations $R_{zz}\mathbf{f} = \mathbf{r}_s$ is

$$\mathbf{f}_\tau = \mathbf{f}_{\tau-1} - \mu_\tau \nabla J(\mathbf{f}_{\tau-1}),$$

for some step-size parameters $\mu_\tau > 0$.

The adaptive *least-mean-square* (LMS) approach [48] replaces the gradient of $J(\mathbf{f})$ with an approximation of the gradient of the function

$G(\mathbf{f}) = |s_n - \hat{s}_n|^2$, which is $-2(s_n - \hat{s}_n)\mathbf{y}_n$. Since we do not know $s_n$, we replace that term with the estimate $p_n$. The iterative step of the LMS method is

$$\mathbf{f}_\tau = \mathbf{f}_{\tau-1} + \mu_\tau(p_\tau - \mathbf{y}_\tau^\dagger \mathbf{f}_{\tau-1})\mathbf{y}_\tau, \qquad (20.5)$$

for $L \leq \tau \leq N$. Notice that it is the approximate gradient of the function $|s_\tau - \hat{s}_\tau|^2$ that is used at this step, in order to involve all the data $z_0, ..., z_N$ as we iterate from $\tau = L$ to $\tau = N$. We illustrate the use of this method in adaptive interference cancellation.

## 20.10.2   Adaptive Interference Cancellation (AIC)

Adaptive interference cancellation (AIC) [161] is used to suppress a dominant noise component $v_n$ in the discrete sequence $z_n = s_n + v_n$. It is assumed that we have available a good estimate $q_n$ of $v_n$. The main idea is to switch the roles of signal and noise in the adaptive LMS method and design a filter to estimate $v_n$. Once we have that estimate, we subtract it from $z_n$ to get our estimate of $s_n$.

In the role of $z_n$ we use

$$q_n = v_n + \epsilon_n,$$

where $\epsilon_n$ denotes a low-level error component. In the role of $p_n$, we take $z_n$, which is approximately $v_n$, since the signal $s_n$ is much lower than the noise $v_n$. Then, $\mathbf{y}_n^\dagger = (q_n, q_{n-1}, ..., q_{n-L})$. The iterative step used to find the filter $\mathbf{f}$ is then

$$\mathbf{f}_\tau = \mathbf{f}_{\tau-1} + \mu_\tau(z_\tau - \mathbf{y}_\tau^\dagger \mathbf{f}_{\tau-1})\mathbf{y}_\tau,$$

for $L \leq \tau \leq N$. When the iterative process has converged to $\mathbf{f}$, we take as our estimate of $s_n$

$$\hat{s}_n = z_n - \sum_{k=0}^{L} f_k q_{n-k}.$$

It has been suggested that this procedure be used in computerized tomography to correct artifacts due to patient motion [66].

## 20.10.3   Recursive Least Squares (RLS)

An alternative to the LMS method is to find the least-squares solution of the system of $N - L + 1$ linear equations

$$p_n = \sum_{k=0}^{L} f_k z_{n-k},$$

for $L \leq n \leq N$. The *recursive least squares* (RLS) method is a recursive approach to solving this system.

For $L \leq \tau \leq N$ let $Z_\tau$ be the matrix whose rows are $\mathbf{y}_n^\dagger$ for $n = L, ..., \tau$, $\mathbf{p}_\tau^T = (p_L, p_{L+1}, ..., p_\tau)$ and $Q_\tau = Z_\tau^\dagger Z_\tau$. The least-squares solution we seek is

$$\mathbf{f} = Q_N^{-1} Z_N^\dagger \mathbf{p}_N.$$

**Ex. 20.4** *Show that $Q_\tau = Q_{\tau-1} + \mathbf{y}_\tau \mathbf{y}_\tau^\dagger$, for $L < \tau \leq N$.*

**Ex. 20.5** *Use the matrix-inversion identity in Equation (20.1) to write $Q_\tau^{-1}$ in terms of $Q_{\tau-1}^{-1}$.*

**Ex. 20.6** *Using the previous exercise, show that the desired least-squares solution $\mathbf{f}$ is $\mathbf{f} = \mathbf{f}_N$, where, for $L \leq \tau \leq N$ we let*

$$\mathbf{f}_\tau = \mathbf{f}_{\tau-1} + \left( \frac{p_\tau - \mathbf{y}_\tau^\dagger \mathbf{f}_{\tau-1}}{1 + \mathbf{y}_\tau^\dagger Q_{\tau-1}^{-1} \mathbf{y}_\tau} \right) Q_{\tau-1}^{-1} \mathbf{y}_\tau.$$

Comparing this iterative step with that given by Equation (20.5), we see that the former gives an explicit value for $\mu_\tau$ and uses $Q_{\tau-1}^{-1} \mathbf{y}_\tau$ instead of $\mathbf{y}_\tau$ as the direction vector for the iterative step. The RMS iteration produces a more accurate estimate of the FCWF than does the LMS method, but requires more computation.

# Chapter 21

## Matrix Theory

## 21.1 Chapter Summary

Matrices and their algebraic properties play an ever-increasing role in signal processing. In this chapter we outline the most important of these properties. The notation associated with matrix and vector algebra is designed to reduce the number of things we have to think about as we perform our calculations. This notation can be extended to multi-variable calculus, as we also show in this chapter.

## 21.2   Matrix Inverses

A square matrix $A$ is said to have inverse $A^{-1}$ provided that

$$AA^{-1} = A^{-1}A = I,$$

where $I$ is the identity matrix. The 2 by 2 matrix $A = \begin{bmatrix} a & b \\ c & d \end{bmatrix}$ has an inverse

$$A^{-1} = \frac{1}{ad - bc} \begin{bmatrix} d & -b \\ -c & a \end{bmatrix}$$

whenever the *determinant* of $A$, $\det(A) = ad - bc$ is not zero. More generally, associated with every complex square matrix is the complex number called its determinant, which is obtained from the entries of the matrix using formulas that can be found in any text on linear algebra. The significance of the determinant is that the matrix is invertible if and only if its determinant is not zero. This is of more theoretical than practical importance, since no computer can tell when a number is precisely zero. A matrix $A$ that is not square cannot have an inverse, but does have a *pseudo-inverse*, which is found using the singular-value decomposition.

## 21.3   Basic Linear Algebra

In this section we discuss systems of linear equations, Gaussian elimination, and the notions of basic and non-basic variables.

### 21.3.1   Bases and Dimension

The notions of a basis and of linear independence are fundamental in linear algebra. Let $V$ be a vector space.

**Definition 21.1** *A collection of vectors $\{u^1, ..., u^N\}$ in $V$ is* linearly independent *if there is no choice of scalars $\alpha_1, ..., \alpha_N$, not all zero, such that*

$$0 = \alpha_1 u^1 + ... + \alpha_N u^N.$$

**Definition 21.2** *The* span *of a collection of vectors $\{u^1, ..., u^N\}$ in $V$ is the set of all vectors $x$ that can be written as linear combinations of the $u^n$; that is, for which there are scalars $c_1, ..., c_N$, such that*

$$x = c_1 u^1 + ... + c_N u^N.$$

**Definition 21.3** *A collection of vectors* $\{w^1, ..., w^N\}$ *in* $\mathcal{V}$ *is called a* spanning set *for a subspace* $S$ *if the set* $S$ *is their span.*

**Definition 21.4** *A collection of vectors* $\{u^1, ..., u^N\}$ *in* $\mathcal{V}$ *is called a* basis *for a subspace* $S$ *if the collection is linearly independent and* $S$ *is their span.*

**Definition 21.5** *A collection of vectors* $\{u^1, ..., u^N\}$ *in an inner product space* $\mathcal{V}$ *is called* orthonormal *if* $\|u^n\|_2 = 1$, *for all* $n$, *and* $\langle u^m, u^n \rangle = 0$, *for* $m \neq n$.

Suppose that $S$ is a subspace of $\mathcal{V}$, that $\{w^1, ..., w^N\}$ is a spanning set for $S$, and $\{u^1, ..., u^M\}$ is a linearly independent subset of $S$. Beginning with $w^1$, we augment the set $\{u^1, ..., u^M\}$ with $w^j$ if $w^j$ is not in the span of the $u^m$ and the $w^k$ previously included. At the end of this process, we have a linearly independent spanning set, and therefore, a basis, for $S$ (Why?). Similarly, beginning with $w^1$, we remove $w^j$ from the set $\{w^1, ..., w^N\}$ if $w^j$ is a linear combination of the $w^k$, $k = 1, ..., j - 1$. In this way we obtain a linearly independent set that spans $S$, hence another basis for $S$. The following lemma will allow us to prove that all bases for a subspace $S$ have the same number of elements.

**Lemma 21.1** *Let* $W = \{w^1, ..., w^N\}$ *be a spanning set for a subspace* $S$ *in* $\mathbb{R}^I$, *and* $V = \{v^1, ..., v^M\}$ *a linearly independent subset of* $S$. *Then* $M \leq N$.

**Proof:** Suppose that $M > N$. Let $B_0 = \{w^1, ..., w^N\}$. To obtain the set $B_1$, form the set $C_1 = \{v^1, w^1, ..., w^N\}$ and remove the first member of $C_1$ that is a linear combination of members of $C_1$ that occur to its left in the listing; since $v^1$ has no members to its left, it is not removed. Since $W$ is a spanning set, $v^1$ is a linear combination of the members of $W$, so that some member of $W$ is a linear combination of $v^1$ and the members of $W$ that precede it in the list; remove the first member of $W$ for which this is true.

We note that the set $B_1$ is a spanning set for $S$ and has $N$ members. Having obtained the spanning set $B_k$, with $N$ members and whose first $k$ members are $v^k, ..., v^1$, we form the set $C_{k+1} = B_k \cup \{v^{k+1}\}$, listing the members so that the first $k + 1$ of them are $\{v^{k+1}, v^k, ..., v^1\}$. To get the set $B_{k+1}$ we remove the first member of $C_{k+1}$ that is a linear combination of the members to its left; there must be one, since $B_k$ is a spanning set, and so $v^{k+1}$ is a linear combination of the members of $B_k$. Since the set $V$ is linearly independent, the member removed is from the set $W$. Continuing in this fashion, we obtain a sequence of spanning sets $B_1, ..., B_N$, each with $N$ members. The set $B_N$ is $B_N = \{v^1, ..., v^N\}$ and $v^{N+1}$ must then be a linear combination of the members of $B_N$, which contradicts the linear independence of $V$. ∎

**Corollary 21.1** *Every basis for a subspace S has the same number of elements.*

**Ex. 21.1** *Let $W = \{w^1, ..., w^N\}$ be a spanning set for a subspace S in $\mathbb{R}^I$, and $V = \{v^1, ..., v^M\}$ a linearly independent subset of S. Let A be the matrix whose columns are the $v^m$, B the matrix whose columns are the $w^n$. Show that there is an N by M matrix C such that $A = BC$. Prove Lemma 21.1 by showing that, if $M > N$, then there is a non-zero vector x with $Cx = Ax = 0$.*

**Definition 21.6** *The dimension of a subspace S is the number of elements in any basis.*

**Lemma 21.2** *For any matrix A, the maximum number of linearly independent rows equals the maximum number of linearly independent columns.*

**Proof:** Suppose that $A$ is an $I$ by $J$ matrix, and that $K \leq J$ is the maximum number of linearly independent columns of $A$. Select $K$ linearly independent columns of $A$ and use them as the $K$ columns of an $I$ by $K$ matrix $U$. Since every column of $A$ must be a linear combination of these $K$ selected ones, there is a $K$ by $J$ matrix $M$ such that $A = UM$. From $A^T = M^T U^T$ we conclude that every column of $A^T$ is a linear combination of the $K$ columns of the matrix $M^T$. Therefore, there can be at most $K$ linearly independent columns of $A^T$. ∎

**Definition 21.7** *The rank of A is the maximum number of linearly independent rows or of linearly independent columns of A.*

### 21.3.2 Systems of Linear Equations

Consider the system of three linear equations in five unknowns given by

$$x_1 + 2x_2 + 2x_4 + x_5 = 0$$
$$-x_1 - x_2 + x_3 + x_4 = 0$$
$$x_1 + 2x_2 - 3x_3 - x_4 - 2x_5 = 0.$$

This system can be written in matrix form as $Ax = 0$, with $A$ the coefficient matrix

$$A = \begin{bmatrix} 1 & 2 & 0 & 2 & 1 \\ -1 & -1 & 1 & 1 & 0 \\ 1 & 2 & -3 & -1 & -2 \end{bmatrix},$$

and $x = (x_1, x_2, x_3, x_4, x_5)^T$. Applying Gaussian elimination to this system, we obtain a second, simpler, system with the same solutions:

$$x_1 - 2x_4 + x_5 = 0$$
$$x_2 + 2x_4 = 0$$
$$x_3 + x_4 + x_5 = 0.$$

From this simpler system we see that the variables $x_4$ and $x_5$ can be freely chosen, with the other three variables then determined by this system of equations. The variables $x_4$ and $x_5$ are then independent, the others dependent. The variables $x_1, x_2$ and $x_3$ are then called *basic variables*. To obtain a basis of solutions we can let $x_4 = 1$ and $x_5 = 0$, obtaining the solution $x = (2, -2, -1, 1, 0)^T$, and then choose $x_4 = 0$ and $x_5 = 1$ to get the solution $x = (-1, 0, -1, 0, 1)^T$. Every solution to $Ax = 0$ is then a linear combination of these two solutions. Notice that which variables are basic and which are non-basic is somewhat arbitrary, in that we could have chosen as the non-basic variables any two whose columns are independent.

Having decided that $x_4$ and $x_5$ are the non-basic variables, we can write the original matrix $A$ as $A = \begin{bmatrix} B & N \end{bmatrix}$, where $B$ is the square invertible matrix

$$B = \begin{bmatrix} 1 & 2 & 0 \\ -1 & -1 & 1 \\ 1 & 2 & -3 \end{bmatrix},$$

and $N$ is the matrix

$$N = \begin{bmatrix} 2 & 1 \\ 1 & 0 \\ -1 & -2 \end{bmatrix}.$$

With $x_B = (x_1, x_2, x_3)^T$ and $x_N = (x_4, x_5)^T$ we can write

$$Ax = Bx_B + Nx_N = 0,$$

so that

$$x_B = -B^{-1}Nx_N.$$

## 21.3.3   Real and Complex Systems of Linear Equations

A system $Ax = b$ of linear equations is called a *complex system*, or a *real system* if the entries of $A$, $x$ and $b$ are complex, or real, respectively. For any matrix $A$, we denote by $A^T$ and $A^\dagger$ the transpose and conjugate transpose of $A$, respectively.

Any complex system can be converted to a real system in the following way. A complex matrix $A$ can be written as $A = A_1 + iA_2$, where $A_1$ and

$A_2$ are real matrices and $i = \sqrt{-1}$. Similarly, $x = x^1 + ix^2$ and $b = b^1 + ib^2$, where $x^1, x^2, b^1$ and $b^2$ are real vectors. Denote by $\tilde{A}$ the real matrix

$$\tilde{A} = \begin{bmatrix} A_1 & -A_2 \\ A_2 & A_1 \end{bmatrix},$$

by $\tilde{x}$ the real vector

$$\tilde{x} = \begin{bmatrix} x^1 \\ x^2 \end{bmatrix},$$

and by $\tilde{b}$ the real vector

$$\tilde{b} = \begin{bmatrix} b^1 \\ b^2 \end{bmatrix}.$$

Then $x$ satisfies the system $Ax = b$ if and only if $\tilde{x}$ satisfies the system $\tilde{A}\tilde{x} = \tilde{b}$.

**Definition 21.8** *A square matrix $A$ is* symmetric *if $A^T = A$ and* Hermitian *if $A^\dagger = A$.*

**Definition 21.9** *A non-zero vector $x$ is said to be an* eigenvector *of the square matrix $A$ if there is a scalar $\lambda$ such that $Ax = \lambda x$. Then $\lambda$ is said to be an* eigenvalue *of $A$.*

If $x$ is an eigenvector of $A$ with eigenvalue $\lambda$, then the matrix $A - \lambda I$ has no inverse, so its determinant is zero; here $I$ is the identity matrix with ones on the main diagonal and zeros elsewhere. Solving for the roots of the determinant is one way to calculate the eigenvalues of $A$. For example, the eigenvalues of the Hermitian matrix

$$B = \begin{bmatrix} 1 & 2+i \\ 2-i & 1 \end{bmatrix}$$

are $\lambda = 1 + \sqrt{5}$ and $\lambda = 1 - \sqrt{5}$, with corresponding eigenvectors $u = (\sqrt{5}, 2 - i)^T$ and $v = (\sqrt{5}, i - 2)^T$, respectively. Then $\tilde{B}$ has the same eigenvalues, but both with multiplicity two. Finally, the associated eigenvectors of $\tilde{B}$ are

$$\begin{bmatrix} u^1 \\ u^2 \end{bmatrix},$$

and

$$\begin{bmatrix} -u^2 \\ u^1 \end{bmatrix},$$

for $\lambda = 1 + \sqrt{5}$, and

$$\begin{bmatrix} v^1 \\ v^2 \end{bmatrix},$$

and

$$\begin{bmatrix} -v^2 \\ v^1 \end{bmatrix},$$

for $\lambda = 1 - \sqrt{5}$.

## 21.4  Solutions of Under-determined Systems of Linear Equations

Suppose that $A\mathbf{x} = \mathbf{b}$ is a consistent linear system of $M$ equations in $N$ unknowns, where $M < N$. Then there are infinitely many solutions. A standard procedure in such cases is to find that solution $\mathbf{x}$ having the smallest norm

$$||\mathbf{x}|| = \sqrt{\sum_{n=1}^{N} |x_n|^2}.$$

As we shall see shortly, the *minimum-norm* solution of $A\mathbf{x} = \mathbf{b}$ is a vector of the form $\mathbf{x} = A^\dagger \mathbf{z}$, where $A^\dagger$ denotes the conjugate transpose of the matrix $A$. Then $A\mathbf{x} = \mathbf{b}$ becomes $AA^\dagger \mathbf{z} = \mathbf{b}$. Typically, $(AA^\dagger)^{-1}$ will exist, and we get $\mathbf{z} = (AA^\dagger)^{-1}\mathbf{b}$, from which it follows that the minimum-norm solution is $\mathbf{x} = A^\dagger(AA^\dagger)^{-1}\mathbf{b}$. When $M$ and $N$ are not too large, forming the matrix $AA^\dagger$ and solving for $\mathbf{z}$ is not prohibitively expensive and time-consuming. However, in image processing the vector $\mathbf{x}$ is often a vectorization of a two-dimensional (or even three-dimensional) image and $M$ and $N$ can be on the order of tens of thousands or more. The ART algorithm gives us a fast method for finding the minimum-norm solution without computing $AA^\dagger$; see [84] and [42].

We begin by proving that the minimum-norm solution of $A\mathbf{x} = \mathbf{b}$ has the form $\mathbf{x} = A^\dagger \mathbf{z}$ for some $M$-dimensional complex vector $\mathbf{z}$.

Let the *null space* of the matrix $A$ be all $N$-dimensional complex vectors $\mathbf{w}$ with $A\mathbf{w} = \mathbf{0}$. If $A\mathbf{x} = \mathbf{b}$ then $A(\mathbf{x} + \mathbf{w}) = \mathbf{b}$ for all $\mathbf{w}$ in the null space of $A$. If $\mathbf{x} = A^\dagger \mathbf{z}$ and $\mathbf{w}$ is in the null space of $A$, then

$$
\begin{aligned}
||\mathbf{x} + \mathbf{w}||^2 &= ||A^\dagger \mathbf{z} + \mathbf{w}||^2 = (A^\dagger \mathbf{z} + \mathbf{w})^\dagger (A^\dagger \mathbf{z} + \mathbf{w}) \\
&= (A^\dagger \mathbf{z})^\dagger (A^\dagger \mathbf{z}) + (A^\dagger \mathbf{z})^\dagger \mathbf{w} + \mathbf{w}^\dagger (A^\dagger \mathbf{z}) + \mathbf{w}^\dagger \mathbf{w} \\
&= ||A^\dagger \mathbf{z}||^2 + (A^\dagger \mathbf{z})^\dagger \mathbf{w} + \mathbf{w}^\dagger (A^\dagger \mathbf{z}) + ||\mathbf{w}||^2 \\
&= ||A^\dagger \mathbf{z}||^2 + ||\mathbf{w}||^2,
\end{aligned}
$$

since

$$\mathbf{w}^\dagger (A^\dagger \mathbf{z}) = (A\mathbf{w})^\dagger \mathbf{z} = \mathbf{0}^\dagger \mathbf{z} = 0$$

and

$$(A^\dagger \mathbf{z})^\dagger \mathbf{w} = \mathbf{z}^\dagger A\mathbf{w} = \mathbf{z}^\dagger \mathbf{0} = 0.$$

Therefore, $||\mathbf{x} + \mathbf{w}|| = ||A^\dagger \mathbf{z} + \mathbf{w}|| > ||A^\dagger \mathbf{z}|| = ||\mathbf{x}||$ unless $\mathbf{w} = \mathbf{0}$.

**Ex. 21.2** *Show that if* $\mathbf{z} = (z_1, ..., z_N)^T$ *is a column vector with complex entries and* $H = H^\dagger$ *is an $N$ by $N$ Hermitian matrix with complex entries* *then the quadratic form* $\mathbf{z}^\dagger H\mathbf{z}$ *is a real number. Show that the quadratic form* $\mathbf{z}^\dagger H\mathbf{z}$ *can be calculated using only real numbers. Let* $\mathbf{z} = \mathbf{x} + i\mathbf{y}$, *with* $\mathbf{x}$ *and* $\mathbf{y}$ *real vectors and let* $H = A + iB$, *where* $A$ *and* $B$ *are real matrices. Then show that* $A^T = A$, $B^T = -B$, $\mathbf{x}^T B\mathbf{x} = 0$ *and finally,*

$$\mathbf{z}^\dagger H\mathbf{z} = \begin{bmatrix} \mathbf{x}^T & \mathbf{y}^T \end{bmatrix} \begin{bmatrix} A & -B \\ B & A \end{bmatrix} \begin{bmatrix} \mathbf{x} \\ \mathbf{y} \end{bmatrix}.$$

*Use the fact that* $\mathbf{z}^\dagger H\mathbf{z}$ *is real for every vector* $\mathbf{z}$ *to conclude that the eigenvalues of* $H$ *are real.*

---

## 21.5    Eigenvalues and Eigenvectors

Given $N$ by $N$ complex matrix $A$, we say that a complex number $\lambda$ is an *eigenvalue* of $A$ if there is a nonzero vector $\mathbf{u}$ with $A\mathbf{u} = \lambda\mathbf{u}$. The column vector $\mathbf{u}$ is then called an *eigenvector* of $A$ associated with eigenvalue $\lambda$; clearly, if $\mathbf{u}$ is an eigenvector of $A$, then so is $c\mathbf{u}$, for any constant $c \neq 0$. If $\lambda$ is an eigenvalue of $A$, then the matrix $A - \lambda I$ fails to have an inverse, since $(A - \lambda I)\mathbf{u} = \mathbf{0}$ but $\mathbf{u} \neq \mathbf{0}$. If we treat $\lambda$ as a variable and compute the determinant of $A - \lambda I$, we obtain a polynomial of degree $N$ in $\lambda$. Its roots $\lambda_1, ..., \lambda_N$ are then the eigenvalues of $A$. If $||\mathbf{u}||^2 = \mathbf{u}^\dagger\mathbf{u} = 1$ then $\mathbf{u}^\dagger A\mathbf{u} = \lambda\mathbf{u}^\dagger\mathbf{u} = \lambda$.

It can be shown that it is possible to find a set of $N$ mutually orthogonal eigenvectors of the Hermitian matrix $H$; call them $\{\mathbf{u}^1, ..., \mathbf{u}^N\}$. The matrix $H$ can then be written as

$$H = \sum_{n=1}^{N} \lambda_n \mathbf{u}^n (\mathbf{u}^n)^\dagger,$$

a linear superposition of the *dyad* matrices $\mathbf{u}^n (\mathbf{u}^n)^\dagger$. We can also write $H = ULU^\dagger$, where $U$ is the matrix whose $n$th column is the column vector $\mathbf{u}^n$ and $L$ is the diagonal matrix with the eigenvalues down the main diagonal and zero elsewhere.

The matrix $H$ is invertible if and only if none of the $\lambda$ are zero and its inverse is

$$H^{-1} = \sum_{n=1}^{N} \lambda_n^{-1} \mathbf{u}^n (\mathbf{u}^n)^\dagger.$$

We also have $H^{-1} = UL^{-1}U^\dagger$.

A Hermitian matrix $Q$ is said to be nonnegative definite (positive definite) if all the eigenvalues of $Q$ are nonnegative (positive). The matrix $Q$ is a nonnegative-definite matrix if and only if there is another matrix $C$ such that $Q = C^\dagger C$. Since the eigenvalues of $Q$ are nonnegative, the diagonal matrix $L$ has a square root, $\sqrt{L}$. Using the fact that $U^\dagger U = I$, we have

$$Q = ULU^\dagger = U\sqrt{L}U^\dagger U\sqrt{L}U^\dagger;$$

we then take $C = U\sqrt{L}U^\dagger$, so $C^\dagger = C$. Then $\mathbf{z}^\dagger Q\mathbf{z} = \mathbf{z}^\dagger C^\dagger C\mathbf{z} = ||C\mathbf{z}||^2$, so that $Q$ is positive definite if and only if $C$ is invertible.

**Ex. 21.3** *Let $A$ be an $M$ by $N$ matrix with complex entries. View $A$ as a linear function with domain $\mathbb{C}^N$, the space of all $N$-dimensional complex column vectors, and range contained within $\mathbb{C}^M$, via the expression $A(\mathbf{x}) = A\mathbf{x}$. Suppose that $M > N$. The range of $A$, denoted $R(A)$, cannot be all of $\mathbb{C}^M$. Show that every vector $\mathbf{z}$ in $\mathbb{C}^M$ can be written uniquely in the form $\mathbf{z} = A\mathbf{x} + \mathbf{w}$, where $A^\dagger\mathbf{w} = \mathbf{0}$. Show that $||\mathbf{z}||^2 = ||A\mathbf{x}||^2 + ||\mathbf{w}||^2$, where $||\mathbf{z}||^2$ denotes the square of the norm of $\mathbf{z}$. Hint: If $\mathbf{z} = A\mathbf{x} + \mathbf{w}$ then consider $A^\dagger\mathbf{z}$. Assume $A^\dagger A$ is invertible.*

## 21.6 Vectorization of a Matrix

When the complex $M$ by $N$ matrix $A$ is stored in the computer it is usually *vectorized*; that is, the matrix

$$A = \begin{bmatrix} A_{11} & A_{12} & \dots & A_{1N} \\ A_{21} & A_{22} & \dots & A_{2N} \\ \cdot & & & \\ \cdot & & & \\ \cdot & & & \\ A_{M1} & A_{M2} & \dots & A_{MN} \end{bmatrix}$$

becomes

$$\mathbf{vec}(A) = (A_{11}, A_{21}, ..., A_{M1}, A_{12}, A_{22}, ..., A_{M2}, ..., A_{MN})^T.$$

**Ex. 21.4 (a)** *Show that the complex dot product $\mathbf{vec}(A) \cdot \mathbf{vec}(B) = \mathbf{vec}(B)^\dagger \mathbf{vec}(A)$ can be obtained by*

$$\mathbf{vec}(A) \cdot \mathbf{vec}(B) = \text{trace}(AB^\dagger) = tr(AB^\dagger),$$

*where, for a square matrix $C$, trace $(C)$ means the sum of the entries along the main diagonal of $C$. We can therefore use the trace to define an inner product between matrices: $< A, B > = \text{trace}(AB^\dagger)$.*

**(b)** *Show that* trace $(AA^\dagger) \geq 0$ *for all* $A$, *so that we can use the trace to define a norm on matrices:* $\|A\|^2 = $ trace $(AA^\dagger)$.

**Ex. 21.5** *Let* $B = ULD^\dagger$ *be an* $M$ *by* $N$ *matrix in diagonalized form; that is,* $L$ *is an* $M$ *by* $N$ *diagonal matrix with entries* $\lambda_1, ..., \lambda_K$ *on its main diagonal, where* $K = \min(M, N)$, *and* $U$ *and* $V$ *are square matrices. Let the n-th column of* $U$ *be denoted* $\mathbf{u}^n$ *and similarly for the columns of* $V$. *Such a diagonal decomposition occurs in the singular value decomposition (SVD). Show that we can write*

$$B = \lambda_1 \mathbf{u}^1 (\mathbf{v}^1)^\dagger + ... + \lambda_K \mathbf{u}^K (\mathbf{v}^K)^\dagger.$$

If $B$ is an $N$ by $N$ Hermitian matrix, then we can take $U = V$ and $K = M = N$, with the columns of $U$ the eigenvectors of $B$, normalized to have Euclidean norm equal to one, and the $\lambda_n$ to be the eigenvalues of $B$. In this case we may also assume that $U$ is a *unitary* matrix; that is, $UU^\dagger = U^\dagger U = I$, where $I$ denotes the identity matrix.

---

## 21.7    The Singular Value Decomposition of a Matrix

We have just seen that an $N$ by $N$ Hermitian matrix $H$ can be written in terms of its eigenvalues and eigenvectors as $H = ULU^\dagger$ or as

$$H = \sum_{n=1}^{N} \lambda_n \mathbf{u}^n (\mathbf{u}^n)^\dagger.$$

The *singular value decomposition* (SVD) is a similar result that applies to any rectangular matrix. It is an important tool in image compression and pseudo-inversion.

### 21.7.1    The SVD

Let $C$ be any $N$ by $K$ complex matrix. In presenting the SVD of $C$ we shall assume that $K \geq N$; the SVD of $C^\dagger$ will come from that of $C$. Let $A = C^\dagger C$ and $B = CC^\dagger$; we assume, reasonably, that $B$, the smaller of the two matrices, is invertible, so all the eigenvalues $\lambda_1, ..., \lambda_N$ of $B$ are positive. Then, write the eigenvalue/eigenvector decomposition of $B$ as $B = ULU^\dagger$.

**Ex. 21.6** *Show that the nonzero eigenvalues of* $A$ *and* $B$ *are the same.*

Let $V$ be the $K$ by $K$ matrix whose first $N$ columns are those of the matrix $C^\dagger U L^{-1/2}$ and whose remaining $K - N$ columns are any mutually orthogonal norm-one vectors that are all orthogonal to each of the first $N$ columns. Let $M$ be the $N$ by $K$ matrix with diagonal entries $M_{nn} = \sqrt{\lambda_n}$ for $n = 1, ..., N$ and whose remaining entries are zero. The nonzero entries of $M$, $\sqrt{\lambda_n}$, are called the *singular values* of $C$. The *singular value decomposition* (SVD) of $C$ is $C = U M V^\dagger$. The SVD of $C^\dagger$ is $C^\dagger = V M^T U^\dagger$.

**Ex. 21.7** *Show that* $U M V^\dagger = C$.

Using the SVD of $C$ we can write

$$C = \sum_{n=1}^{N} \sqrt{\lambda_n} \mathbf{u}^n (\mathbf{v}^n)^\dagger,$$

where $\mathbf{v}^n$ denotes the $n$th column of the matrix $V$.

In image processing, matrices such as $C$ are used to represent discrete two-dimensional images, with the entries of $C$ corresponding to the grey level or color at each pixel. It is common to find that most of the $N$ singular values of $C$ are nearly zero, so that $C$ can be written approximately as a sum of far fewer than $N$ dyads; this is SVD image compression.

## 21.7.2   An Application in Space Exploration

The *Galileo* was deployed from the space shuttle *Atlantis* on October 18, 1989. After a detour around Venus and back past Earth to pick up gravity-assisted speed, *Galileo* headed for Jupiter. Its mission included a study of Jupiter's moon Europa, and the plan was to send back one high-resolution photo per minute, at a rate of 134 KB per second, via a huge high-gain antenna, one with a high degree of directionality that can transmit most of the limited signal energy in a narrow beam. When the time came to open the antenna, it stuck. Without the pictures, the mission would be a failure.

There was a much smaller *low-gain* antenna on board, but the best transmission rate was going to be ten bits per second, and the directionality was much less. All that could be done from earth was to reprogram an old on-board computer to compress the pictures prior to transmission. The problem was that pictures could be taken much faster than they could be transmitted to earth; some way to store them prior to transmission was key. The original designers of the software had long since retired, but the engineers figured out a way to introduce state-of-the-art image compression algorithms into the computer. It happened that there was an ancient reel-to-reel storage device on board that was there only to serve as a backup for storing atmospheric data. Using this device and the compression methods, the engineers saved the mission [5].

### 21.7.3 Pseudo-Inversion

If $N \neq K$ then $C$ cannot have an inverse; it does, however, have a *pseudo-inverse*, $C^* = V M^* U^\dagger$, where $M^*$ is the matrix obtained from $M$ by taking the inverse of each of its nonzero entries and leaving the remaining zeros the same. The pseudo-inverse of $C^\dagger$ is

$$(C^\dagger)^* = (C^*)^\dagger = U(M^*)^T V^\dagger = U(M^\dagger)^* V^\dagger.$$

Some important properties of the pseudo-inverse are the following:

1. $CC^*C = C$,

2. $C^*CC^* = C^*$,

3. $(C^*C)^\dagger = C^*C$,

4. $(CC^*)^\dagger = CC^*$.

The pseudo-inverse of an arbitrary $I$ by $J$ matrix $G$ can be used in much the same way as the inverse of nonsingular matrices to find approximate or exact solutions of systems of equations $Gx = d$. The following examples illustrate this point.

**Ex. 21.8** *If $I > J$ the system $Gx = d$ probably has no exact solution. Show that whenever $G^\dagger G$ is invertible the pseudo-inverse of $G$ is $G^* = (G^\dagger G)^{-1} G^\dagger$ so that the vector $x = G^* d$ is the least-squares approximate solution.*

**Ex. 21.9** *If $I < J$ the system $Gx = d$ probably has infinitely many solutions. Show that whenever the matrix $GG^\dagger$ is invertible the pseudo-inverse of $G$ is $G^* = G^\dagger(GG^\dagger)^{-1}$, so that the vector $x = G^* d$ is the exact solution of $Gx = d$ closest to the origin; that is, it is the minimum-norm solution.*

---

## 21.8 Singular Values of Sparse Matrices

In image reconstruction from projections the $M$ by $N$ matrix $A$ is usually quite large and often $\epsilon$-sparse; that is, most of its elements do not exceed $\epsilon$ in absolute value, where $\epsilon$ denotes a small positive quantity. In transmission tomography each column of $A$ corresponds to a single pixel in the digitized image, while each row of $A$ corresponds to a line segment through the object, along which an x-ray beam has traveled. The entries of a given row of $A$ are nonzero only for those columns whose associated pixel lies on that line segment; clearly, most of the entries of any given row of $A$ will then be zero. In emission tomography the $I$ by $J$ nonnegative matrix $P$ has entries $P_{ij} \geq 0$; for each detector $i$ and pixel $j$, $P_{ij}$ is the probability that an emission at the $j$th pixel will be detected at the $i$th detector. When a detection is recorded at the $i$th detector, we want the likely source of the emission to be one of only a small number of pixels. For single photon emission tomography (SPECT), a lead collimator is used to permit detection of only those photons approaching the detector straight on. In positron emission tomography (PET), coincidence detection serves much the same purpose. In both cases the probabilities $P_{ij}$ will be zero (or nearly zero) for most combinations of $i$ and $j$. Such matrices are called *sparse* (or *almost sparse*). We discuss now a convenient estimate for the largest singular value of an almost sparse matrix $A$, which, for notational convenience only, we take to be real.

In [40] it was shown that if $A$ is normalized so that each row has length one, then the spectral radius of $A^T A$, which is the square of the largest singular value of $A$ itself, does not exceed the maximum number of nonzero elements in any column of $A$. A similar upper bound on $\rho(A^T A)$ can be obtained for non-normalized, $\epsilon$-sparse $A$.

Let $A$ be an $M$ by $N$ matrix. For each $n = 1, ..., N$, let $s_n > 0$ be the number of nonzero entries in the $n$th column of $A$, and let $s$ be the maximum of the $s_n$. Let $G$ be the $M$ by $N$ matrix with entries

$$G_{mn} = A_{mn} / \left( \sum_{l=1}^{N} s_l A_{ml}^2 \right)^{1/2}.$$

Lent has shown that the eigenvalues of the matrix $G^T G$ do not exceed one [107]. This result suggested the following proposition, whose proof was given in [40].

**Proposition 21.1** *Let $A$ be an $M$ by $N$ matrix. For each $m = 1, ..., M$ let $\nu_m = \sum_{n=1}^{N} A_{mn}^2 > 0$. For each $n = 1, ..., N$ let $\sigma_n = \sum_{m=1}^{M} e_{mn} \nu_m$, where $e_{mn} = 1$ if $A_{mn} \neq 0$ and $e_{mn} = 0$ otherwise. Let $\sigma$ denote the maximum of the $\sigma_n$. Then the eigenvalues of the matrix $A^T A$ do not exceed $\sigma$. If $A$ is normalized so that the Euclidean length of each of its rows is one, then the eigenvalues of $A^T A$ do not exceed $s$, the maximum number of nonzero elements in any column of $A$.*

**Proof:** For simplicity, we consider only the normalized case; the proof for the more general case is similar.

Let $A^T A \mathbf{v} = c\mathbf{v}$ for some nonzero vector $\mathbf{v}$. We show that $c \leq s$. We have $AA^T A\mathbf{v} = cA\mathbf{v}$ and so $\mathbf{w}^T AA^T \mathbf{w} = \mathbf{v}^T A^T AA^T A\mathbf{v} = c\mathbf{v}^T A^T A\mathbf{v} = c\mathbf{w}^T \mathbf{w}$, for $\mathbf{w} = A\mathbf{v}$. Then, with $e_{mn} = 1$ if $A_{mn} \neq 0$ and $e_{mn} = 0$ otherwise, we have

$$
\begin{aligned}
\left( \sum_{m=1}^{M} A_{mn} w_m \right)^2 &= \left( \sum_{m=1}^{M} A_{mn} e_{mn} w_m \right)^2 \\
&\leq \left( \sum_{m=1}^{M} A_{mn}^2 w_m^2 \right) \left( \sum_{m=1}^{M} e_{mn}^2 \right) \\
&= \left( \sum_{m=1}^{M} A_{mn}^2 w_m^2 \right) s_j \leq \left( \sum_{m=1}^{M} A_{mn}^2 w_m^2 \right) s.
\end{aligned}
$$

Therefore,

$$
\mathbf{w}^T AA^T \mathbf{w} = \sum_{n=1}^{N} \left( \sum_{m=1}^{M} A_{mn} w_m \right)^2 \leq \sum_{n=1}^{N} \left( \sum_{m=1}^{M} A_{mn}^2 w_m^2 \right) s,
$$

and

$$
\begin{aligned}
\mathbf{w}^T AA^T \mathbf{w} &= c \sum_{m=1}^{M} w_m^2 = c \sum_{m=1}^{M} w_m^2 \left( \sum_{n=1}^{N} A_{mn}^2 \right) \\
&= c \sum_{m=1}^{M} \sum_{n=1}^{N} w_m^2 A_{mn}^2.
\end{aligned}
$$

The result follows immediately. ∎

If we normalize $A$ so that its rows have length one, then the trace of the matrix $AA^T$ is $\mathrm{tr}(AA^T) = M$, which is also the sum of the eigenvalues of $A^T A$. Consequently, the maximum eigenvalue of $A^T A$ does not exceed $M$; Proposition 21.1 improves that upper bound considerably, if $A$ is sparse and so $s \ll M$.

In image reconstruction from projection data that includes scattering we often encounter matrices $A$ most of whose entries are small, if not exactly zero. A slight modification of the proof provides us with a useful upper bound for $L$, the largest eigenvalue of $A^T A$, in such cases. Assume that the rows of $A$ have length one. For $\epsilon > 0$ let $s$ be the largest number of entries in any column of $A$ whose magnitudes exceed $\epsilon$. Then we have

$$
L \leq s + MN\epsilon^2 + 2\epsilon(MNs)^{1/2}.
$$

The proof of this result is similar to that for Proposition 21.1.

## 21.9 Matrix and Vector Differentiation

As we saw previously, the least-squares approximate solution of $Ax = b$ is a vector $\hat{x}$ that minimizes the function $||Ax - b||$. In our discussion of band-limited extrapolation we showed that, for any nonnegative-definite matrix $Q$, the vector having norm one that maximizes the quadratic form $x^\dagger Qx$ is an eigenvector of $Q$ associated with the largest eigenvalue. In the chapter on best linear unbiased optimization we seek a matrix that minimizes a certain function. All of these examples involve what we can call *matrix-vector differentiation*, that is, the differentiation of a function with respect to a matrix or a vector. The gradient of a function of several variables is a well-known example and we begin there. Since there is some possibility of confusion, we adopt the notational convention that boldfaced symbols, such as $\mathbf{x}$, indicate a column vector, while $x$ denotes a scalar.

## 21.10 Differentiation with Respect to a Vector

Let $\mathbf{x} = (x_1, ..., x_N)^T$ be an $N$-dimensional real column vector. Let $z = f(\mathbf{x})$ be a real-valued function of the entries of $\mathbf{x}$. The derivative of $z$ with respect to $\mathbf{x}$, also called the *gradient* of $z$, is the column vector

$$\frac{\partial z}{\partial \mathbf{x}} = \mathbf{a} = (a_1, ..., a_N)^T$$

with entries

$$a_n = \frac{\partial z}{\partial x_n}.$$

**Ex. 21.10** *Let* $\mathbf{y}$ *be a fixed real column vector and* $z = f(\mathbf{x}) = \mathbf{y}^T\mathbf{x}$. *Show that*

$$\frac{\partial z}{\partial \mathbf{x}} = \mathbf{y}.$$

**Ex. 21.11** *Let* $Q$ *be a real symmetric nonnegative-definite matrix, and let* $z = f(\mathbf{x}) = \mathbf{x}^T Q\mathbf{x}$. *Show that the gradient of this quadratic form is*

$$\frac{\partial z}{\partial \mathbf{x}} = 2Q\mathbf{x}.$$

*Hint: Write* $Q$ *as a linear combination of dyads involving the eigenvectors.*

**Ex. 21.12** *Let* $z = ||A\mathbf{x} - \mathbf{b}||^2$. *Show that*

$$\frac{\partial z}{\partial \mathbf{x}} = 2A^T A\mathbf{x} - 2A^T \mathbf{b}.$$

*Hint: Use* $z = (A\mathbf{x} - \mathbf{b})^T (A\mathbf{x} - \mathbf{b})$.

We can also consider the second derivative of $z = f(\mathbf{x})$, which is the *Hessian matrix* of $z$

$$H = \frac{\partial^2 z}{\partial \mathbf{x}^2} = \nabla^2 f(\mathbf{x})$$

with entries

$$H_{mn} = \frac{\partial^2 z}{\partial x_m \partial x_n}.$$

If the entries of the vector $\mathbf{z} = (z_1, ..., z_M)^T$ are real-valued functions of the vector $\mathbf{x}$, the derivative of $\mathbf{z}$ is the matrix whose $m$th column is the derivative of the real-valued function $z_m$. This matrix is usually called the *Jacobian matrix* of $\mathbf{z}$. If $M = N$ the determinant of the Jacobian matrix is the *Jacobian*.

**Ex. 21.13** *Suppose* $(u, v) = (u(x, y), v(x, y))$ *is a change of variables from the Cartesian* $(x, y)$ *coordinate system to some other* $(u, v)$ *coordinate system. Let* $\mathbf{x} = (x, y)^T$ *and* $\mathbf{z} = (u(\mathbf{x}), v(\mathbf{x}))^T$.

**(a)** *Calculate the Jacobian for the rectangular coordinate system obtained by rotating the* $(x, y)$ *system through an angle of* $\theta$.

**(b)** *Calculate the Jacobian for the transformation from the* $(x, y)$ *system to polar coordinates.*

---

## 21.11    Differentiation with Respect to a Matrix

Now we consider real-valued functions $z = f(A)$ of a real matrix $A$. As an example, for square matrices $A$ we have

$$z = f(A) = \text{trace}\,(A) = \sum_{n=1}^{N} A_{nn},$$

the sum of the entries along the main diagonal of $A$.

The derivative of $z = f(A)$ is the matrix

$$\frac{\partial z}{\partial A} = B$$

whose entries are

$$B_{mn} = \frac{\partial z}{\partial A_{mn}}.$$

**Ex. 21.14** *Show that the derivative of* trace $(A)$ *is* $B = I$, *the identity matrix.*

**Ex. 21.15** *Show that the derivative of* $z = $ trace $(DAC)$ *with respect to* $A$ *is*

$$\frac{\partial z}{\partial A} = D^T C^T.$$

Consider the function $f$ defined for all $J$ by $J$ positive-definite symmetric matrices by

$$f(Q) = -\log\det(Q).$$

**Proposition 21.2** *The gradient of* $f(Q)$ *is* $g(Q) = -Q^{-1}$.

**Proof:** Let $\Delta Q$ be symmetric. Let $\gamma_j$, for $j = 1, 2, ..., J$, be the eigenvalues of the symmetric matrix $Q^{-1/2}(\Delta Q)Q^{-1/2}$. These $\gamma_j$ are then real and are also the eigenvalues of the matrix $Q^{-1}(\Delta Q)$. We shall consider $\|\Delta Q\|$ small, so we may safely assume that $1 + \gamma_j > 0$.

Note that

$$\langle Q^{-1}, \Delta Q \rangle = \sum_{j=1}^{J} \gamma_j,$$

since the trace of any square matrix is the sum of its eigenvalues. Then we have

$$f(Q + \Delta Q) - f(Q) = -\log\det(Q + \Delta Q) + \log\det(Q)$$

$$= -\log\det(I + Q^{-1}(\Delta Q)) = -\sum_{j=1}^{J} \log(1 + \gamma_j).$$

From the submultiplicativity of the Frobenius norm we have

$$\|Q^{-1}(\Delta Q)\|/\|Q^{-1}\| \le \|\Delta Q\| \le \|Q^{-1}(\Delta Q)\| \|Q\|.$$

Therefore, taking the limit as $\|\Delta Q\|$ goes to zero is equivalent to taking the limit as $\|\gamma\|$ goes to zero, where $\gamma$ is the vector whose entries are the $\gamma_j$.

To show that $g(Q) = -Q^{-1}$ note that

$$\limsup_{\|\Delta Q\| \to 0} \frac{f(Q + \Delta Q) - f(Q) - \langle -Q^{-1}, \Delta Q \rangle}{\|\Delta Q\|}$$

$$= \limsup_{\|\Delta Q\| \to 0} \frac{|-\log \det(Q + \Delta Q) + \log \det(Q) + \langle Q^{-1}, \Delta Q \rangle|}{\|\Delta Q\|}$$

$$\leq \limsup_{\|\gamma\| \to 0} \frac{\sum_{j=1}^{J} |\log(1 + \gamma_j) - \gamma_j|}{\|\gamma\|/\|Q^{-1}\|}$$

$$\leq \|Q^{-1}\| \sum_{j=1}^{J} \lim_{\gamma_j \to 0} \frac{\gamma_j - \log(1 + \gamma_j)}{|\gamma_j|} = 0.$$

∎

We note in passing that the derivative of $\det(DAC)$ with respect to $A$ is the matrix $\det(DAC)(A^{-1})^T$.

Although the trace is not independent of the order of the matrices in a product, it is independent of cyclic permutation of the factors:

$$\text{trace}(ABC) = \text{trace}(CAB) = \text{trace}(BCA).$$

Therefore, the trace is independent of the order for the product of two matrices:

$$\text{trace}(AB) = \text{trace}(BA).$$

From this fact we conclude that

$$\mathbf{x}^T \mathbf{x} = \text{trace}(\mathbf{x}^T \mathbf{x}) = \text{trace}(\mathbf{x}\mathbf{x}^T).$$

If $\mathbf{x}$ is a random vector with correlation matrix

$$R = E(\mathbf{x}\mathbf{x}^T),$$

then

$$E(\mathbf{x}^T \mathbf{x}) = E(\text{trace}(\mathbf{x}\mathbf{x}^T)) = \text{trace}(E(\mathbf{x}\mathbf{x}^T)) = \text{trace}(R).$$

**Ex. 21.16** *Let $z = \text{trace}(A^T C A)$. Show that the derivative of $z$ with respect to the matrix $A$ is*

$$\frac{\partial z}{\partial A} = CA + C^T A.$$

*Therefore, if $C = Q$ is symmetric, then the derivative is $2QA$.*

We have restricted the discussion here to real matrices and vectors. It often happens that we want to optimize a real quantity with respect to a complex vector. We can rewrite such quantities in terms of the real and imaginary parts of the complex values involved, to reduce everything to the real case just considered. For example, let $Q$ be a Hermitian matrix; then the quadratic form $\mathbf{k}^\dagger Q \mathbf{k}$ is real, for any complex vector $\mathbf{k}$. As we saw in Exercise 21.2, we can write the quadratic form entirely in terms of real matrices and vectors.

If $w = u + iv$ is a complex number with real part $u$ and imaginary part $v$, the function $z = f(w) = |w|^2$ is real-valued. The derivative of $z = f(w)$ with respect to the complex variable $w$ does not exist. When we write $z = u^2 + v^2$, we consider $z$ as a function of the real vector $\mathbf{x} = (u, v)^T$. The derivative of $z$ with respect to $\mathbf{x}$ is the vector $(2u, 2v)^T$.

Similarly, when we consider the real quadratic form $\mathbf{k}^\dagger Q \mathbf{k}$, we view each of the complex entries of the $N$ by 1 vector $\mathbf{k}$ as two real numbers forming a two-dimensional real vector. We then differentiate the quadratic form with respect to the $2N$ by 1 real vector formed from these real and imaginary parts. If we turn the resulting $2N$ by 1 real vector back into an $N$ by 1 complex vector, we get $2Q\mathbf{k}$ as the derivative; so, it appears as if the formula for differentiating in the real case carries over to the complex case.

## 21.12 Eigenvectors and Optimization

We can use these results concerning differentiation with respect to a vector to show that eigenvectors solve certain optimization problems.

Consider the problem of maximizing the quadratic form $\mathbf{x}^\dagger Q \mathbf{x}$, subject to $\mathbf{x}^\dagger \mathbf{x} = 1$; here the matrix $Q$ is Hermitian, positive-definite, so that all of its eigenvalues are positive. We use the Lagrange-multiplier approach, with the Lagrangian

$$L(\mathbf{x}, \lambda) = \mathbf{x}^\dagger Q \mathbf{x} - \lambda \mathbf{x}^\dagger \mathbf{x},$$

where the scalar variable $\lambda$ is the Lagrange multiplier. We differentiate $L(\mathbf{x}, \lambda)$ with respect to $\mathbf{x}$ and set the result equal to zero, obtaining

$$2Q\mathbf{x} - 2\lambda\mathbf{x} = 0,$$

or

$$Q\mathbf{x} = \lambda\mathbf{x}.$$

Therefore, $\mathbf{x}$ is an eigenvector of $Q$ and $\lambda$ is its eigenvalue. Since

$$\mathbf{x}^\dagger Q \mathbf{x} = \lambda \mathbf{x}^\dagger \mathbf{x} = \lambda,$$

we conclude that $\lambda = \lambda_1$, the largest eigenvalue of $Q$, and $\mathbf{x} = \mathbf{u}^1$, a norm-one eigenvector associated with $\lambda_1$.

Now consider the problem of maximizing $\mathbf{x}^\dagger Q \mathbf{x}$, subject to $\mathbf{x}^\dagger \mathbf{x} = 1$, and $\mathbf{x}^\dagger \mathbf{u}^1 = 0$. The Lagrangian is now

$$L(\mathbf{x}, \lambda, \alpha) = \mathbf{x}^\dagger Q \mathbf{x} - \lambda \mathbf{x}^\dagger \mathbf{x} - \alpha \mathbf{x}^\dagger \mathbf{u}^1.$$

Differentiating with respect to the vector $\mathbf{x}$ and setting the result equal to zero, we find that

$$2Q\mathbf{x} - 2\lambda \mathbf{x} - \alpha \mathbf{u}^1 = 0,$$

or

$$Q\mathbf{x} = \lambda \mathbf{x} + \beta \mathbf{u}^1,$$

for $\beta = \alpha/2$. But, we know that

$$(\mathbf{u}^1)^\dagger Q \mathbf{x} = \lambda (\mathbf{u}^1)^\dagger \mathbf{x} + \beta (\mathbf{u}^1)^\dagger \mathbf{u}^1 = \beta,$$

and

$$(\mathbf{u}^1)^\dagger Q \mathbf{x} = (Q\mathbf{u}^1)^\dagger \mathbf{x} = \lambda_1 (\mathbf{u}^1)^\dagger \mathbf{x} = 0,$$

so $\beta = 0$ and we have

$$Q\mathbf{x} = \lambda \mathbf{x}.$$

Since

$$\mathbf{x}^\dagger Q \mathbf{x} = \lambda,$$

we conclude that $\mathbf{x}$ is a norm-one eigenvector of $Q$ associated with the second-largest eigenvalue, $\lambda = \lambda_2$.

Continuing in this fashion, we can show that the norm-one eigenvector of $Q$ associated with the $n$th largest eigenvalue $\lambda_n$ maximizes the quadratic form $\mathbf{x}^\dagger Q \mathbf{x}$, subject to the constraints $\mathbf{x}^\dagger \mathbf{x} = 1$ and $\mathbf{x}^\dagger \mathbf{u}^m = 0$, for $m = 1, 2, ..., n - 1$.

# Chapter 22

## Compressed Sensing

## 22.1   Chapter Summary

Large amounts of data are often redundant and methods for compressing these data sets play an increasingly important role in a number of applications. The basic idea is to find ways to expand the data vector as a superposition of known vectors, so that only a few of the coefficients are nonzero. Much of the research in this field goes under the names *compressed sensing* and *compressed sampling* (CS) [67]. The key notion in CS is *sparseness*. The JPEG technology uses such an approach to represent images as a superposition of sinusoids and wavelets. For applications such as medical imaging, CS provides a means of reducing radiation dosage to the patient without sacrificing image quality. An important aspect of CS is finding sparse solutions of underdetermined systems of linear equations, which can often be accomplished by one-norm minimization. The best reference on CS to date is probably [16].

## 22.2  An Overview

In this section we "compress" Justin Romberg's article [133] that captures well the essence of compressed sensing and compressed sampling.

In classical data compression, the data vector is first transformed into a superposition of known basis "signals." If this basis is well chosen, then most of the information in the data will be concentrated in a few terms with relatively large coefficients; the representation of the data is then said to be *sparse* with respect to the chosen basis. The data compression is then achieved by discarding the terms with relatively small coefficients. For example, vectorized digital photographs often have a sparse representation with respect to a wavelet basis that measures intensity at different scales. In this traditional approach, a great deal of data is obtained by sampling at a very high rate, and then applying a transform and selection process to produce a much smaller vector of important coefficients. This procedure of gathering a large amount of data just to produce a much smaller vector of coefficients, seems wasteful. Compressed sensing attempts to avoid this wastefulness by integrating the compression step into the sampling process itself.

Normally, *sampling* means recording the values of an analog signal at some discrete set of points. Instead, CS devices provide initial data consisting of "correlations" that is, matched-filter values, between the signal and a set of known test signals. The big question is: How do we select this set of test signals? One might think that it is best to use as the test signals the members of the basis with respect to which the data is sparsely represented. Certainly, a small number of correlations would suffice to capture the signal, since the representation is known to be sparse. But we don't know which basis members are the important ones, and we would have to obtain a large number of correlations to find out which ones are the important ones, defeating the purpose of reducing the sampling effort. The solution, surprisingly, is to select the test signals at random, making the test signals quite unlike the basis vectors that produced the sparse representation.

At the final step, an algorithm is then applied to extract the desired information from this smaller set of correlations. Now we seek a solution that is consistent with the sampled data and also sparse, with respect to the original basis. With $b$ the vector of correlations, $x$ a vector of coefficients in the sparse-representation basis, and $A$ the matrix describing the linear transformation, we seek a maximally sparse solution of $Ax = b$. Finding such a maximally sparse solution is not easy; it is an NP-hard problem. It has been discovered that finding the minimum one-norm solution is often a reasonable substitute, which means that the computation can be converted to a linear-programming problem.

## 22.3   Compressed Sensing

The objective in CS is to exploit sparseness to reconstruct a vector $f$ in $\mathbb{R}^J$ from relatively few linear functional measurements [67].

Let $U = \{u^1, u^2, ..., u^J\}$ and $V = \{v^1, v^2, ..., v^J\}$ be two orthonormal bases for $\mathbb{R}^J$, with all members of $\mathbb{R}^J$ represented as column vectors. For $i = 1, 2, ..., J$, let

$$\mu_i = \max_{1 \le j \le J} \{|\langle u^i, v^j \rangle|\}$$

and

$$\mu(U, V) = \max_{1 \le i \le J} \mu_i.$$

We know from Cauchy's Inequality that

$$|\langle u^i, v^j \rangle| \le 1,$$

and from Parseval's Equation

$$\sum_{j=1}^J |\langle u^i, v^j \rangle|^2 = ||u^i||^2 = 1.$$

Therefore, we have

$$\frac{1}{\sqrt{J}} \le \mu(U, V) \le 1.$$

The quantity $\mu(U, V)$ is the *coherence* measure of the two bases; the closer $\mu(U, V)$ is to the lower bound of $\frac{1}{\sqrt{J}}$, the more *incoherent* the two bases are. We give an example of incoherent bases for $\mathbb{C}^J$.

Let $U = \{u^1, u^2, ..., u^J\}$ be the usual orthonormal basis for $\mathbb{C}^J$, where all the entries of $u^j$ are zero, except that $u_j^j = 1$. Let $V = \{v^1, v^2, ..., v^J\}$ be the Fourier basis, with the $k$th entry of $v^j$ given by

$$v_k^j = \frac{1}{\sqrt{J}} e^{2\pi i k j / J}.$$

Then it is easy to show that $\mu(U, V) = \frac{1}{\sqrt{J}}$. Clearly, each vector $u^j$ has a maximally sparse representation in the $U$ basis, but not in the $V$ basis. Similarly, each $v^j$ has a maximally sparse representation in the $V$ basis, but not in the $U$ basis. When $J$ is large, we may well want to estimate the index $j$ from the measurement of relatively few coefficients of $u^j$ in the $V$-basis representation. This is compressed sampling.

Let $f$ be a fixed member of $\mathbb{R}^J$; we expand $f$ in the $V$ basis as

$$f = x_1 v^1 + x_2 v^2 + ... + x_J v^J.$$

We say that the coefficient vector $x = (x_1, ..., x_J)$ is $s$-sparse if $s$ is the number of nonzero $x_j$.

If $s$ is small, most of the $x_j$ are zero, but since we do not know which ones these are, we would have to compute all the linear functional values

$$x_j = \langle f, v^j \rangle$$

to recover $f$ exactly. In fact, the smaller $s$ is, the harder it would be to learn anything from randomly selected $x_j$, since most would be zero. The idea in CS is to obtain measurements of $f$ with members of a different orthonormal basis, which we call the $U$ basis. If the members of $U$ are very much like the members of $V$, then nothing is gained. But, if the members of $U$ are quite unlike the members of $V$, then each inner product measurement

$$y_i = \langle f, u^i \rangle = f^T u^i$$

should tell us something about $f$. If the two bases are sufficiently inco-herent, then relatively few $y_i$ values should tell us quite a bit about $f$. Specifically, we have the following result due to Candès and Romberg [46]: suppose the coefficient vector $x$ for representing $f$ in the $V$ basis is $s$-sparse. Select uniformly randomly $I \leq J$ members of the $U$ basis and compute the measurements $y_i = \langle f, u^i \rangle$. Then, if $I$ is sufficiently large, it is highly prob-able that $z = x$ also solves the problem of minimizing the one-norm

$$||z||_1 = |z_1| + |z_2| + ... + |z_J|,$$

subject to the conditions

$$y_i = \langle g, u^i \rangle = g^T u^i,$$

for those $M$ randomly selected $u^i$, where

$$g = z_1 v^1 + z_2 v^2 + ... + z_J v^J.$$

The smaller $\mu(U, V)$ is, the smaller the $I$ is permitted to be without reduc-ing the probability of perfect reconstruction.

---

## 22.4   Sparse Solutions

Suppose that $A$ is a real $I$ by $J$ matrix, with $I < J$, and that the linear system $Ax = b$ has infinitely many solutions. For any vector $x$, we define the *support* of $x$ to be the subset $S$ of $\{1, 2, ..., J\}$ consisting of those $j$ for which the entries $x_j \neq 0$. For any under-determined system $Ax = b$, there will, of course, be at least one solution of minimum support, that is, for which $|S|$, the size of the support set $S$, is minimum. However, finding such a maximally sparse solution requires combinatorial optimization, and is known to be computationally difficult. It is important, therefore, to have a computationally tractable method for finding maximally sparse solutions. The discussion in this section is based on [16].

## 22.4.1 Maximally Sparse Solutions

Consider the problem $P_0$: among all solutions $x$ of the consistent system $b = Ax$, find one, call it $\hat{x}$, that is maximally sparse, that is, has the minimum number of nonzero entries. Obviously, there will be at least one such solution having minimal support, but finding one, however, is a combinatorial optimization problem and is generally NP-hard. For notational convenience, we denote by $\|x\|_0$ the number of nonzero entries of $x$.

There are two basic questions concerning the problem $P_0$:

1. Can uniqueness of the solution be claimed? Under what conditions?

2. If a candidate for the solution is available, is there a simple test to determine if it is, in fact, a solution?

**Definition 22.1** *Let $A$ be an $I$ by $J$ matrix, with $I < J$. The* spark *of $A$ is the smallest number of linearly dependent columns.*

We denote the spark of $A$ by $sp(A)$. The definition of the spark of $A$ is superficially similar to that of the rank of $A$, but the spark is a more difficult quantity to calculate. Notice that, if we change the word "columns" to "rows" in the definition, we may get a different number. For example, the 5 by 6 matrix

$$A = \begin{bmatrix} 1 & 0 & 0 & 1 & 0 & 0 \\ 0 & 1 & 0 & 0 & 1 & 0 \\ 0 & 0 & 1 & 0 & 0 & 1 \\ 0 & 0 & 0 & 0 & 0 & 0 \\ 0 & 0 & 0 & 0 & 0 & 0 \end{bmatrix}$$

has a rank of 3 and a spark of 2, although the smallest number of linearly dependent rows is 1. The rank of the matrix

$$B = \begin{bmatrix} 1 & 0 & 0 & 0 & 0 & 1 \\ 0 & 1 & 0 & 0 & 0 & 1 \\ 0 & 0 & 1 & 0 & 0 & 1 \\ 0 & 0 & 0 & 1 & 0 & 1 \\ 0 & 0 & 0 & 0 & 1 & 1 \end{bmatrix}$$

is 5, and the spark is 6, while the rank of the matrix

$$C = \begin{bmatrix} 1 & 0 & 0 & 0 & 0 & 1 \\ 0 & 1 & 0 & 0 & 0 & 0 \\ 0 & 0 & 1 & 0 & 0 & 0 \\ 0 & 0 & 0 & 1 & 0 & 0 \\ 0 & 0 & 0 & 0 & 1 & 0 \end{bmatrix}$$

is 5 and its spark is 2. The spark is an important notion when seeking sparse solutions of $Ax = b$, where we assume that $I < J$. The spark is not defined for matrices with $I \geq J$.

The following theorem is not difficult to prove.

**Theorem 22.1** *If $Ax = b$ and $\|x\|_0 < sp(A)/2$, then $x$ solves $P_0$.*

Unfortunately, calculating the spark of a matrix is typically more difficult than solving $P_0$. There is a simpler way, fortunately. We denote by $a_k$ the $k$th column of the matrix $A$.

**Definition 22.2** *The* mutual coherence *of the matrix $A$ is*

$$\mu(A) = \max_{1 \leq k,j \leq m, k \neq j} \frac{|a_k^\dagger a_j|}{\|a_k\|_2 \|a_j\|_2}.$$

The matrix $A$ is said to have *nearly incoherent columns* if $\mu(A)$ is nearly equal to zero. If $A$ were square and orthogonal, then we would have $\mu(A) = 0$. However, we are assuming that $A$ is $I$ by $J$, with $I < J$, so that $\mu(A) > 0$. The following lemma is helpful.

**Lemma 22.1** *For any matrix $A$ we have*

$$sp(A) \geq 1 + \frac{1}{\mu(A)}.$$

As a consequence, we get the following theorem.

**Theorem 22.2** *If $Ax = b$ and*

$$\|x\|_0 < \frac{1}{2}\left(1 + \frac{1}{\mu(A)}\right),$$

*then $x$ solves $P_0$.*

## 22.4.2   Minimum One-Norm Solutions

A more tractable problem is to seek a *minimum one-norm* solution, that is, we can solve the problem $P_1$: minimize

$$||x||_1 = \sum_{j=1}^{J} |x_j|,$$

subject to $Ax = b$. Let $x^*$ be a solution of $P_1$. Problem $P_1$ can be formulated as a linear programming problem, so is more easily solved. The big questions are: when does $P_1$ have a unique solution $x^*$, and when is $x^* = \hat{x}$? The problem $P_1$ will have a unique solution if and only if $A$ is such that the one-norm satisfies

$$||x^*||_1 < ||x^* + v||_1,$$

for all nonzero $v$ in the null space of $A$. We have the following theorem.

**Theorem 22.3** *If $A$ is $I$ by $J$, with full rank and $I < J$, and $Ax = b$, with*

$$||x||_0 < \frac{1}{2}\left(1 + \frac{1}{\mu(A)}\right),$$

*then $x$ solves both $P_0$ and $P_1$.*

## 22.4.3   Minimum One-Norm as an LP Problem

The entries of $x$ need not be nonnegative, so the problem is not yet a linear programming problem. Let

$$B = \begin{bmatrix} A & -A \end{bmatrix},$$

and consider the linear programming problem of minimizing the function

$$c^T z = \sum_{j=1}^{2J} z_j,$$

subject to the constraints $z \geq 0$, and $Bz = b$. Let $z^*$ be the solution. We write

$$z^* = \begin{bmatrix} u^* \\ v^* \end{bmatrix}.$$

Then, as we shall see, $x^* = u^* - v^*$ minimizes the one-norm, subject to $Ax = b$.

First, we show that $u_j^* v_j^* = 0$, for each $j$. If, say, there is a $j$ such that $0 < v_j^* < u_j^*$, then we can create a new vector $z$ by replacing the old $u_j^*$ with $u_j^* - v_j^*$ and the old $v_j^*$ with zero, while maintaining $Bz = b$. But then, since $u_j^* - v_j^* < u_j^* + v_j^*$, it follows that $c^T z < c^T z^*$, which is a contradiction. Consequently, we have $\|x^*\|_1 = c^T z^*$.

Now we select any $x$ with $Ax = b$. Write $u_j = x_j$, if $x_j \geq 0$, and $u_j = 0$, otherwise. Let $v_j = u_j - x_j$, so that $x = u - v$. Then let

$$z = \begin{bmatrix} u \\ v \end{bmatrix}.$$

Then $b = Ax = Bz$, and $c^T z = \|x\|_1$. Consequently,

$$\|x^*\|_1 = c^T z^* \leq c^T z = \|x\|_1,$$

and $x^*$ must be a minimum one-norm solution.

### 22.4.4   Why the One-Norm?

When a system of linear equations $Ax = b$ is under-determined, we can find the *minimum-two-norm solution* that minimizes the square of the two-norm,

$$\|x\|_2^2 = \sum_{j=1}^{J} x_j^2,$$

subject to $Ax = b$. One drawback to this approach is that the two-norm penalizes relatively large values of $x_j$ much more than the smaller ones, so tends to provide non-sparse solutions. Alternatively, we may seek the solution for which the one-norm,

$$\|x\|_1 = \sum_{j=1}^{J} |x_j|,$$

is minimized. The one-norm still penalizes relatively large entries $x_j$ more than the smaller ones, but much less than the two-norm does. As a result, it often happens that the minimum one-norm solution actually solves $P_0$ as well.

### 22.4.5   Comparison with the PDFT

The PDFT approach to solving the under-determined system $Ax = b$ is to select weights $w_j \geq 0$ and then to find the solution $\tilde{x}$ that minimizes the weighted two-norm given by

$$\sum_{j \in S} |x_j|^2 w_j,$$

where $S$ is the support set of $w$, meaning that $S$ is the set of all $j$ for which $w_j > 0$. Our intention is to select weights $w_j$ so that $w_j^{-1}$ is reasonably close to $|x_j^*|$. Consider, therefore, what happens when $S$ is the support set of $x^*$ and $w_j^{-1} = |x_j^*|$ for $j \in S$. We show that $\tilde{x}$ is also a minimum-one-norm solution.

To see why this is true, note that, for any $x$ supported on $S$, we have

$$\|x\|_1 = \sum_{j \in S} |x_j| = \sum_{j \in S} \frac{|x_j|}{\sqrt{|x_j^*|}} \sqrt{|x_j^*|}$$

$$\leq \sqrt{\sum_{j \in S} \frac{|x_j|^2}{|x_j^*|}} \sqrt{\sum_{j \in S} |x_j^*|}.$$

Therefore,

$$\|\tilde{x}\|_1 = \sum_{j \in S} |\tilde{x}_j| \leq \sqrt{\sum_{j \in S} \frac{|\tilde{x}_j|^2}{|x_j^*|}} \sqrt{\sum_{j \in S} |x_j^*|}$$

$$\leq \sqrt{\sum_{j \in S} \frac{|x_j^*|^2}{|x_j^*|}} \sqrt{\sum_{j \in S} |x_j^*|}$$

$$= \sum_{j \in S} |x_j^*| = \|x^*\|_1.$$

Therefore, $\tilde{x}$ also minimizes the one-norm.

## 22.4.6   Iterative Reweighting

Let $x$ be the truth. Generally, we want each weight $w_j$ to be a good prior estimate of the reciprocal of $|x_j|$. Because we do not yet know $x$, we may take a sequential-optimization approach, beginning with weights $w_j^0 > 0$, finding the PDFT solution using these weights, then using this PDFT solution to get a (we hope!) better choice for the weights, and so on. This sequential approach was successfully implemented in the early 1980's by Michael Fiddy and his students [74].

In [47], the same approach is taken, but with respect to the one-norm. Since the one-norm still penalizes larger values disproportionately, balance can be achieved by minimizing a weighted-one-norm, with weights close to the reciprocals of the $|x_j|$. Again, not yet knowing $x$, they employ a sequential approach, using the previous minimum-weighted-one-norm solution to obtain the new set of weights for the next minimization. At each step of the sequential procedure, the previous reconstruction is used to estimate the true support of the desired solution.

It is interesting to note that an on-going debate among users of the PDFT concerns the nature of the prior weighting. Does $w_j$ approximate

$|x_j|^{-1}$ or $|x_j|^{-2}$? This is close to the issue treated in [47], the use of a weight in the minimum-one-norm approach.

It should be noted again that finding a sparse solution is not usually the goal in the use of the PDFT, but the use of the weights has much the same effect as using the one-norm to find sparse solutions. To the extent that the weights approximate the entries of $\hat{x}$, their use reduces the penalty associated with the larger entries of an estimated solution.

## 22.5   Why Sparseness?

One obvious reason for wanting sparse solutions of $Ax = b$ is that we have prior knowledge that the desired solution is sparse. Such a problem arises in signal analysis from Fourier-transform data. In other cases, such as in the reconstruction of locally constant signals, it is not the signal itself, but its discrete derivative, that is sparse.

### 22.5.1   Signal Analysis

Suppose that our signal $f(t)$ is known to consist of a small number of complex exponentials, so that $f(t)$ has the form

$$f(t) = \sum_{j=1}^{J} a_j e^{i\omega_j t},$$

for some small number of frequencies $\omega_j$ in the interval $[0, 2\pi)$. For $n = 0, 1, ..., N - 1$, let $f_n = f(n)$, and let $f$ be the vector in $\mathbb{C}^N$ with entries $f_n$; we assume that $J$ is much smaller than $N$. The discrete (vector) Fourier transform of $f$ is the vector $F$ having the entries

$$F_k = \frac{1}{\sqrt{N}} \sum_{n=0}^{N-1} f_n e^{2\pi ikn/N},$$

for $k = 0, 1, ..., N - 1$; we write $F = Ef$, where $E$ is the $N$ by $N$ matrix with entries $E_{kn} = \frac{1}{\sqrt{N}} e^{2\pi ikn/N}$. If $N$ is large enough, we may safely assume that each of the $\omega_j$ is equal to one of the frequencies $2\pi ik$ and that the vector $F$ is $J$-sparse. The question now is: How many values of $f(n)$ do we need to calculate in order to be sure that we can recapture $f(t)$ exactly? We have the following theorem [45]:

**Theorem 22.4** *Let $N$ be prime. Let $S$ be any subset of $\{0, 1, ..., N - 1\}$ with $|S| \geq 2J$. Then the vector $F$ can be uniquely determined from the measurements $f_n$ for $n$ in $S$.*

We know that
$$f = E^\dagger F,$$
where $E^\dagger$ is the conjugate transpose of the matrix $E$. The point here is that, for any matrix $R$ obtained from the identity matrix $I$ by deleting $N - |S|$ rows, we can recover the vector $F$ from the measurements $Rf$.

If $J$ is not prime, then the assertion of the theorem may not hold, since we can have $j = 0 \bmod J$, without $j = 0$. However, the assertion remains valid for most sets of $J$ frequencies and most subsets $S$ of indices; therefore, with high probability, we can recover the vector $F$ from $Rf$. Note the similarity between this and Prony's method.

Note that the matrix $E$ is *unitary*, that is, $E^\dagger E = I$, and, equivalently, the columns of $E$ form an orthonormal basis for $\mathbb{C}^N$. The data vector is

$$b = Rf = RE^\dagger F.$$

In this example, the vector $f$ is not sparse, but can be represented sparsely in a particular orthonormal basis, namely as $f = E^\dagger F$, using a sparse vector $F$ of coefficients. The *representing basis* then consists of the columns of the matrix $E^\dagger$. The measurements pertaining to the vector $f$ are the values $f_n$, for $n$ in $S$. Since $f_n$ can be viewed as the inner product of $f$ with $\delta^n$, the $n$th column of the identity matrix $I$, that is,

$$f_n = \langle \delta^n, f \rangle,$$

the columns of $I$ provide the so-called *sampling basis*. With $A = RE^\dagger$ and $x = F$, we then have
$$Ax = b,$$
with the vector $x$ sparse. It is important for what follows to note that the matrix $A$ is random, in the sense that we choose which rows of $I$ to use to form $R$.

## 22.5.2 Locally Constant Signals

Suppose now that the function $f(t)$ is locally constant, its graph consisting of some number of horizontal lines. We discretize the function $f(t)$ to get the vector $f = (f(0), f(1), ..., f(N-1))^T$. The discrete derivative vector is $g = (g_1, g_2, ..., g_{N-1})^T$, with

$$g_n = f(n) - f(n-1).$$

Since $f(t)$ is locally constant, the vector $g$ is sparse. The data we will have will not typically be values $f(n)$. The goal will be to recover $f$ from $M$ linear functional values pertaining to $f$, where $M$ is much smaller than $N$. We shall assume, from now on, that we have measured, or can estimate, the value $f(0)$.

Our $M$ by 1 data vector $d$ consists of measurements pertaining to the vector $f$:

$$d_m = \sum_{n=0}^{N-1} H_{mn} f_n,$$

for $m = 1, ..., M$, where the $H_{mn}$ are known. We can then write

$$d_m = f(0)\left(\sum_{n=0}^{N-1} H_{mn}\right) + \sum_{k=1}^{N-1}\left(\sum_{n=k}^{N-1} H_{mn}\right) g_k.$$

Since $f(0)$ is known, we can write

$$b_m = d_m - f(0)\left(\sum_{n=0}^{N-1} H_{mn}\right) = \sum_{k=1}^{N-1} A_{mk} g_k,$$

where

$$A_{mk} = \sum_{n=k}^{J} H_{mn}.$$

The problem is then to find a sparse solution of $Ax = g$. As in the previous example, we often have the freedom to select the linear functions, that is, the values $H_{mn}$, so the matrix $A$ can be viewed as random.

### 22.5.3 Tomographic Imaging

The reconstruction of tomographic images is an important aspect of medical diagnosis, and one that combines aspects of both of the previous examples. The data one obtains from the scanning process can often be interpreted as values of the Fourier transform of the desired image; this is precisely the case in magnetic-resonance imaging, and approximately true for x-ray transmission tomography, positron-emission tomography (PET) and single-photon emission tomography (SPECT). The images one encounters in medical diagnosis are often approximately locally constant, so the associated array of discrete partial derivatives will be sparse. If this sparse derivative array can be recovered from relatively few Fourier-transform values, then the scanning time can be reduced.

We turn now to the more general problem of compressed sampling.

---

## 22.6 Compressed Sampling

Our goal is to recover the vector $f = (f_1, ..., f_J)^T$ from $I$ linear functional values of $f$, where $I$ is much less than $J$. In general, this is not possible without prior information about the vector $f$. In compressed sampling, the prior information concerns the sparseness of either $f$ itself, or another vector linearly related to $f$.

Let $U$ and $V$ be unitary $J$ by $J$ matrices, so that the column vectors of both $U$ and $V$ form orthonormal bases for $\mathbb{C}^J$. We shall refer to the bases associated with $U$ and $V$ as the *sampling basis* and the *representing basis*, respectively. The first objective is to find a unitary matrix $V$ so that $f = Vx$, where $x$ is sparse. Then we want to find a second unitary matrix $U$ such that, when an $I$ by $J$ matrix $R$ is obtained from $U$ by deleting rows, the sparse vector $x$ can be determined from the data $b = RVx = Ax$. Theorems in compressed sensing describe properties of the matrices $U$ and $V$ such that, when $R$ is obtained from $U$ by a random selection of the rows of $U$, the vector $x$ will be uniquely determined, with high probability, as the unique solution that minimizes the one-norm.

# Chapter 23

## Probability

## 23.1  Chapter Summary

In this chapter we review a few important results from the theory of probability.

## 23.2 Independent Random Variables

Let $X_1, ..., X_N$ be $N$ independent real random variables with the same mean (that is, expected value) $\mu$ and same variance $\sigma^2$. The main consequence of independence is that $E(X_i X_j) = E(X_i)E(X_j) = \mu^2$ for $i \neq j$. Then, it is easily shown that the *sample average*

$$\bar{X} = N^{-1} \sum_{n=1}^{N} X_n$$

has $\mu$ for its mean and $\sigma^2/N$ for its variance.

**Ex. 23.1** *Prove these two assertions.*

## 23.3 Maximum Likelihood Parameter Estimation

Suppose that the random variable $X$ has a probability density function $p(x; \theta)$, where $\theta$ is an unknown parameter. A common problem in statistics is to estimate $\theta$ from independently sampled values of $X$, say $x_1, ..., x_N$. A frequently used approach is to maximize the function of $\theta$ given by

$$L(\theta) = L(\theta; x_1, ..., x_N) = \prod_{n=1}^{N} p(x_n; \theta).$$

The function $L(\theta)$ is the *likelihood function* and a value of $\theta$ maximizing $L(\theta)$ is a *maximum likelihood estimate*. We give two examples of maximum likelihood (ML) estimation.

### 23.3.1 An Example: The Bias of a Coin

Let $\theta$ in the interval $[0, 1]$ be the unknown probability of success on one trial of a binomial distribution (a coin flip, for example), so that the probability of $k$ successes in $N$ trials is $L(\theta; k, N) = \frac{N!}{k!(N-k)!} \theta^k (1 - \theta)^{N-k}$, for $k = 0, 1, ..., N$. If we have observed $N$ trials and have recorded $k$ successes, we can estimate $\theta$ by selecting that $\hat{\theta}$ for which $L(\theta, k, N)$ is maximized as a function of $\theta$.

**Ex. 23.2** *Show that, for the binomial case described above, the maximum likelihood estimate of $\theta$ is $\hat{\theta} = k/N$.*

### 23.3.2 Estimating a Poisson Mean

A random variable $X$ taking on only nonnegative integer values is said to have the *Poisson distribution* with parameter $\lambda > 0$ if, for each nonnegative integer $k$, the probability $p_k$ that $X$ will take on the value $k$ is given by

$$p_k = e^{-\lambda}\lambda^k/k!.$$

**Ex. 23.3** *Show that the sequence $\{p_k\}_{k=0}^{\infty}$ sums to one.*

**Ex. 23.4** *Show that the expected value $E(X)$ is $\lambda$, where the expected value in this case is*

$$E(X) = \sum_{k=0}^{\infty} kp_k.$$

**Ex. 23.5** *Show that the variance of $X$ is also $\lambda$, where the variance of $X$ in this case is*

$$\mathrm{var}(X) = \sum_{k=0}^{\infty}(k-\lambda)^2 p_k.$$

**Ex. 23.6** *Show that the ML estimate of $\lambda$ based on $N$ independent samples is the sample mean.*

---

## 23.4 Independent Poisson Random Variables

Let $Z_1, ..., Z_N$ be independent Poisson random variables with expected value $E(Z_n) = \lambda_n$. Let $\mathbf{Z}$ be the random vector with $Z_n$ as its entries, $\lambda$ the vector whose entries are the $\lambda_n$, and $\lambda_+ = \sum_{n=1}^{N} \lambda_n$. Then the probability function for $\mathbf{Z}$ is

$$f(\mathbf{Z}|\lambda) = \prod_{n=1}^{N} \lambda_n^{z_n} \exp(-\lambda_n)/z_n! = \exp(-\lambda_+) \prod_{n=1}^{N} \lambda_n^{z_n}/z_n! \, .$$

Now let $Y = \sum_{n=1}^{N} Z_n$. Then, the probability function for $Y$ is

$$\mathrm{Prob}(Y = y) = \mathrm{Prob}(Z_1 + ... + Z_N = y) = \sum_{z_1 + ... z_N = y} \exp(-\lambda_+) \prod_{n=1}^{N} \lambda_n^{z_n}/z_n! \, .$$

But, as we shall see shortly, $Y$ is a Poisson random variable with $E(Y) = \lambda_+$, since we have

$$\sum_{z_1 + ... z_N = y} \exp(-\lambda_+) \prod_{n=1}^{N} \lambda_n^{z_n}/z_n! = \exp(-\lambda_+)\lambda_+^y/y! \, . \tag{23.1}$$

When we observe an instance of $y$, we can consider the conditional distribution $f(\mathbf{Z}|\lambda, y)$ of $\{Z_1, ..., Z_N\}$, subject to $y = Z_1 + ... + Z_N$. We have

$$f(\mathbf{Z}|\lambda, y) = \frac{y!}{z_1!...z_N!}\left(\frac{\lambda_1}{\lambda_+}\right)^{z_1}...\left(\frac{\lambda_N}{\lambda_+}\right)^{z_N}.$$

This is a *multinomial distribution*. Given $y$ and $\lambda$, the conditional expected value of $Z_n$ is then $E(Z_n|\lambda, y) = y\lambda_n/\lambda_+$. To see why Equation (23.1) is true, we discuss the multinomial distribution.

---

## 23.5   The Multinomial Distribution

When we expand the quantity $(a_1 + ... + a_N)^y$, we obtain a sum of terms, each of the form $a_1^{z_1}...a_N^{z_N}$, with $z_1 + ... + z_N = y$. How many terms of the same form are there? There are $N$ variables. We are to select $z_n$ of type $n$, for each $n = 1, ..., N$, to get $y = z_1 + ... + z_N$ factors. Imagine $y$ blank spaces, to be filled in by various factor types as we do the selection. We select $z_1$ of these blanks and mark them $a_1$, for type one. We can do that in $\binom{y}{z_1}$ ways. We then select $z_2$ of the remaining blank spaces and enter $a_2$ in them; we can do this in $\binom{y-z_1}{z_2}$ ways. Continuing in this way, we find that we can select the $N$ factor types in

$$\binom{y}{z_1}\binom{y-z_1}{z_2}...\binom{y-(z_1+...+z_{N-2})}{z_{N-1}}$$

ways, or in

$$\frac{y!}{z_1!(y-z_1)!}...\frac{(y-(z_1+...+z_{N-2}))!}{z_{N-1}!(y-(z_1+...+z_{N-1}))!} = \frac{y!}{z_1!...z_N!}.$$

This tells us in how many different sequences the factor types can be selected. Applying this, we get the multinomial theorem:

$$(a_1 + ... + a_N)^y = \sum_{z_1+...+z_N=y}\frac{y!}{z_1!...z_N!}a_1^{z_1}...a_N^{z_N}.$$

Select $a_n = \lambda_n/\lambda_+$. Then,

$$1 = 1^y = \left(\frac{\lambda_1}{\lambda_+} + ... + \frac{\lambda_N}{\lambda_+}\right)^y$$

$$= \sum_{z_1+...+z_N=y}\frac{y!}{z_1!...z_N!}\left(\frac{\lambda_1}{\lambda_+}\right)^{z_1}...\left(\frac{\lambda_N}{\lambda_+}\right)^{z_N}.$$

From this we get

$$\sum_{z_1+\dots z_N=y} \exp(-\lambda_+) \prod_{n=1}^{N} \lambda_n^{z_n}/z_n! = \exp(-\lambda_+)\lambda_+^{y}/y! \ .$$

---

## 23.6    Characteristic Functions

The Fourier transform shows up in probability theory in the guise of the *characteristic function* of a random variable. The characteristic function is related to, but more general than, the moment-generating function and serves much the same purposes.

A real-valued random variable $X$ is said to have the probability density function (pdf) $f(x)$ if, for any interval $[a,b]$, the probability that $X$ takes its value within this interval is given by the integral $\int_a^b f(x)dx$. To be a pdf, $f(x)$ must be nonnegative and $\int_{-\infty}^{\infty} f(x)dx = 1$. The *characteristic function* of $X$ is then

$$F(\omega) = \int_{-\infty}^{\infty} f(x)e^{ix\omega}dx.$$

The formulas for differentiating the Fourier transform are quite useful in determining the moments of a random variable.

The *expected value* of $X$ is

$$E(X) = \int_{-\infty}^{\infty} xf(x)dx,$$

and for any real-valued function $g(x)$ the expected value of the random variable $g(X)$ is

$$E(g(X)) = \int_{-\infty}^{\infty} g(x)f(x)dx.$$

The $n$th moment of $X$ is

$$E(X^n) = \int_{-\infty}^{\infty} x^n f(x)dx;$$

the *variance* of $X$ is then $\text{var}(X) = E(X^2) - E(X)^2$. It follows, therefore, that the $n$th moment of the random variable $X$ is given by

$$E(X^n) = (i)^n F^{(n)}(0).$$

If we have $N$ real-valued random variables $X_1, ..., X_N$, their *joint probability density function* is $f(x_1, ..., x_N) \geq 0$ having the property that, for any intervals $[a_1, b_1], ..., [a_N, b_N]$, the probability that $X_n$ takes its value within $[a_n, b_n]$, for each $n$, is given by the multiple integral

$$\int_{a_1}^{b_1} \cdots \int_{a_N}^{b_N} f(x_1, ..., x_N) dx_1 \cdots dx_N.$$

The joint moments are then

$$E(X_1^{m_1} \cdots X_N^{m_N}) = \int_{-\infty}^{\infty} \cdots \int_{-\infty}^{\infty} x_1^{m_1} \cdots x_N^{m_N} f(x_1, ..., x_N) dx_1 \cdots dx_N.$$

The joint moments can be calculated by evaluating at zero the partial derivatives of the characteristic function of the joint pdf.

The random variables are said to be *independent* if

$$f(x_1, ..., x_N) = f(x_1) \cdots f(x_N),$$

where, in keeping with the convention used in the probability literature, $f(x_n)$ denotes the pdf of the random variable $X_n$.

If $X$ and $Y$ are independent random variables with probability density functions $f(x)$ and $g(y)$, then the probability density function for the random variable $Z = X + Y$ is $(f * g)(z)$, the convolution of $f$ and $g$. To see this, we first calculate the cumulative distribution function

$$H(z) = \text{Prob}\,(X + Y \leq z),$$

which is

$$H(z) = \int_{x=-\infty}^{+\infty} \int_{y=-\infty}^{z-x} f(x)g(y)dydx.$$

Using the change of variable $t = x + y$, we get

$$H(z) = \int_{x=-\infty}^{+\infty} \int_{t=-\infty}^{z} f(x)g(t - x)dtdx.$$

The pdf for the random variable $Z$ is $h(z) = H'(z)$, the derivative of $H(z)$. Differentiating the inner integral with respect to $z$, we obtain

$$h(z) = \int_{x=-\infty}^{+\infty} f(x)g(z - x)dx;$$

therefore, $h(z) = (f*g)(z)$. It follows that the characteristic function for the random variable $Z = X + Y$ is the product of the characteristic functions for $X$ and $Y$.

## 23.7 Gaussian Random Variables

A real-valued random variable $X$ is called *Gaussian* or *normal* with mean $\mu$ and variance $\sigma^2$ if its probabilty density function (pdf) is

$$f(x) = \frac{1}{\sigma\sqrt{2\pi}} \exp\left(-\frac{(x-\mu)^2}{2\sigma^2}\right).$$

In the statistical literature a normal random variable is *standard* if its mean is $\mu = 0$ and its variance is $\sigma^2 = 1$.

### 23.7.1 Gaussian Random Vectors

Suppose now that $Z_1, ..., Z_N$ are independent standard normal random variables. Then, their joint pdf is the function

$$f(z_1, ..., z_N) = \prod_{n=1}^{N} \frac{1}{\sqrt{2\pi}} \exp\left(-\frac{1}{2}z_n^2\right) = \frac{1}{(\sqrt{2\pi})^N} \exp\left(-\frac{1}{2}(z_1^2 + ... + z_N^2)\right).$$

By taking linear combinations of these random variables, we can obtain a new set of normal random variables that are no longer independent. For each $m = 1, ..., M$ let

$$X_m = \sum_{n=1}^{N} A_{mn} Z_n.$$

Then $E(X_m) = 0$.

The *covariance matrix* associated with the $X_m$ is the matrix $R$ with entries $R_{mn} = E(X_m X_n)$, $m, n = 1, 2, ..., M$. We have

$$E(X_m X_n) = \sum_{k=1}^{N} A_{mk} \sum_{j=1}^{N} A_{nj} E(Z_k Z_j).$$

Since the $Z_n$ are independent with mean zero, we have $E(Z_k Z_j) = 0$ for $k \neq j$ and $E(Z_k^2) = 1$. Therefore,

$$E(X_m X_n) = \sum_{k=1}^{N} A_{mk} A_{nk},$$

and the covariance matrix is $R = AA^T$.

Writing $\mathbf{X} = (X_1, ..., X_M)^T$ and $\mathbf{Z} = (Z_1, ..., Z_N)^T$, we have $\mathbf{X} = A\mathbf{Z}$, where $A$ is the $M$ by $N$ matrix with entries $A_{mn}$. Using the standard formulas for changing variables, we find that the joint pdf for the random variables $X_1, ..., X_M$ is

$$f(x_1, ..., x_M) = \frac{1}{\sqrt{\det(R)}} \frac{1}{(\sqrt{2\pi})^N} \exp\left(-\frac{1}{2}\mathbf{x}^T R^{-1}\mathbf{x}\right),$$

with $\mathbf{x} = (x_1, ..., x_N)^T$. For the remainder of this chapter, we limit the discussion to the case of $M = N = 2$ and use the notation $X_1 = X$, $X_2 = Y$ and $f(x_1, x_2) = f(x, y)$. We also let $\rho = E(XY)/\sigma_1\sigma_2$.

The two-dimensional FT of the function $f(x, y)$, the characteristic function of the Gaussian random vector $\mathbf{X}$, is

$$F(\alpha, \beta) = \exp\left(-\frac{1}{2}(\sigma_1^2\alpha^2 + \sigma_2^2\beta^2 + 2\sigma_1\sigma_2\rho\alpha\beta)\right).$$

**Ex. 23.7** *Use partial derivatives of $F(\alpha, \beta)$ to show that $E(X^2Y^2) = 2\sigma_1^2\sigma_2^2\rho^2$.*

**Ex. 23.8** *Show that $E(X^2Y^2) = E(X^2)E(Y^2) + 2E(XY)^2$.*

### 23.7.2 Complex Gaussian Random Variables

Let $X$ and $Y$ be independent real Gaussian random variables with means $\mu_x$ and $\mu_y$, respectively, and common variance $\sigma^2$. Then $W = X + iY$ is a *complex Gaussian random variable* with mean $\mu_w = E(W) = \mu_x + i\mu_y$ and variance $\sigma_w^2 = 2\sigma^2$.

The results of Exercise 23.7 extend to complex Gaussian random variables $W$ and $V$. In the complex case we have

$$E(|V|^2|W|^2) = E(|V|^2)E(|W|^2) + |E(V\overline{W})|^2.$$

This is important in optical image processing, where it is called the *Hanbury–Brown Twiss effect* and provides the basis for intensity interferometry [78]. The main point is that we can obtain magnitude information about $E(V\overline{W})$, but not phase information, by measuring the correlation between the magnitudes of $V$ and $W$; that is, we learn something about $E(V\overline{W})$ from intensity measurements. Since we have only the magnitude of $E(V\overline{W})$, we then have a *phase problem*.

### 23.8  Using *A Priori* Information

We know that to get information out we need to put information in; but how to do it is the problem. One approach that is quite popular within the image-reconstruction community is the use of statistical Bayesian methods and maximum *a posteriori* (MAP) estimation.

---

## 23.9 Conditional Probabilities and Bayes' Rule

Suppose that $A$ and $B$ are two events with positive probabilities $P(A)$ and $P(B)$, respectively. The *conditional probability* of $B$, given $A$, is defined to be $P(B|A) = P(A \cap B)/P(A)$. It follows that Bayes' Rule holds:

$$P(A|B) = P(B|A)P(A)/P(B).$$

To illustrate the use of this rule, we consider the following example.

### 23.9.1 An Example of Bayes' Rule

Suppose that, in a certain town, 10 percent of the adults over 50 have diabetes. The town doctor correctly diagnoses those with diabetes as having the disease 95 percent of the time. In two percent of the cases he incorrectly diagnoses those not having the disease as having it. Let $D$ mean that the patient has diabetes, $N$ that the patient does not have the disease, $A$ that a diagnosis of diabetes is made, and $B$ that a diagnosis of diabetes is not made. The probability that he will diagnose a given adult as having diabetes is given by the rule of total probability:

$$P(A) = P(A|D)P(D) + P(A|N)P(N).$$

In this example, we obtain $P(A) = 0.113$. Now suppose a patient receives a diagnosis of diabetes. What is the probability that this diagnosis is correct? In other words, what is $P(D|A)$? For this we use Bayes' Rule:

$$P(D|A) = P(A|D)P(D)/P(A),$$

which turns out to be 0.84.

### 23.9.2 Using Prior Probabilities

So far nothing is controversial. The fun begins when we attempt to broaden the use of Bayes' Rule to ascribe *a priori* probabilities to quantities that are not random. The example used originally by Thomas Bayes in the eighteenth century is as follows. Imagine a billiard table with a line drawn across it parallel to its shorter side, cutting the table into two rectangular regions, the nearer called A and the farther B. Balls are tossed on to the table, coming to rest in either of the two regions. Suppose that we are told only that after $N$ such tosses $n$ of the balls ended up in region A. What is the probability that the next ball will end up in region A?

At first it would seem that we cannot answer this question unless we are told the probability of any ball ending up in region A; Bayes argues differently, however. Let $A$ be the event that a ball comes to rest in region A, and let $P(A) = x$ be the unknown probability of coming to rest in region A; we may consider $x$ to be the relative area of region A, although this is not necessary. Let $D$ be the event that $n$ out of $N$ balls end up in A. Then,

$$P(D|x) = \binom{N}{n} x^n (1-x)^{N-n}.$$

Bayes then adopts the view that the horizontal line on the table was randomly positioned so that the unknown $x$ can be treated as a random variable. Using Bayes' Rule, we have

$$P(x|D) = P(D|x)P(x)/P(D),$$

where $P(x)$ is the probability density function (pdf) of the random variable $x$, which Bayes takes to be uniform over the interval $[0, 1]$. Therefore, we have

$$P(x|D) = c \binom{N}{n} x^n (1-x)^{N-n},$$

where $c$ is chosen so as to make $P(x|D)$ a pdf.

**Ex. 23.9** *Use integration by parts or the Beta function to show that*

$$\binom{N}{n} \int_0^1 x^n (1-x)^{N-n} dx = 1/(N+1),$$

*and*

$$\binom{N+1}{n+1} \int_0^1 x^{n+1} (1-x)^{N-n} dx = 1/(N+2)$$

*for $n = 0, 1, ..., N$.*

From the exercise we can conclude that $c = N + 1$; therefore we have the pdf $P(x|D)$. Now we want to estimate $x$ itself. One way to do this is to calculate the expected value of this pdf, which, according to the exercise, is $(n+1)/(N+2)$. So even though we do not know $x$, we can reasonably say $(n+1)/(N+2)$ is the probability that the next ball will end up in region A, given the behavior of the previous $N$ balls.

There is a second way to estimate $x$; we can find the value of $x$ for which the pdf reaches its maximum. A quick calculation shows this value to be $n/N$. This estimate of $x$ is not the same as the one we calculated using the expected value but they are close for large $N$.

What is controversial here is the decision to treat the positioning of the line as a random act, whereby $x$ becomes a random variable, as well as the selection of a specific pdf to govern the random variable $x$. Even if $x$ were a random variable, we do not necessarily know its pdf. Bayes takes the pdf to be uniform over $[0, 1]$, more as an expression of ignorance than of knowledge. It is this broader use of prior probabilities that is generally known as *Bayesian methods* and not the use of Bayes' Rule itself.

## 23.10    Maximum *A Posteriori* Estimation

Bayesian methods provide us with an alternative to maximum likelihood parameter estimation. Suppose that a random variable (or vector) $Z$ has the pdf $f(z; \theta)$, where $\theta$ is a parameter. When we hold $z$ fixed and view $f(z; \theta)$ as a function only of $\theta$, it is called the *likelihood function*. Having observed an instance of $Z$, call it $z$, we can estimate the parameter $\theta$ by selecting that value for which the likelihood function $f(z; \theta)$ has its maximum. This is the *maximum likelihood* (ML) estimator. Alternatively, suppose that we treat $\theta$ itself as one value of a random variable $\Theta$ having its own pdf, say $g(\theta)$. Then, Bayes' Rule says that the conditional pdf of $\Theta$, given $z$, is

$$g(\theta|z) = f(z; \theta)g(\theta)/f(z),$$

where

$$f(z) = \int f(z; \theta)g(\theta)d\theta.$$

The maximum *a posteriori* (MAP) estimate of $\theta$ is the one for which the function $g(\theta|z)$ is maximized. Taking logs and ignoring terms that do not involve $\theta$, we find that the MAP estimate of $\theta$ maximizes the function $\log f(z; \theta) + \log g(\theta)$.

Because the ML estimate maximizes $\log f(z; \theta)$, the MAP estimate is viewed as involving a *penalty term* $\log g(\theta)$ missing in the ML approach. This penalty function is based on the prior pdf $g(\theta)$. We have flexibility in selecting $g(\theta)$ and often choose $g(\theta)$ in a way that expresses our prior knowledge of the parameter $\theta$.

## 23.11    MAP Reconstruction of Images

In emission tomography the parameter $\theta$ is actually a vectorized image that we wish to reconstruct and the observed data constitute $z$. Our prior knowledge about $\theta$ may be that the true image is near some prior estimate, say $\rho$, of the correct answer, in which case $g(\theta)$ is selected to peak at $\rho$ [105]. Frequently our prior knowledge of $\theta$ is that the image it represents is nearly constant locally, except for edges. Then $g(\theta)$ is designed to weight more heavily the locally constant images and less heavily the others [82, 85, 106, 89, 109].

## 23.12    Penalty-Function Methods

The so-called *penalty function* that appears in the MAP approach comes from a prior pdf for $\theta$. This suggests more general methods that involve a penalty function term that does not necessarily emerge from Bayes' Rule [34]. Such methods are well-known in optimization. We are free to estimate $\theta$ as the maximizer of a suitable objective function whether or not that function is a posterior probability. Using penalty-function methods permits us to avoid the controversies that accompany Bayesian methods.

## 23.13    Basic Notions

The *covariance* between two complex-valued random variables $x$ and $y$ is

$$\text{cov}_{xy} = E((x - E(x))(\overline{y - E(y)})),$$

and the *correlation coefficient* is

$$\rho_{xy} = \text{cov}_{xy} / \sqrt{E(|x - E(x)|^2)} \sqrt{E(|y - E(y)|^2)}.$$

The two random variables are said to be *uncorrelated* if and only if $\rho_{xy} = 0$. The *covariance matrix* of a random vector $\mathbf{v}$ is the matrix $Q$ whose entries are the covariances of all the pairs of entries of $\mathbf{v}$. The vector $\mathbf{v}$ is said to be *uncorrelated* if $Q$ is diagonal; otherwise, we call $\mathbf{v}$ *correlated*. If the expected value of each of the entries of $\mathbf{v}$ is zero, we also have $Q = E(\mathbf{v}\mathbf{v}^\dagger)$. We saw in our discussion of the BLUE that when the noise vector $\mathbf{v}$ is correlated we need to employ the covariance matrix to obtain the best linear unbiased estimator.

---

## 23.14  Generating Correlated Noise Vectors

We can obtain an $N$ by 1 correlated-noise random vector $\mathbf{v}$ as follows. Select a positive integer $K$, an arbitrary $N$ by $K$ matrix $C$, and $K$ independent standard normal random variables $z_1, ..., z_K$; that is, their means are equal to zero and their variances are equal to one. Then let $\mathbf{z}$ be the random vector with entries $z_k$. Define $\mathbf{v} = C\mathbf{z}$. Then, we have $E(\mathbf{v}) = \mathbf{0}$ and $E(\mathbf{vv}^\dagger) = CC^\dagger = Q$. In fact, for the Gaussian case, this is the only way to obtain a correlated Gaussian random vector. The matrix $C$ producing the covariance matrix $Q$ is not unique.

---

## 23.15  Covariance Matrices

In order for $Q$ to be a covariance matrix, it is necessary and sufficient that it be Hermitian and nonnegative-definite; that is, $Q^\dagger = Q$ and the eigenvalues of $Q$ are nonnegative. Given any such $Q$, we can create an $N$ by 1 noise vector $\mathbf{v}$ having $Q$ as its covariance matrix using the eigenvalue/eigenvector decomposition of $Q$. Then, taking $U$ to be the matrix whose columns are the orthonormal eigenvectors of $Q$ and $L$ the diagonal matrix whose diagonal entries are $\lambda_n$, $n = 1, ..., N$, the eigenvalues of $Q$, we have $Q = ULU^\dagger$. For convenience, we assume that $\lambda_1 \geq \lambda_2 \geq ... \geq \lambda_N > 0$. Let $\mathbf{z}$ be a random $N$ by 1 vector whose entries are independent, standard normal random variables, and let $C = U\sqrt{L}U^\dagger$, the Hermitian square root of $Q$. Then, $\mathbf{v} = C\mathbf{z}$ has $Q$ for its covariance matrix.

If we write this $\mathbf{v}$ as

$$\mathbf{v} = (U\sqrt{L}U^\dagger)\mathbf{z} = U(\sqrt{L}U^\dagger\mathbf{z}) = U\mathbf{p},$$

then $\mathbf{p} = \sqrt{L}U^\dagger\mathbf{z}$ is uncorrelated; $E(\mathbf{pp}^\dagger) = L$.

## 23.16    Principal Component Analysis

We can write the vector $\mathbf{v} = U\mathbf{p}$ as

$$\mathbf{v} = \sum_{n=1}^{N} p_n \mathbf{u}^n,$$

so that the entries of $\mathbf{v}$ are

$$v_m = \sum_{n=1}^{N} U_{mn} p_n \qquad\qquad (23.2)$$

where $\mathbf{u}^n = (u_{1,n}, ..., u_{N,n})^T$ is the eigenvector of $Q$ associated with eigenvalue $\lambda_n$. Since the variance of $p_n$ is $\lambda_n$, Equation (23.2) decomposes the vector $\mathbf{v}$ into components of decreasing strength. The terms in the sum corresponding to the smaller indices describe most of $\mathbf{v}$; they are the *principal components* of $\mathbf{v}$. Each $p_n$ is a linear combination of the entries of $\mathbf{v}$, and *principal component analysis* consists of finding these uncorrelated linear combinations that best describe the correlated entries of $\mathbf{v}$. The representation $\mathbf{v} = U\mathbf{p}$ expresses $\mathbf{v}$ as a linear combination of orthonormal vectors with uncorrelated coefficients. This is analogous to the *Karhunen-Loève expansion* for stochastic processes [2].

Principal component analysis has as its goal the approximation of the covariance matrix $Q = E(\mathbf{v}\mathbf{v}^\dagger)$ by nonnegative-definite matrices of lower rank. A related area is *factor analysis*, which attempts to describe the $N$ by $N$ covariance matrix $Q$ as $Q = AA^\dagger + D$, where $A$ is some $N$ by $J$ matrix, for some $J < N$, and $D$ is diagonal. Factor analysis attempts to account for the correlated components of $Q$ using the lower-rank matrix $AA^\dagger$. Underlying this is a model for the random vector $\mathbf{v}$:

$$\mathbf{v} = A\mathbf{x} + \mathbf{w},$$

where both $\mathbf{x}$ and $\mathbf{w}$ are uncorrelated. The entries of the random vector $\mathbf{x}$ are the *common factors* that affect each entry of $\mathbf{v}$ while those of $\mathbf{w}$ are the *special factors*, each associated with a single entry of $\mathbf{v}$. Factor analysis plays an increasingly prominent role in signal and image processing [17] as well as in the social sciences.

In [151] Gil Strang points out that, from a linear algebra standpoint, factor analysis raises some questions. As his example shows, the representation of $Q$ as $Q = AA^\dagger + D$ is not unique. The matrix $Q$ does not uniquely determine the size of the matrix $A$:

$$Q = \begin{bmatrix} 1 & .74 & .24 & .24 \\ .74 & 1 & .24 & .24 \\ .24 & .24 & 1 & .74 \\ .24 & .24 & .74 & 1 \end{bmatrix} = \begin{bmatrix} .7 & .5 \\ .7 & .5 \\ .7 & -.5 \\ .7 & -.5 \end{bmatrix} \begin{bmatrix} .7 & .7 & .7 & .7 \\ .5 & .5 & -.5 & -.5 \end{bmatrix} + .26I$$

and

$$Q = \begin{bmatrix} .6 & \sqrt{.38} & 0 \\ .6 & \sqrt{.38} & 0 \\ .4 & 0 & \sqrt{.58} \\ .4 & 0 & \sqrt{.58} \end{bmatrix} \begin{bmatrix} .6 & .6 & .4 & .4 \\ \sqrt{.38} & \sqrt{.38} & 0 & 0 \\ 0 & 0 & \sqrt{.58} & \sqrt{.58} \end{bmatrix} .26I.$$

It is also possible to represent $Q$ with different diagonal components $D$.

# Chapter 24

# Using the Wave Equation

## 24.1  Chapter Summary

In this chapter we demonstrate how the problem of Fourier-transform estimation from sampled data arises in the processing of measurements obtained by sampling electromagnetic- or acoustic-field fluctuations, as in radar or sonar. We continue the discussion, begun in Chapter 9, of plane-wave solutions of the wave equation. To illustrate the use of non-plane-wave solutions we consider the problem of detecting a source of acoustic energy in a shallow-water environment.

## 24.2  The Wave Equation

In many areas of remote sensing, what we measure are the fluctuations in time of an electromagnetic or acoustic field. Such fields are described mathematically as solutions of certain partial differential equations, such as the *wave equation*. A function $u(x, y, z, t)$ is said to satisfy the *three-dimensional wave equation* if

$$u_{tt} = c^2(u_{xx} + u_{yy} + u_{zz}) = c^2 \nabla^2 u,$$

where $u_{tt}$ denotes the second partial derivative of $u$ with respect to the time variable $t$ twice and $c > 0$ is the (constant) speed of propagation. More complicated versions of the wave equation permit the speed of propagation $c$ to vary with the spatial variables $x, y, z$, but we shall not consider that here.

Using the method of *separation of variables*, we start with solutions $u(t, x, y, z)$ having the simple form

$$u(t, x, y, z) = f(t)g(x, y, z).$$

Inserting this separated form into the wave equation, we get

$$f''(t)g(x, y, z) = c^2 f(t)\nabla^2 g(x, y, z)$$

or

$$f''(t)/f(t) = c^2 \nabla^2 g(x, y, z)/g(x, y, z).$$

The function on the left is independent of the spatial variables, while the one on the right is independent of the time variable; consequently, they must both equal the same constant, which we denote $-\omega^2$. From this we have two separate equations,

$$f''(t) + \omega^2 f(t) = 0, \tag{24.1}$$

and

$$\nabla^2 g(x, y, z) + \frac{\omega^2}{c^2} g(x, y, z) = 0. \tag{24.2}$$

Equation (24.2) is the *Helmholtz equation*.

Equation (24.1) has for its solutions the functions $f(t) = \cos(\omega t)$ and $\sin(\omega t)$, or, in complex form, the complex exponential functions $f(t) = e^{i\omega t}$ and $f(t) = e^{-i\omega t}$. Functions $u(t, x, y, z) = f(t)g(x, y, z)$ with such time dependence are called *time-harmonic* solutions.

In three-dimensional spherical coordinates with $r = \sqrt{x^2 + y^2 + z^2}$ a radial function $u(r, t)$ satisfies the wave equation if

$$u_{tt} = c^2 \left( u_{rr} + \frac{2}{r} u_r \right).$$

Radial solutions to the wave equation have the property that at any fixed time the value of $u$ is the same for all the points on a sphere centered at the origin; the curves of constant value of $u$ are these spheres, for each fixed time.

Suppose that at time $t = 0$ the function $h(r, 0)$ is zero except for $r$ near zero; that is, initially, there is a localized disturbance centered at the origin. As time passes that disturbance spreads out spherically. When the radius of a sphere is very large, the surface of the sphere appears planar, to an observer on that surface, who is said then to be in the *far field*. This motivates the study of solutions of the wave equation that are constant on planes; the so-called *plane-wave solutions*.

We simplify the situation by assuming that all the plane-wave solutions are associated with the same frequency, $\omega$. In the continuous superposition model, the field is a superposition of plane waves;

$$u(\mathbf{s}, t) = e^{i\omega t} \int f(\mathbf{k}) e^{i\mathbf{k} \cdot \mathbf{s}} d\mathbf{k}.$$

Our measurements at the sensor locations $\mathbf{s}_m$ give us the values

$$F(\mathbf{s}_m) = \int f(\mathbf{k}) e^{i\mathbf{k} \cdot \mathbf{s}_m} d\mathbf{k},$$

for $m = 1, ..., M$. The data are then Fourier transform values of the complex function $f(\mathbf{k})$; $f(\mathbf{k})$ is defined for all three-dimensional real vectors $\mathbf{k}$, but is zero, in theory, at least, for those $\mathbf{k}$ whose squared length $||\mathbf{k}||^2$ is not equal to $\omega^2/c^2$. Our goal is then to estimate $f(\mathbf{k})$ from finitely many values of its Fourier transform. Since each $\mathbf{k}$ is a normal vector for its plane-wave field component, determining the value of $f(\mathbf{k})$ will tell us the strength of the plane-wave component coming from the direction $\mathbf{k}$.

The collection of sensors at the spatial locations $\mathbf{s}_m$, $m = 1, ..., M$, is called *an array*, and the size of the array, in units of the wavelength $\lambda = 2\pi c/\omega$, is called the *aperture* of the array. Generally, the larger the aperture the better, but what is a large aperture for one value of $\omega$ will be a smaller aperture for a lower frequency. The book by Haykin [88] is a useful reference, as is the review paper by Wright, Pridham, and Kay [164].

In some applications the sensor locations are essentially arbitrary, while in others their locations are carefully chosen. Sometimes, the sensors are collinear, as in sonar towed arrays. Let's look more closely at the collinear case.

We assume now that the sensors are equi-spaced along the $x$-axis, at locations $(m\Delta, 0, 0)$, $m = 1, ..., M$, where $\Delta > 0$ is the sensor spacing; such an arrangement is called a *uniform line array*. This setup was illustrated in Figure 9.1 in Chapter 9. Our data is then

$$F_m = F(\mathbf{s}_m) = F((m\Delta, 0, 0)) = \int f(\mathbf{k}) e^{im\Delta \mathbf{k} \cdot (1,0,0)} d\mathbf{k}.$$

Since $\mathbf{k} \cdot (1, 0, 0) = \frac{\omega}{c} \cos \theta$, for $\theta$ the angle between the vector $\mathbf{k}$ and the $x$-axis, we see that there is some ambiguity now; we cannot distinguish the cone of vectors that have the same $\theta$. It is common then to assume that the wavevectors $\mathbf{k}$ have no $z$-component and that $\theta$ is the angle between two vectors in the $x, y$-plane, the so-called *angle of arrival*. The *wavenumber* variable $k = \frac{\omega}{c} \cos \theta$ lies in the interval $[-\frac{\omega}{c}, \frac{\omega}{c}]$, and we imagine that $f(\mathbf{k})$ is now $f(k)$, defined for $|k| \leq \frac{\omega}{c}$. The Fourier transform of $f(k)$ is $F(s)$, a function of a single real variable $s$. Our data is then viewed as the values $F(m\Delta)$, for $m = 1, ..., M$. Since the function $f(k)$ is zero for $|k| > \frac{\omega}{c}$, the Nyquist spacing in $s$ is $\frac{\pi c}{\omega}$, which is $\frac{\lambda}{2}$, where $\lambda = \frac{2\pi c}{\omega}$ is the wavelength.

To avoid aliasing, which now means mistaking one direction of arrival for another, we need to select $\Delta \le \frac{\lambda}{2}$. When we have oversampled, so that $\Delta < \frac{\lambda}{2}$, the interval $[-\frac{\omega}{c}, \frac{\omega}{c}]$, the so-called *visible region*, is strictly smaller than the interval $[-\frac{\pi}{\Delta}, \frac{\pi}{\Delta}]$. If the model of propagation is accurate, all the signal component plane waves will correspond to wavenumbers $k$ in the visible region and the background noise will also appear as a superposition of such propagating plane waves. In practice, there can be components in the noise that appear to come from wavenumbers $k$ outside of the visible region; this means these components of the noise are not due to distant sources propagating as plane waves, but, perhaps, to sources that are in the *near field*, or localized around individual sensors, or coming from the electronics within the sensors.

Using the relation $\lambda\omega = 2\pi c$, we can calculate the Nyquist spacing for any particular case of plane-wave array processing. For electromagnetic waves the propagation speed is the speed of light, which we shall take here to be $c = 3 \times 10^8$ meters per second. The wavelength $\lambda$ for gamma rays is around one Angstrom, which is $10^{-10}$ meters; for x-rays it is about one millimicron, or $10^{-9}$ meters. The visible spectrum has wavelengths that are a little less than one micron, that is, $10^{-6}$ meters. Shortwave radio has wavelength around one millimeter; broadcast radio has a $\lambda$ running from about 10 meters to 1000 meters, while the so-called long radio waves can have wavelengths several thousand meters long. At the one extreme it is impractical (if not physically impossible) to place individual sensors at the Nyquist spacing of fractions of microns, while at the other end, managing to place the sensors far enough apart is the challenge.

In ocean acoustics it is usually assumed that the speed of propagation of sound is around 1500 meters per second, although deviations from this *ambient sound speed* are significant and since they are caused by such things as temperature differences in the ocean, can be used to estimate these differences. At around the frequency $\omega = 50$ Hz, we find sound generated by man-made machinery, such as motors in vessels, with higher frequency harmonics sometimes present also; at other frequencies the main sources of acoustic energy may be wind-driven waves or whales. The wavelength for 50 Hz is $\lambda = 30$ meters; sonar will typically operate both above and below this wavelength. It is sometimes the case that the array of sensors is fixed in place, so what may be Nyquist spacing for 50 Hz will be oversampling for 20 Hz.

It is often the case that we are primarily interested in the values $|f(\mathbf{k})|$, not the complex values $f(\mathbf{k})$. Since the Fourier transform of the function $|f(\mathbf{k})|^2$ is the autocorrelation function obtained by convolving the function $F$ with $\overline{F}$, we can mimic the approach used earlier for power spectrum estimation to find $|f(\mathbf{k})|$. We can now employ the nonlinear methods such as Burg's MEM and Capon's maximum-likelihood method.

In array processing, as in other forms of signal and image processing, we want to remove the noise and enhance the information-bearing component, the signal. To do this we need some idea of the statistical behavior of the noise, we need a physically accurate description of what the signals probably look like, and we need a way to use this information. Much of our discussion up to now has been about the many ways in which such prior information can be incorporated in linear and nonlinear procedures. We have not said much about the important issue of the sensitivity of these methods to mismatch; that is, what happens when our physical model is wrong or the statistics of the noise is not what we thought it was? We did note earlier how Burg's MEM resolves closely spaced sinusoids when the background is white noise, but when the noise is correlated, MEM can degrade rapidly.

Even when the physical model and noise statistics are reasonably accurate, slight errors in the hardware can cause rapid degradation of the processor. Sometimes acoustic signal processing is performed with sensors that are designed to be expendable and are therefore less expensive and more prone to errors than more permanent equipment. Knowing what a sensor has received is important, but so is knowing when it received it. Slight phase errors caused by the hardware can go unnoticed when the data is processed in one manner, but can ruin the performance of another method.

The information we seek is often stored redundantly in the data and hardware errors may harm only some of these storage locations, making robust processing still possible. As we saw in our discussion of eigenvector methods, information about the frequencies of the complex exponential components of the signal are stored in the roots of the polynomials obtained from some of the eigenvectors. In [28] it was demonstrated that, in the presence of correlated noise background, phase errors distort the roots of some of these polynomials more than others; robust estimation of the frequencies is still possible if the stable roots are interrogated.

We have focused here exclusively on plane-wave propagation, which results when the source is far enough away from the sensors and the speed of propagation is constant. In many important applications these conditions are violated, and different versions of the wave equation are needed, which have different solutions. For example, sonar signal processing in environments such as shallow channels, in which some of the sound reaches the sensors only after interacting with the ocean floor or the surface, requires more complicated parameterized models for solutions of the appropriate wave equation. Lack of information about the depth and nature of the bottom can also cause errors in the signal processing. In some cases it is possible to use acoustic energy from known sources to determine the needed information.

Array signal processing can be done in *passive* or *active* mode. In passive mode the energy is either reflected off of or originates at the object of interest: the moon reflects sunlight, while ships generate their own noise. In the active mode the object of interest does not generate or reflect enough energy by itself, so the energy is generated by the party doing the sensing: active sonar is sometimes used to locate quiet vessels, while radar is used to locate planes in the sky or to map the surface of the earth. Near-earth asteroids are initially detected by passive optical observation, as small dots of reflected sunlight; once detected, they are then imaged by active radar to determine their size, shape, rotation and such.

Previously we considered the array processing problem in the context of plane-wave propagation. When the environment is more complicated, the wave equation must be modified to reflect the physics of the situation and the signal processing modified to incorporate that physics. A good example of such modification is provided by acoustic signal processing in shallow water, the subject of the rest of this chapter.

---

## 24.3   The Shallow-Water Case

In the shallow-water situation the acoustic energy from the source interacts with the surface and with the bottom of the channel, prior to being received by the sensors. The nature of this interaction is described by the wave equation in cylindrical coordinates. The deviation from the ambient pressure is the function $p(t, \mathbf{s}) = p(t, r, z, \theta)$, where $\mathbf{s} = (r, z, \theta)$ is the spatial vector variable, $r$ is the range, $z$ the depth, and $\theta$ the bearing angle in the horizontal. We assume a single frequency, $\omega$, so that

$$p(t, \mathbf{s}) = e^{i\omega t} g(r, z, \theta).$$

We shall assume cylindrical symmetry to remove the $\theta$ dependence; in many applications the bearing is essentially known or limited by the environment or can be determined by other means. The sensors are usually positioned in a vertical array in the channel, with the top of the array taken to be the origin of the coordinate system and positive $z$ taken to mean positive depth below the surface. We shall also assume that there is a single source of acoustic energy located at range $r_s$ and depth $z_s$.

To simplify a bit, we assume here that the sound speed $c = c(z)$ does not change with range, but only with depth, and that the channel has constant depth and density. Then, the Helmholtz equation for the function $g(r, z)$ is

$$\nabla^2 g(r, z) + [\omega/c(z)]^2 g(r, z) = 0.$$

The Laplacian is

$$\nabla^2 g(r, z) = g_{rr}(r, z) + \frac{1}{r} g_r(r, z) + g_{zz}(r, z).$$

We separate the variables once again, writing

$$g(r, z) = f(r)u(z).$$

Then, the range function $f(r)$ must satisfy the differential equation

$$f''(r) + \frac{1}{r} f'(r) = -\alpha f(r),$$

and the depth function $u(z)$ satisfies the differential equation

$$u''(z) + k(z)^2 u(z) = \alpha u(z),$$

where $\alpha$ is a separation constant and

$$k(z)^2 = [\omega/c(z)]^2.$$

Taking $\lambda^2 = \alpha$, the range equation becomes

$$f''(r) + \frac{1}{r} f'(r) + \lambda^2 f(r) = 0,$$

which is Bessel's equation, with Hankel-function solutions. The depth equation becomes

$$u''(z) + (k(z)^2 - \lambda^2)u(z) = 0,$$

which is of Sturm-Liouville type. The boundary conditions pertaining to the surface and the channel bottom will determine the values of $\lambda$ for which a solution exists.

To illustrate the way in which the boundary conditions become involved, we consider two examples.

---

## 24.4 The Homogeneous-Layer Model

We assume now that the channel consists of a single homogeneous layer of water of constant density, constant depth $d$, and constant sound speed $c$. We impose the following boundary conditions:

1. Pressure-release surface: $u(0) = 0$;

2. Rigid bottom: $u'(d) = 0$.

With $\gamma^2 = (k^2 - \lambda^2)$, we get $\cos(\gamma d) = 0$, so the permissible values of $\lambda$ are

$$\lambda_m = (k^2 - [(2m-1)\pi/2d]^2)^{1/2}, \ m = 1, 2, \dots.$$

The normalized solutions of the depth equation are now

$$u_m(z) = \sqrt{2/d}\sin(\gamma_m z),$$

where

$$\gamma_m = \sqrt{k^2 - \lambda_m^2} = (2m-1)\pi/2d, \ m = 1, 2, \dots.$$

For each $m$ the corresponding function of the range satisfies the differential equation

$$f''(r) + \frac{1}{r}f'(r) + \lambda_m^2 f(r),$$

which has solution $H_0^{(1)}(\lambda_m r)$, where $H_0^{(1)}$ is the zeroth order Hankel-function solution of Bessel's equation. The asymptotic form for this function is

$$\pi i H_0^{(1)}(\lambda_m r) = \sqrt{2\pi/\lambda_m r}\exp\left(-i\left(\lambda_m r + \frac{\pi}{4}\right)\right).$$

It is this asymptotic form that is used in practice. Note that when $\lambda_m$ is complex with a negative imaginary part, there will be a decaying exponential in this solution, so this term will be omitted in the signal processing.

Having found the range and depth functions, we write $g(r, z)$ as a superposition of these elementary products, called the *modes*:

$$g(r, z) = \sum_{m=1}^{M} A_m H_0^{(1)}(\lambda_m r)u_m(z),$$

where $M$ is the number of propagating modes free of decaying exponentials. The $A_m$ can be found from the original Helmholtz equation; they are

$$A_m = (i/4)u_m(z_s),$$

where $z_s$ is the depth of the source of the acoustic energy. Notice that the depth of the source also determines the strength of each mode in this superposition; this is described by saying that the source has *excited* certain modes and not others.

The eigenvalues $\lambda_m$ of the depth equation will be complex when

$$k = \frac{\omega}{c} < \frac{(2m-1)\pi}{2d}.$$

If $\omega$ is below the *cut-off frequency* $\frac{\pi c}{2d}$, then all the $\lambda_m$ are complex and there are no propagating modes $(M = 0)$. The number of propagating modes is

$$M = \frac{1}{2} + \frac{\omega d}{\pi c},$$

which is $\frac{1}{2}$ plus the depth of the channel in units of half-wavelengths.

This model for shallow-water propagation is helpful in revealing a number of the important aspects of modal propagation, but is of limited practical utility. A more useful and realistic model is the *Pekeris waveguide*.

---

## 24.5   The Pekeris Waveguide

Now we assume that the water column has constant depth $d$, sound speed $c$, and density $b$. Beneath the water is an infinite half-space with sound speed $c' > c$, and density $b'$. Figure 24.1 illustrates the situation.

Using the new depth variable $v = \frac{\omega z}{c}$, the depth equation becomes

$$u''(v) + \lambda^2 u(v) = 0, \text{ for } 0 \leq v \leq \frac{\omega d}{c},$$

and

$$u''(v) + \left( \left( \frac{c}{c'} \right)^2 - 1 + \lambda^2 \right) u(v) = 0, \text{ for } \frac{\omega d}{c} < v.$$

To have a solution, $\lambda$ must satisfy the equation

$$\tan(\lambda \omega d/c) = -(\lambda b/b')/\sqrt{1 - \left( \frac{c}{c'} \right)^2 - \lambda^2},$$

with

$$1 - \left( \frac{c}{c'} \right)^2 - \lambda^2 \geq 0.$$

The *trapped modes* are those whose corresponding $\lambda$ satisfies

$$1 \geq 1 - \lambda^2 \geq \left( \frac{c}{c'} \right)^2.$$

The eigenfunctions are

$$u_m(v) = \sin(\lambda_m v), \text{ for } 0 \leq v \leq \frac{\omega d}{c}$$

and

$$u_m(v) = \exp\left( -v\sqrt{1 - \left( \frac{c}{c'} \right)^2 - \lambda^2} \right), \text{ for } \frac{\omega d}{c} < v.$$

Although the Pekeris model has its uses, it still may not be realistic enough in some cases and more complicated propagation models will be needed.

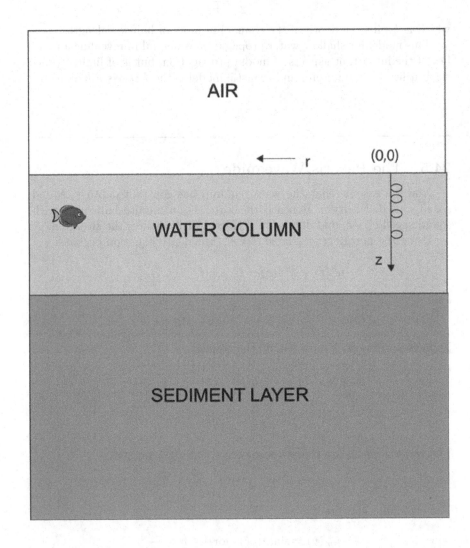

**FIGURE 24.1**: The Pekeris Model.

## 24.6 The General Normal-Mode Model

Regardless of the model by which the modal functions are determined, the general *normal-mode expansion* for the $\theta$-independent case is

$$g(r, z) = \sum_{m=1}^{M} u_m(z) s_m(r, z_s),$$

where $M$ is the number of propagating modes and $s_m(r, z_s)$ is the *modal amplitude* containing all the information about the source of the sound.

### 24.6.1 Matched-Field Processing

In plane-wave array processing we write the acoustic field as a superposition of plane-wave fields and try to find the corresponding amplitudes. This can be done using a matched filter, although high-resolution methods can also be used. In the matched-filter approach, we fix a wavevector and then match the data with the vector that describes what we would have received at the sensors had there been but a single plane wave present corresponding to that fixed wavevector; we then repeat for other fixed wavevectors. In more complicated acoustic environments, such as normal-mode propagation in shallow water, we write the acoustic field as a superposition of fields due to sources of acoustic energy at individual points in range and depth and then seek the corresponding amplitudes. Once again, this can be done using a matched filter.

In matched-field processing we fix a particular range and depth and compute what we would have received at the sensors had the acoustic field been generated solely by a single source at that location. We then match the data with this computed vector. We repeat this process for many different choices of range and depth, obtaining a function of $r$ and $z$ showing the likely locations of actual sources. As in the plane-wave case, high-resolution nonlinear methods can also be used.

As in the plane-wave case, the performance of our processing methods can be degraded by incorrect description of the environment, as well as by phase errors and the like introduced by the hardware [28]. Once again, it is necessary to seek out those locations within the data where the information we seek is less disturbed by such errors [32, 33].

Good sources for more information concerning matched-field processing are the book by Tolstoy [155] and the papers [4], [18], [70], [91], [92], [139], [140], [141], [154], and [165].

# Chapter 25

## Reconstruction in Hilbert Space

## 25.1   Chapter Summary

In many of the examples we have considered in this book, the data has been finitely many linear-functional values of the function of interest. In this chapter we consider this problem from a purely mathematical perspective. We take the function of interest to be a member of a Hilbert space, and use best approximation to solve the problem.

## 25.2   The Basic Problem

We want to reconstruct a function $f : \mathbb{R}^D \to \mathbb{C}$ from finitely many linear-functional values pertaining to that function. For example, we may want to reconstruct $f$ from values $f(x_n)$ of $f$ itself, or from Fourier-transform values $F(\gamma_n)$. We adopt the view that $f$ is a member of some infinite-dimensional Hilbert space $H$ with inner product $\langle \cdot, \cdot \rangle$, and the data values are

$$g_n = \langle f, h^n \rangle, \tag{25.1}$$

for $n = 1, ..., N$, where the $h^n$ are known members of the Hilbert space. For example, suppose $f(x)$ is supported on the interval $[a, b]$ and we have

Fourier-transform data,

$$g_n = F(\gamma_n) = \int_a^b f(x)e^{i\gamma_n x}dx = \langle f, e^n \rangle = \int_a^b f(x)\overline{e^n(x)}dx,$$

where $e^n(x) = e^{-i\gamma_n x}$. Because there are infinitely many solutions to our problem, we need some approach that singles out one solution. The most common approach is to select the estimate $\hat{f}$ of $f$ that minimizes the norm $\|\hat{f}\| = \sqrt{\langle \hat{f}, \hat{f} \rangle}$, subject to $\hat{f}$ satisfying Equation (25.1); that is,

$$g_n = \langle \hat{f}, h^n \rangle. \tag{25.2}$$

We know that every element $f$ of $H$ can be written uniquely as

$$f = \sum_{m=1}^N a_m h^m + u,$$

where $\langle u, h^n \rangle = 0$, for $n = 1, ..., N$. We may reasonably conclude from this that the probing or measuring of the function $f$ that resulted in our data is incapable of telling us anything about $u$, so that we have no choice but to take the finite sum as our estimate of $f$. We then solve the system of linear equations

$$g_n = \sum_{m=1}^N a_m \langle h^m, h^n \rangle,$$

for the $a_m$. In the case of Fourier-transform data, this approach leads to the DFT estimator. This argument has been offered several times by researchers who should know better. There is a flaw in this argument that we can exploit to obtain better estimates of $f$. To illustrate the point, we consider the problem of reconstructing $f(x)$ from Fourier-transform data.

---

## 25.3   Fourier-Transform Data

Suppose $f(x)$ is zero outside the interval $[a, b]$ and our data are the values $F(\gamma_n)$, $n = 1, ..., N$, where $F(\gamma)$ is the Fourier transform of $f(x)$. It is reasonable to suppose that $f(x)$ is a member of the Hilbert space $L^2(a, b)$ and the inner product is

$$\langle f, g \rangle = \int_a^b f(x)\overline{g(x)}dx. \tag{25.3}$$

With $h^n(x) = e^n(x) = e^{-i\gamma_n x}$, we have

$$g_n = F(\gamma_n) = \langle f, e^n \rangle.$$

But there are other inner products that we can use to represent the data. Suppose that $p(x)$ is a bounded positive function on $[a, b]$, bounded away from zero, with $w(x) = p(x)^{-1}$, and we define a new inner product on $L^2(a, b)$ by

$$\langle f, g \rangle_w = \int_a^b f(x)\overline{g(x)}w(x)dx. \tag{25.4}$$

Then we can represent tht data as

$$g_n = \int_a^b f(x)\overline{e^n(x)p(x)}w(x)dx = \langle f, t^n \rangle_w,$$

with

$$t^n(x) = e^n(x)p(x).$$

Arguing just as in the previous section, we may claim that the only reasonable estimator of $f(x)$ is in the span of the functions $t^n(x)$, since we know that $f(x)$ can be written uniquely as

$$f(x) = \sum_{m=1}^N b_m t^m(x) + v(x),$$

where

$$\langle v, t^n \rangle_w = 0,$$

for $n = 1, ..., N$. The resulting estimator is

$$\hat{f}(x) = p(x) \sum_{m=1}^N b_m e^{i\gamma_m x},$$

where the coefficients $b_m$ are found by forcing $\hat{f}(x)$ to be consistent with the inner product data; that is, the $b_m$ solve the system of linear equations

$$g_n = \langle f, t^n \rangle_w = \sum_{m=1}^N b_m \int_a^b p(x)e^{i(\gamma_m - \gamma_n)x}dx.$$

The point we are making here is that, even after we have decided which Hilbert space to use, $L^2(a, b)$ in this example, there will still be infinitely many inner products that can be chosen to represent the data, and therefore, infinitely many estimators of $f(x)$, each one arguably the right choice.

## 25.4 The General Case

Let $H$ be our chosen ambient Hilbert space, which contains $f$, with given inner product $\langle \cdot, \cdot \rangle$. Let $T : H \to H$ be a continuous, linear, invertible operator. The *adjoint* of $T$, with respect to the original inner product, is $T^\dagger$, defined by

$$\langle Tf, g \rangle = \langle f, T^\dagger g \rangle.$$

Define the $T$-inner product to be

$$\langle f, g \rangle_T = \langle T^{-1}f, T^{-1}g \rangle.$$

The adjoint of $T$, with respect to the $T$-inner product, is $T^*$, defined by

$$\langle Tf, g \rangle_T = \langle f, T^*g \rangle_T.$$

**Ex. 25.1** *Prove that $T^*T = TT^\dagger$, so that*

$$T^* = TT^\dagger T^{-1}.$$

Then the data is

$$g_n = \langle f, h^n \rangle = \langle Tf, Th^n \rangle_T = \langle f, T^*Th^n \rangle_T = \langle f, TT^\dagger h^n \rangle_T.$$

Now we consider the reconstruction problem within the Hilbert space endowed with the $T$-inner product.

With this new inner product, the minimum-norm estimate of $f$ is

$$\hat{f} = \sum_{m=1}^{N} c_m TT^\dagger h^m,$$

with

$$g_n = \langle \hat{f}, TT^\dagger h^n \rangle_T = \sum_{m=1}^{N} c_m \langle TT^\dagger h^m, TT^\dagger h^n \rangle_T,$$

or

$$g_n = \sum_{m=1}^{N} c_m \langle T^\dagger h^m, T^\dagger h^n \rangle.$$

With $G$ the Gram matrix with entries

$$G_{m,n} = \langle T^\dagger h^m, T^\dagger h^n \rangle,$$

we have to solve the system of linear equations

$$g_n = \sum_{m=1}^{N} G_{m,n} c_m,$$

for $n = 1, ..., N$.

## 25.5   Some Examples

In this section we illustrate the general case with two examples.

### 25.5.1   Choosing the Inner Product

If the function $f(x)$ to be estimated is support-limited to the interval $[a, b]$, it is reasonable to assume that $f(x)$ is a member of $L^2(a, b)$, with the inner product given by Equation (25.3). In this case, the operator $T$ is just the identity operator. The minimum-norm estimator associated with this usual inner product has the form

$$\hat{f}(x) = \sum_{m=1}^{N} a_m h^m(x).$$

As we saw in the case of Fourier-transform data, there may be other inner products on $L^2(a, b)$ that lead to better estimates of $f(x)$; in particular, the inner product given by Equation (25.4) permits us to incorporate prior information about the function $|f(x)|$ in the estimate. The minimum-norm estimate associated with this inner product has the form

$$\hat{f}(x) = p(x) \sum_{m=1}^{N} b_m h^m(x).$$

In this case, the linear operator $T$ is defined by

$$T(f)(x) = \sqrt{p(x)} f(x).$$

In both cases, the coefficients are determined by making the estimator consistent with the data; that is, by satisfying Equation (25.2).

### 25.5.2   Choosing the Hilbert Space

We even have a choice to make in the selection of the Hilbert space itself. Suppose we know that $f(x)$ is really zero outside the smaller interval $[c, d] \subseteq [a, b]$. We can select as $H$ the space $L^2[c, d]$, or perhaps the closed subspace of all members of $L^2[a, b]$ that are zero outside $[c, d]$. If we take the view that, once we have changed the inner product we have already changed the Hilbert space, then there are still more Hilbert spaces we may use.

## 25.6   Summary

The flaw in the original argument presented in the first section is that it assumes that the function $f(x)$ is a member of only one Hilbert space, with only one inner product and norm to be dealt with, and that the linear-functional data must be represented using this single inner product. The minimum-norm solution is determined, once we settle on a particular Hilbert space and inner product, but we have a great deal of choice in selecting these. This is the stage at which we can incorporate prior knowledge to improve our estimator of $f(x)$.

# Chapter 26

# Some Theory of Fourier Analysis

## 26.1   Chapter Summary

In this appendix we survey, without proofs, some of the basic theorems concerning Fourier series and Fourier transforms. The discussion here is taken largely from [134] and [51]. There are many books, such as [80], that the reader interested in further details may consult. The book [101] is a delightful, if unconventional, journey through the theory and applications of Fourier analysis.

## 26.2   Fourier Series

Let $f : [-L, L] \to \mathbb{C}$. The Fourier series associated with the function $f$ is

$$f(x) \approx \sum_{n=-\infty}^{\infty} c_n e^{i\frac{n\pi}{L}x},$$

with

$$c_n = \frac{1}{2L} \int_{-L}^{L} f(x) e^{-i\frac{n\pi}{L}x} dx.$$

The $N$th partial sum is defined to be

$$S_N(x) = \sum_{n=-N}^{N} c_n e^{i\frac{n\pi}{L}x}.$$

Convergence of the Fourier series involves the behavior of the sequence $\{S_N(x)\}$ as $N \to \infty$.

It is known that, even if $f$ can be extended to a $2L$-periodic function that is everywhere continuous, there can be values of $x$, even a non-denumerable and everywhere dense set of $x$, at which the Fourier series fails to converge. However, it was shown by Carleson in 1966 that, under these conditions on $f$, the series will converge to $f$ almost everywhere; that is, except on a set of Lebesgue measure zero.

We can't expect $S_N(x)$ to converge to $f(x)$ for all $x$, since, if $f(x)$ and $g(x)$ differ at only finitely many points, they have the same associated Fourier series. If both $f$ and $g$ are continuous and $2L$-periodic, and the Fourier coefficients are the same, must $f = g$? The answer is yes, because of Fejer's Theorem.

Instead of considering $S_N(x)$, we consider

$$\sigma_N(x) = \frac{1}{N+1}(S_0(x) + S_1(x) + ... + S_N(x)).$$

We have the following theorem.

**Theorem 26.1 (Fejer's Theorem)** *Let $f$ have a continuous $2L$-periodic extension. Then the sequence $\{\sigma_N(x)\}$ converges to $f(x)$ uniformly.*

**Corollary 26.1** *If $f$ and $g$ both have continuous $2L$-periodic extensions and their Fourier coefficients agree, then $f = g$.*

**Theorem 26.2** *If $f$ has a continuous $2L$-periodic extension, then*

$$\lim_{N \to \infty} \int_{-L}^{L} |f(x) - S_N(x)|^2 dx = 0,$$

*and*

$$\frac{1}{2L} \int_{-L}^{L} |f(x)|^2 dx = \sum_{n=-\infty}^{\infty} |c_n|^2.$$

**Definition 26.1** *The function $f$ is said to be* Lipschitz continuous, *or just* Lipschitz, *at $x$ if there are constants $M > 0$ and $\delta > 0$ such that $|x - y| < \delta$ implies $|f(x) - f(y)| < |x - y|$.*

**Theorem 26.3** *If $f$ is Lipschitz at $x$ then $S_N(x) \to f(x)$.*

**Corollary 26.2** *If $f$ is differentiable at $x$, then $S_N(x) \to f(x)$.*

**Proof:** Since $f$ is differentiable at $x$ it is also Lipschitz at $x$. ∎

## 26.3   Fourier Transforms

In previous chapters it was our practice to treat the basic formulas for a Fourier-transform pair,

$$F(\gamma) = \int f(x)e^{i\gamma x}dx, \tag{26.1}$$

and

$$f(x) = \frac{1}{2\pi} \int F(\gamma)e^{-i\gamma x}d\gamma, \tag{26.2}$$

as formal expressions, rather than as universally valid statements. Theorems concerning the validity of these expressions must always include assumptions about the properties of $f$ and $F$, and about the nature of the integrals involved.

In the theory of Riemann integration the two symbols

$$\int_a^{+\infty} f(x)dx \tag{26.3}$$

and

$$\lim_{b\to+\infty} \int_a^b f(x)dx \tag{26.4}$$

are equivalent; in the theory of Lebesgue integration they are different. In the Lebesgue theory, the integral in Equation (26.3) involves two approximations done simultaneously; we approximate the function $f$ by a sequence of step functions, while at the same time extending the domain of the step functions to infinity. In Equation (26.4) the two limiting processes are done sequentially; first approximate $f$ by step functions on $[a, b]$ to get the integral, and then take the limit, as $b$ approaches infinity. For example, the function $f(x) = \frac{\sin x}{x}$ on $[0, +\infty)$ is not Lebesgue integrable, since its positive and negative parts are not separately Lebesgue integrable, but the Rieman integral is

$$\int_0^{+\infty} \frac{\sin x}{x}dx = \frac{\pi}{2},$$

which can be shown using the theory of residues.

**Definition 26.2** *Let $1 \le p < +\infty$. A function $f : \mathbb{R} \to \mathbb{C}$ is said to be in the class $L^p$ if $f$ is measurable in the sense of Lebesgue and the Lebesgue integral*

$$\int_{-\infty}^{\infty} |f(x)|^p dx$$

*is finite. Functions f in $L^1$ are said to be* absolutely integrable; *functions f in $L^2$ are* square integrable.

If $f$ is in $L^1$, then the integral in Equation (26.1) exists for all $\gamma$ and defines a bounded, continuous function on the whole of $\mathbb{R}$. If, in addition, the function $F$ is in $L^1$, then the integral in Equation (26.2) also exists for all $x$ and defines a bounded, continuous function that is equal, almost everywhere, to the original $f$. In general, however, $F$ need not be a member of $L^1$, and more complicated efforts are needed to give meaning to Equation (26.2).

If $f$ is in $L^2$, then the limit

$$F(\gamma) = \lim_{A \to +\infty} \left( \int_{-A}^{A} f(x) e^{i\gamma x} dx \right)$$

exists, in the $L^2$ sense, and defines the Fourier transform of $f$ as a member of $L^2$. In addition, the limit

$$f(x) = \lim_{A \to +\infty} \left( \frac{1}{2\pi} \int_{-A}^{A} F(\gamma) e^{-i\gamma x} d\gamma \right)$$

also exists, in the $L^2$ sense, and provides the inversion formula.

In order for the spaces $L^1$ and $L^2$ to be complete as metric spaces, the members of $L^1$ and $L^2$ are not individual functions, but equivalence classes of functions. Two functions $f$ and $g$ are *equivalent* if the function $f - g$ is equal to zero, except possibly on a set of measure zero. However, we shall continue to speak of the members of these spaces as functions.

## 26.4 Functions in the Schwartz Class

As we just discussed, the integrals in Equations (26.1) and (26.2) may have to be interpreted carefully if they are to be applied to fairly general classes of functions $f(x)$ and $F(\gamma)$. In this section we describe a class of functions for which these integrals can be defined.

If both $f(x)$ and $F(\gamma)$ are measurable and absolutely integrable then both functions are continuous. To illustrate some of the issues involved, we consider the functions in the Schwartz class [80]

A function $f(x)$ is said to be in the *Schwartz class*, or to be a *Schwartz function*, if $f(x)$ is infinitely differentiable and

$$|x|^m f^{(n)}(x) \to 0,$$

as $|x| \to +\infty$. Here $f^{(n)}(x)$ denotes the $n$th derivative of $f(x)$. An example of a Schwartz function is $f(x) = e^{-x^2}$, with Fourier transform $F(\gamma) = \sqrt{\pi}e^{-\gamma^2/4}$. The following proposition tells us that Schwartz functions are absolutely integrable on the real line, and so the Fourier transform is well defined.

**Proposition 26.1** *If $f(x)$ is a Schwartz function, then*

$$\int_{-\infty}^{\infty} |f(x)|dx < +\infty.$$

**Proof:** There is a constant $M > 0$ such that $|x|^2|f(x)| \leq 1$, for $|x| \geq M$. Then

$$\int_{-\infty}^{\infty} |f(x)|dx \leq \int_{-M}^{M} |f(x)|dx + \int_{|x|\geq M} |x|^{-2}dx < +\infty.$$

∎

If $f(x)$ is a Schwartz function, then so is its Fourier transform. To prove the Fourier Inversion Formula it is sufficient to show that

$$f(0) = \int_{-\infty}^{\infty} F(\gamma)d\gamma/2\pi.$$

Write

$$f(x) = f(0)e^{-x^2} + (f(x) - f(0)e^{-x^2}) = f(0)e^{-x^2} + g(x). \quad (26.5)$$

Then $g(0) = 0$, so $g(x) = xh(x)$, where $h(x) = g(x)/x$ is also a Schwartz function. Then the Fourier transform of $g(x)$ is the derivative of the Fourier transform of $h(x)$; that is,

$$G(\gamma) = H'(\gamma).$$

The function $H(\gamma)$ is a Schwartz function, so it goes to zero at the infinities. Computing the Fourier transform of both sides of Equation (26.5), we obtain

$$F(\gamma) = f(0)\sqrt{\pi}e^{-\gamma^2/4} + H'(\gamma).$$

Therefore,

$$\int_{-\infty}^{\infty} F(\gamma)d\gamma = 2\pi f(0) + H(+\infty) - H(-\infty) = 2\pi f(0).$$

To prove the Fourier Inversion Formula, we let $K(\gamma) = F(\gamma)e^{-ix_0\gamma}$, for fixed $x_0$. Then the inverse Fourier transform of $K(\gamma)$ is $k(x) = f(x + x_0)$, and therefore

$$\int_{-\infty}^{\infty} K(\gamma)d\gamma = 2\pi k(0) = 2\pi f(x_0). \quad (26.6)$$

## 26.5  Generalized Fourier Series

Let $\mathcal{H}$ be a Hilbert space, with inner product $\langle \cdot, \cdot \rangle$, and $\{\phi^1, \phi^2, ...\}$ an orthonormal basis for $\mathcal{H}$. Let $f$ be a member of $\mathcal{H}$. Then there are unique coefficients $c_1, c_2, ...$ such that the generalized Fourier series converges to $f$; that is,

$$f(x) = \sum_{n=1}^{\infty} c_n \phi^n(x).$$

The coefficients are given by

$$c_n = \langle f, \phi^n \rangle.$$

Let the $N$th partial sum of the series be

$$S_N(x) = \sum_{n=1}^{N} c_n \phi^n(x).$$

Then when we say that the series converges to $f$ we mean that

$$\lim_{N \to \infty} \|f - S_n\| = 0.$$

The following exercise shows that the $N$th partial sum is also a best approximation of $f$.

**Ex. 26.1** *Let*

$$T_N(x) = \sum_{n=1}^{N} b_n \phi^n(x),$$

*for an arbitrary selection of the coefficients $b_n$. Show that*

$$\|f - S_N\| \le \|f - T_N\|,$$

*with equality if and only if $b_n = c_n$ for $n = 1, ..., N$.*

## 26.6  Wiener Theory

The study of periodic components of functions is one of the main topics in generalized harmonic analysis [163]. To analyze such functions Norbert Wiener focused on the *autocorrelation function* of $f$, defined by

$$r_f(\tau) = \lim_{T \to +\infty} \frac{1}{2T} \int_{-T}^{T} f(t) \overline{f(t - \tau)} dt.$$

For example, let

$$f(t) = \sum_{n=1}^{N} a_n e^{i\omega_n t}.$$

Then we have

$$r_f(\tau) = \sum_{n=1}^{N} |a_n|^2 e^{i\omega_n \tau},$$

and

$$|a_n|^2 = \lim_{T \to +\infty} \frac{1}{2T} \int_{-T}^{T} r_f(\tau) e^{-i\omega_n \tau} d\tau.$$

Notice that the Fourier transform of $r_f(\tau)$ is

$$R_f(\omega) = \sum_{n=1}^{N} |a_n|^2 \delta(\omega - \omega_n),$$

the power spectrum of the function $f$. In order to avoid involving delta functions, Wiener takes a different approach to analyzing the spectrum of $f$.

In general, whenever the function $r_f(\tau)$ exists, the *integrated spectrum* of $f$ is the function

$$S(\omega) = \int_{-\infty}^{\infty} r_f(\tau) \frac{e^{i\omega\tau} - 1}{i\tau} d\tau.$$

Let's try to make sense of this definition.

Let $G(\gamma) = \chi_{[0,\omega]}(\gamma)$ be the characteristic function of the interval $[0, \omega]$. Then the inverse Fourier transform of $G(\gamma)$ is

$$g(x) = \frac{1}{2\pi} \frac{e^{-i\omega x} - 1}{-ix} = \frac{1}{2\pi} \overline{\left( \frac{e^{i\omega x} - 1}{ix} \right)}.$$

When the Parseval-Plancherel Equation (2.9) holds, we have

$$S(\omega) = 2\pi \int_{-\infty}^{\infty} r_f(\tau) \overline{g(\tau)} d\tau = \int_{0}^{\omega} R_f(\gamma) d\gamma,$$

so that $S'(\omega) = R_f(\omega)$. In such cases, $S(\omega)$ is differentiable, $S'(\omega) = R_f(\omega)$ is nonnegative, and $S'(\omega)$ is the *power spectrum* or *spectral density function* of $f$. When the function $f$ contains periodic components, the function $S(\omega)$ will have discontinuities, which is why Wiener focuses on $S(\omega)$, rather than on $S'(\omega)$.

# Chapter 27

# Reverberation and Echo Cancellation

## 27.1  Chapter Summary

A nice application of Dirac delta-function models is the problem of reverberation and echo cancellation, as discussed in [116]. The received signal is viewed as a filtered version of the original and we want to remove the effects of the filter, thereby removing the echo. This leads to the problem of finding the inverse filter. A version of the echo cancellation problem arises in telecommunications, as discussed in [147] and [148].

## 27.2  The Echo Model

Suppose that $x(t)$ is the original transmitted signal and the received signal is

$$y(t) = x(t) + \alpha x(t - d),$$

where $d > 0$ is the delay present in the echo term. We assume that the echo term is weaker than the original signal, so we make $0 < \alpha < 1$. With the filter function $h(t)$ defined by

$$h(t) = \delta(t) + \alpha\delta(t - d) = \delta(t) + \alpha\delta_d(t), \qquad (27.1)$$

where $\delta_d(t) = \delta(t - d)$, we can write $y(t)$ as the convolution of $x(t)$ and $h(t)$; that is,

$$y(t) = x(t) * h(t).$$

A more general model is used to describe reverberation:

$$h(t) = \sum_{k=0}^{K} \alpha_k \delta(t - d_k),$$

with $\alpha_0 = 1$, $d_0 = 0$, and $d_k > 0$ and $0 < \alpha_k < 1$ for $k = 1, 2, ..., K$.

Our goal is to find a second filter, denoted $h_i(t)$, the inverse of $h(t)$ in Equation (27.1), such that

$$h(t) * h_i(t) = \delta(t),$$

and therefore

$$x(t) = y(t) * h_i(t). \tag{27.2}$$

For now, we use trial and error to find $h_i(t)$; later we shall use the Fourier transform.

---

## 27.3    Finding the Inverse Filter

As a first guess, let us try

$$g_1(t) = \delta(t) - \alpha \delta_d(t).$$

Convolving $g_1(t)$ with $h(t)$, we get

$$h(t) * g_1(t) = \delta(t) * \delta(t) - \alpha^2 \delta_d(t) * \delta_d(t).$$

We need to find out what $\delta_d(t) * \delta_d(t)$ is.

**Ex. 27.1** *Use the sifting property of the Dirac delta and the definition of convolution to show that*

$$\delta_d(t) * \delta_d(t) = \delta_{2d}(t).$$

The Fourier transform of $\delta_d(t)$ is the function $\exp(id\omega)$, so that the Fourier transform of the convolution of $\delta_d(t)$ with itself is the square of $\exp(id\omega)$, or $\exp(i(2d)\omega)$. This tells us again that the convolution of $\delta_d(t)$ with itself is $\delta_{2d}(t)$. Therefore,

$$h(t) * g_1(t) = \delta(t) - \alpha^2 \delta_{2d}(t).$$

We do not quite have what we want, but since $0 < \alpha < 1$, the $\alpha^2$ is much smaller than $\alpha$.

Suppose that we continue down this path, and take for our next guess the filter function $g_2(t)$ given by

$$g_2(t) = \delta(t) - \alpha \delta_d(t) + \alpha^2 \delta_{2d}(t).$$

We then find that

$$h(t) * g_2(t) = \delta(t) + \alpha^3 \delta_{3d}(t);$$

the coefficient is $\alpha^3$ now, which is even smaller, and the delay in the echo term has moved to $3d$. We could continue along this path, but a final solution is beginning to suggest itself.

Suppose that we define

$$g_N(t) = \sum_{n=0}^{N} (-1)^n \alpha^n \delta_{nd}(t).$$

It would then follow that

$$h(t) * g_N(t) = \delta(t) - (-1)^{N+1} \alpha^{N+1} \delta_{(N+1)d}(t).$$

The coefficient $\alpha^{N+1}$ goes to zero and the delay goes to infinity, as $N \to \infty$. This suggests that the inverse filter should be the infinite sum

$$h_i(t) = \sum_{n=0}^{\infty} (-1)^n \alpha^n \delta_{nd}(t). \tag{27.3}$$

Then Equation (27.2) becomes

$$x(t) = y(t) - \alpha y(t-d) + \alpha^2 y(t-2d) - \alpha^3 y(t-3d) + \dots .$$

Obviously, to remove the echo completely in this manner we need infinite memory.

**Ex. 27.2** *Assume that $x(t) = 0$ for $t < 0$. Show that the problem of removing the echo is simpler now.*

---

## 27.4 Using the Fourier Transform

The Fourier transform of the filter function $h(t)$ in Equation (27.1) is

$$H(\omega) = 1 + \alpha \exp(id\omega).$$

If we are to have

$$h(t) * h_i(t) = \delta(t),$$

we must have

$$H(\omega)H_i(\omega) = 1,$$

where $H_i(\omega)$ is the Fourier transform of the inverse filter function $h_i(t)$ that we seek. It follows that

$$H_i(\omega) = (1 + \alpha \exp(id\omega))^{-1}.$$

Recalling the formula for the sum of a geometric progression,

$$1 - r + r^2 - r^3 + \ldots = \frac{1}{1+r},$$

for $|r| < 1$, we find that we can write

$$H_i(\omega) = 1 - \alpha \exp(id\omega) + \alpha^2 \exp(i(2d)\omega) - \alpha^3 \exp(i(3d)\omega) + \ldots,$$

which tells us that $h_i(t)$ is precisely as given in Equation (27.3).

---

## 27.5   The Teleconferencing Problem

In teleconferencing, each separate room is equipped with microphones for transmitting to the other rooms and loudspeakers for broadcasting what the people in the other rooms are saying. For simplicity, consider two rooms, the transmitting room (TR), in which people are currently speaking, and the receiving room (RR), where the people are currently listening to the broadcast from the TR. The RR also has microphones and the problem arises when the signal broadcast into the RR from the TR reaches the microphones in the RR and is broadcast back into the TR. If it reaches the microphones in the TR, it will be re-broadcast to the RR, creating an echo, or worse.

The signal that reaches a microphone in the RR will depend on the signals broadcast into the RR from the TR, as well as on the acoustics of the RR and on the placement of the microphone in the RR; that is, it will be a filtered version of what is broadcast into the RR. The hope is to be able to estimate the filter, generate an approximation of what is about to be re-broadcast, and subtract the estimate prior to re-broadcasting, thereby reducing to near zero what is re-broadcast back to the TR.

In practice, all signals are viewed as discrete time series, and all filters are taken to be *finite impulse response* (FIR) filters. Because the acoustics of the RR are not known a priori, the filter that the RR imposes must be estimated. This is done adaptively, by comparing vectors of samples of the original transmissions with the filtered version that is about to be re-broadcast, as described in [148].

# Bibliography

[1] Anderson, T. (1972) "Efficient estimation of regression coefficients in time series." In *Proc. of Sixth Berkeley Symposium on Mathematical Statistics and Probability, Volume 1: The Theory of Statistics.* Berkeley, CA: University of California Press, pp. 471–482.

[2] Ash, R. and Gardner, M. (1975) *Topics in Stochastic Processes.* Boston: Academic Press.

[3] Auslander, L., Kailath, T., and Mitter, S., eds. (1990) *Signal Processing Part I: Signal Processing Theory*, IMA Volumes in Mathematics and Its Applications, Volume 22. New York: Springer-Verlag.

[4] Baggeroer, A., Kuperman, W., and Schmidt, H. (1988) "Matched field processing: source localization in correlated noise as optimum parameter estimation." *Journal of the Acoustical Society of America* **83**, pp. 571–587.

[5] Benson, M. (2003) "What *Galileo* Saw." In *The New Yorker*; reprinted in [57].

[6] Bertero, M. (1992) "Sampling theory, resolution limits and inversion methods." In [8], pp. 71–94.

[7] Bertero, M. and Boccacci, P. (1998) *Introduction to Inverse Problems in Imaging.* Bristol, UK: Institute of Physics Publishing.

[8] Bertero, M. and Pike, E.R., eds. (1992) *Inverse Problems in Scattering and Imaging.* Malvern Physics Series, Adam Hilger, IOP Publishing, London.

[9] Bertsekas, D.P. (1997) "A new class of incremental gradient methods for least squares problems." *SIAM J. Optim.* **7**, pp. 913–926.

[10] Blackman, R. and Tukey, J. (1959) *The Measurement of Power Spectra.* New York: Dover Publications.

[11] Boggess, A. and Narcowich, F. (2001) *A First Course in Wavelets, with Fourier Analysis.* Englewood Cliffs, NJ: Prentice-Hall.

[12] Born, M. and Wolf, E. (1999) *Principles of Optics: 7th edition.* Cambridge, UK: Cambridge University Press.

[13] Bochner, S. and Chandrasekharan, K. (1949) *Fourier Transforms,* Annals of Mathematical Studies, No. 19. Princeton, NJ: Princeton University Press.

[14] Bolles, E.B. (1997) *Galileo's Commandment: 2,500 Years of Great Science Writing.* New York: W.H. Freeman.

[15] Bracewell, R.C. (1979) "Image reconstruction in radio astronomy." In [90], pp. 81–104.

[16] Bruckstein, A., Donoho, D., and Elad, M. (2009) "From sparse solutions of systems of equations to sparse modeling of signals and images." *SIAM Review* **51(1)**, pp. 34–81.

[17] Bruyant, P., Sau, J., and Mallet, J.J. (1999) "Noise removal using factor analysis of dynamic structures: application to cardiac gated studies." *Journal of Nuclear Medicine* **40(10)**, pp. 1676–1682.

[18] Bucker, H. (1976) "Use of calculated sound fields and matched field detection to locate sound sources in shallow water." *Journal of the Acoustical Society of America* **59**, pp. 368–373.

[19] Burg, J. (1967) "Maximum entropy spectral analysis." Paper presented at the 37th Annual SEG meeting, Oklahoma City, OK.

[20] Burg, J. (1972) "The relationship between maximum entropy spectra and maximum likelihood spectra." *Geophysics* **37**, pp. 375–376.

[21] Burg, J. (1975) *Maximum Entropy Spectral Analysis.* Ph.D. dissertation, Stanford University.

[22] Byrne, C. and Fitzgerald, R. (1979) "A unifying model for spectrum estimation." In *Proceedings of the RADC Workshop on Spectrum Estimation.* October 1979. Rome, NY: Griffiss AFB.

[23] Byrne, C. and Fitzgerald, R. (1982) "Reconstruction from partial information, with applications to tomography." *SIAM J. Applied Math.* **42(4)**, pp. 933–940.

[24] Byrne, C., Fitzgerald, R., Fiddy, M., Hall, T. and Darling, A. (1983) "Image restoration and resolution enhancement." *J. Opt. Soc. Amer.* **73**, pp. 1481–1487.

[25] Byrne, C. and Wells, D. (1983) "Limit of continuous and discrete finite-band Gerchberg iterative spectrum extrapolation." *Optics Letters* **8(10)**, pp. 526–527.

[26] Byrne, C. and Fitzgerald, R. (1984) "Spectral estimators that extend the maximum entropy and maximum likelihood methods." *SIAM J. Applied Math.* **44(2)**, pp. 425–442.

[27] Byrne, C., Levine, B.M., and Dainty, J.C. (1984) "Stable estimation of the probability density function of intensity from photon frequency counts." *JOSA Communications* **1(11)**, pp. 1132–1135.

[28] Byrne, C. and Steele, A. (1985) "Stable nonlinear methods for sensor array processing." *IEEE Transactions on Oceanic Engineering* **OE-10(3)**, pp. 255–259.

[29] Byrne, C. and Wells, D. (1985) "Optimality of certain iterative and non-iterative data extrapolation procedures." *Journal of Mathematical Analysis and Applications* **111(1)**, pp. 26–34.

[30] Byrne, C. and Fiddy, M. (1987) "Estimation of continuous object distributions from Fourier magnitude measurements." *JOSA A* **4**, pp. 412–417.

[31] Byrne, C. and Fiddy, M. (1988) "Images as power spectra; reconstruction as Wiener filter approximation." *Inverse Problems* **4**, pp. 399–409.

[32] Byrne, C., Brent, R., Feuillade, C., and DelBalzo, D (1990) "A stable data-adaptive method for matched-field array processing in acoustic waveguides." *Journal of the Acoustical Society of America* **87(6)**, pp. 2493–2502.

[33] Byrne, C., Frichter, G., and Feuillade, C. (1990) "Sector-focused stability methods for robust source localization in matched-field processing." *Journal of the Acoustical Society of America* **88(6)**, pp. 2843–2851.

[34] Byrne, C. (1993) "Iterative image reconstruction algorithms based on cross-entropy minimization." *IEEE Transactions on Image Processing* **IP-2**, pp. 96–103.

[35] Byrne, C., Haughton, D., and Jiang, T. (1993) "High-resolution inversion of the discrete Poisson and binomial transformations." *Inverse Problems* **9**, pp. 39–56.

[36] Byrne, C. (1995) "Erratum and addendum to 'Iterative image reconstruction algorithms based on cross-entropy minimization'." *IEEE Transactions on Image Processing* **IP-4**, pp. 225–226.

[37] Byrne, C. (1996) "Block-iterative methods for image reconstruction from projections." *IEEE Transactions on Image Processing* **IP-5**, pp. 792–794.

[38] Byrne, C. (1997) "Convergent block-iterative algorithms for image reconstruction from inconsistent data." *IEEE Transactions on Image Processing* **IP-6**, pp. 1296–1304.

[39] Byrne, C. (1998) "Accelerating the EMML algorithm and related iterative algorithms by rescaled block-iterative (RBI) methods." *IEEE Transactions on Image Processing* **IP-7**, pp. 100–109.

[40] Byrne, C. (2002) "Iterative oblique projection onto convex sets and the split feasibility problem." *Inverse Problems* **18**, pp. 441–453.

[41] Byrne, C. (2008) *Applied Iterative Methods.* Wellesley, MA: A K Peters, Publ.

[42] Byrne, C. (2009) *Applied and Computational Linear Algebra: A First Course.* Unpublished text available at my web site.

[43] Byrne, C. (2014) *A First Course in Optimization.* Boca Raton: CRC Press.

[44] Byrne, C. (2014) *Iterative Optimization in Inverse Problems.* Boca Raton: CRC Press.

[45] Candès, E., Romberg, J., and Tao, T. (2006) "Robust uncertainty principles: Exact signal reconstruction from highly incomplete frequency information." *IEEE Transactions on Information Theory* **52(2)**, pp. 489–509.

[46] Candès, E. and Romberg, J. (2007) "Sparsity and incoherence in compressive sampling." *Inverse Problems* **23(3)**, pp. 969–985.

[47] Candès, E., Wakin, M., and Boyd, S. (2007) "Enhancing sparsity by reweighted $l_1$ minimization." Preprint available at http://www.acm.caltech.edu/ emmanuel/publications.html.

[48] Candy, J. (1988) *Signal Processing: The Modern Approach.* New York: McGraw-Hill Publ.

[49] Capon, J. (1969) "High-resolution frequency-wavenumber spectrum analysis." *Proc. of the IEEE* **57**, pp. 1408–1418.

[50] Cederquist, J., Fienup, J., Wackerman, C., Robinson, S., and Kryskowski, D. (1989) "Wave-front phase estimation from Fourier intensity measurements." *Journal of the Optical Society of America A* **6(7)**, pp. 1020–1026.

[51] Champeney, D.C. (1987) *A Handbook of Fourier Theorems.* Cambridge, UK: Cambridge University Press.

[52] Chang, J.-H., Anderson, J.M.M., and Votaw, J.R. (2004) "Regularized image reconstruction algorithms for positron emission tomography." *IEEE Transactions on Medical Imaging* **23(9)**, pp. 1165–1175.

[53] Childers, D., ed. (1978) *Modern Spectral Analysis*. New York: IEEE Press.

[54] Christensen, O. (2003) *An Introduction to Frames and Riesz Bases*. Boston: Birkhäuser.

[55] Chui, C. (1992) *An Introduction to Wavelets*. Boston: Academic Press.

[56] Chui, C. and Chen, G. (1991) *Kalman Filtering, 2nd edition*. Berlin: Springer-Verlag.

[57] Cohen, J., ed. (2010) *The Best of The Best American Science Writing*. New York: Harper-Collins Publ.

[58] Cooley, J. and Tukey, J. (1965) "An algorithm for the machine calculation of complex Fourier series." *Math. Comp.* **19**, pp. 297–301.

[59] Cox, H. (1973) "Resolving power and sensitivity to mismatch of optimum array processors." *Journal of the Acoustical Society of America* **54**, pp. 771–785.

[60] Csiszár, I. (1991) "Why least squares and maximum entropy? An axiomatic approach to inference for linear inverse problems." *The Annals of Statistics* **19(4)**, pp. 2032–2066.

[61] Csiszár, I. and Tusnády, G. (1984) "Information geometry and alternating minimization procedures." *Statistics and Decisions* **Supp. 1**, pp. 205–237.

[62] Dainty, J. C. and Fiddy, M. (1984) "The essential role of prior knowledge in phase retrieval." *Optica Acta* **31**, pp. 325–330.

[63] Daubechies, I. (1988) "Orthogonal bases of compactly supported wavelets." *Commun. Pure Appl. Math.* **41**, pp. 909–996.

[64] Daubechies, I. (1992) *Ten Lectures on Wavelets*. Philadelphia: Society for Industrial and Applied Mathematics.

[65] De Bruijn, N. (1967) "Uncertainty principles in Fourier analysis." In *Inequalties*, O. Shisha, editor, pp. 57–71. Boston: Academic Press.

[66] Dhanantwari, A., Stergiopoulos, S., and Iakovidis, I. (2001) "Correcting organ motion artifacts in x-ray CT medical imaging systems by adaptive processing. I. Theory." *Med. Phys.* **28(8)**, pp. 1562–1576.

[67] Donoho, D. (2006) "Compressed sensing." *IEEE Transactions on Information Theory* **52(4)**, pp. 1289–1306.

[68] Duda, R., Hart, P., and Stork, D. (2001) *Pattern Classification*. New York: John Wiley and Sons, Inc.

[69] Eddington, A. (1927) "The story of Algol." *Stars and Atoms*; reprinted in [14].

[70] Feuillade, C., DelBalzo, D., and Rowe, M. (1989) "Environmental mismatch in shallow-water matched-field processing: geoacoustic parameter variability." *Journal of the Acoustical Society of America* **85**, pp. 2354–2364.

[71] Feynman, R. (1985) *QED: The Strange Theory of Light and Matter.* Princeton, NJ: Princeton University Press.

[72] Feynman, R., Leighton, R., and Sands, M. (1963) *The Feynman Lectures on Physics, Vol. 1*. Boston: Addison-Wesley.

[73] Fiddy, M. (1983) "The phase retrieval problem." In *Inverse Optics*, SPIE Proceedings 413 (A.J. Devaney, editor), pp. 176–181.

[74] Fiddy, M. (2008) *Private communication*.

[75] Fienup, J. (1979) "Space object imaging through the turbulent atmosphere." *Optical Engineering* **18**, pp. 529–534.

[76] Fienup, J. (1982) "Phase retrieval algorithms: a comparison." *Applied Optics* **21**, pp. 2758–2769.

[77] Fienup, J. (1987) "Reconstruction of a complex-valued object from the modulus of its Fourier transform using a support constraint." *Journal of the Optical Society of America A* **4(1)**, pp. 118–123.

[78] Frieden, B. R. (1982) *Probability, Statistical Optics and Data Testing.* Berlin: Springer-Verlag.

[79] Gabor, D. (1946) "Theory of communication." *Journal of the IEE (London)* **93**, pp. 429–457.

[80] Gasquet, C. and Witomski, F. (1998) *Fourier Analysis and Applications.* Berlin: Springer-Verlag.

[81] Gelb, A., ed. (1974) *Applied Optimal Estimation*, written by the technical staff of The Analytic Sciences Corporation. Cambridge, MA: MIT Press.

[82] Geman, S. and Geman, D. (1984) "Stochastic relaxation, Gibbs distributions and the Bayesian restoration of images." *IEEE Transactions on Pattern Analysis and Machine Intelligence* **PAMI-6**, pp. 721–741.

[83] Gerchberg, R. W. (1974) "Super-restoration through error energy reduction." *Optica Acta* **21**, pp. 709–720.

[84] Gordon, R., Bender, R., and Herman, G.T. (1970) "Algebraic reconstruction techniques (ART) for three-dimensional electron microscopy and x-ray photography." *J. Theoret. Biol.* **29**, pp. 471–481.

[85] Green, P. (1990) "Bayesian reconstructions from emission tomography data using a modified EM algorithm." *IEEE Transactions on Medical Imaging* **9**, pp. 84–93.

[86] Groetsch, C. (1999) *Inverse Problems: Activities for Undergraduates.* Washington, DC: The Mathematical Association of America.

[87] Grünbaum, F.A., Helton, J.W., and Khargonekar, P., eds. (1990) *Signal Processing Part II: Control Theory and Applications*, IMA Volumes in Mathematics and Its Applications, Volume 23. New York: Springer-Verlag.

[88] Haykin, S. (1985) *Array Signal Processing.* Englewood Cliffs, NJ: Prentice-Hall.

[89] Hebert, T. and Leahy, R. (1989) "A generalized EM algorithm for 3-D Bayesian reconstruction from Poisson data using Gibbs priors." *IEEE Transactions on Medical Imaging* **8**, pp. 194–202.

[90] Herman, G.T., ed. (1979) *Image Reconstruction from Projections: Topics in Applied Physics, Vol. 32*, Berlin: Springer-Verlag.

[91] Hinich, M. (1973) "Maximum likelihood signal processing for a vertical array." *Journal of the Acoustical Society of America* **54**, pp. 499–503.

[92] Hinich, M. (1979) "Maximum likelihood estimation of the position of a radiating source in a waveguide." *Journal of the Acoustical Society of America* **66**, pp. 480–483.

[93] Hoffman, K. (1962) *Banach Spaces of Analytic Functions.* Englewood Cliffs, NJ: Prentice-Hall.

[94] Hogg, R. and Craig, A. (1978) *Introduction to Mathematical Statistics.* New York: MacMillan.

[95] Hubbard, B. (1998) *The World According to Wavelets.* Natick, MA: A K Peters, Inc.

[96] Johnson, C., Hendriks, E., Berezhnoy, I., Brevdo, E., Hughes, S., Daubechies, I., Li, J., Postma, E., and Wang, J. (2008) "Image Processing for Artist Identification." *IEEE Signal Processing Magazine* **25(4)**, pp. 37–48.

[97] Kaiser, G. (1994) *A Friendly Guide to Wavelets*. Boston: Birkhäuser.

[98] Kalman, R. (1960) "A new approach to linear filtering and prediction problems." *Trans. ASME, J. Basic Eng.* **82**, pp. 35–45.

[99] Katznelson, Y. (1983) *An Introduction to Harmonic Analysis*. New York: John Wiley and Sons, Inc.

[100] Kheifets, A. (2004) *Private communication*.

[101] Körner, T. (1988) *Fourier Analysis*. Cambridge, UK: Cambridge University Press.

[102] Körner, T. (1996) *The Pleasures of Counting*. Cambridge, UK: Cambridge University Press.

[103] Lane, R. (1987) "Recovery of complex images from Fourier magnitude." *Optics Communications* **63(1)**, pp. 6–10.

[104] Lange, K. and Carson, R. (1984) "EM reconstruction algorithms for emission and transmission tomography." *Journal of Computer Assisted Tomography* **8**, pp. 306–316.

[105] Lange, K., Bahn, M. and Little, R. (1987) "A theoretical study of some maximum likelihood algorithms for emission and transmission tomography." *IEEE Trans. Med. Imag.* **MI-6(2)**, pp. 106–114.

[106] Leahy, R., Hebert, T., and Lee, R. (1989) "Applications of Markov random field models in medical imaging." In *Proceedings of the Conference on Information Processing in Medical Imaging*. Berkeley, CA: Lawrence-Berkeley Laboratory.

[107] Lent, A. (1998) *Private communication*.

[108] Levi, A. and Stark, H. (1984) "Image restoration by the method of generalized projections, with application to restoration from magnitude." *J. Opt. Soc. Am. A* **1**, pp. 932–943.

[109] Levitan, E. and Herman, G.T. (1987) "A maximum *a posteriori* probability expectation maximization algorithm for image reconstruction in emission tomography." *IEEE Transactions on Medical Imaging* **6**, pp. 185–192.

[110] Liao, C.-W., Fiddy, M., and Byrne, C. (1997) "Imaging from the zero locations of far-field intensity data." *Journal of the Optical Society of America: A* **14(12)**, pp. 3155–3161.

[111] Lightman, A. (2005) *Discoveries: Great Breakthroughs in 20th Century Science.* New York: Vintage Books.

[112] Lindberg, D. (1992) *The Beginnings of Western Science.* Chicago: University of Chicago Press.

[113] Luenberger, D. (1969) *Optimization by Vector Space Methods.* New York: John Wiley and Sons, Inc.

[114] Magness, T. and McQuire, J. (1962) "Comparison of least squares and minimum variance estimates of regression parameters." *Annals of Mathematical Statistics* **33**, pp. 462–470.

[115] Mallat, S.G. (1989) "A theory of multiresolution signal decomposition: The wavelet representation." *IEEE Transactions on Pattern Analysis and Machine Intelligence* **PAMI-11**, pp. 674–693.

[116] McClellan, J., Schafer, R., and Yoder, M. (2003) *Signal Processing First.* Upper Saddle River, NJ: Prentice Hall, Inc.

[117] Meidunas, E. (2001) *Re-scaled Block Iterative Expectation Maximization Maximum Likelihood (RBI-EMML) Abundance Estimation and Sub-pixel Material Identification in Hyperspectral Imagery.* MS thesis, Department of Electrical Engineering, University of Massachusetts Lowell.

[118] Meyer, Y. (1993) *Wavelets: Algorithms and Applications.* Philadelphia, PA: SIAM Publ.

[119] Muller, R. (2008) *Physics for Future Presidents: The Science Behind the Headlines.* New York: W.W. Norton and Company.

[120] Oppenheim, A. and Schafer, R. (1975) *Digital Signal Processing.* Englewood Cliffs, NJ: Prentice-Hall.

[121] Papoulis, A. (1975) "A new algorithm in spectral analysis and band-limited extrapolation." *IEEE Transactions on Circuits and Systems* **22**, pp. 735–742.

[122] Papoulis, A. (1977) *Signal Analysis.* New York: McGraw-Hill.

[123] Paulraj, A., Roy, R., and Kailath, T. (1986) "A subspace rotation approach to signal parameter estimation." *Proceedings of the IEEE* **74**, pp. 1044–1045.

[124] Pelagotti, A., Del Mastio, A., De Rosa, A., Piva, A. (2008) "Multispectral imaging of paintings." *IEEE Signal Processing Magazine* **25(4)**, pp. 27–36.

[125] Penrose, R. (2007) *The Road to Reality: A Complete Guide to the Laws of the Universe*. New York: Vintage Books.

[126] Pisarenko, V. (1973) "The retrieval of harmonics from a covariance function." *Geoph. J. R. Astrom. Soc.* **30**, pp. 347–366 .

[127] Pižurica, A., Philips, W., Lemahieu, I., and Acheroy, M. (2003) "A versatile wavelet domain noise filtration technique for medical imaging." *IEEE Transactions on Medical Imaging: Special Issue on Wavelets in Medical Imaging* **22**, pp. 323–331.

[128] Poggio, T. and Smale, S. (2003) "The mathematics of learning: dealing with data." *Notices of the American Mathematical Society* **50(5)**, pp. 537–544.

[129] Priestley, M. B. (1981) *Spectral Analysis and Time Series*. Boston: Academic Press.

[130] Prony, G.R.B. (1795) "Essai expérimental et analytique sur les lois de la dilatabilité de fluides élastiques et sur celles de la force expansion de la vapeur de l'alcool, à différentes températures." *Journal de l'Ecole Polytechnique* (Paris) **1(2)**, pp. 24–76.

[131] Qian, H. (1990) "Inverse Poisson transformation and shot noise filtering." *Rev. Sci. Instrum.* **61**, pp. 2088–2091.

[132] Ribés, A., Pillay, R., Schmitt, F., and Lahanier, C. (2008) "Studying that smile." *IEEE Signal Processing Magazine* **25(4)**, pp. 14–26.

[133] Romberg, J. (2007) "Compressed sensing creates a buzz at ICIAM '07." *SIAM NEWS*, October 2007, p. 7.

[134] Rudin, W. (1953) *Principles of Mathematical Analysis*. New York: McGraw-Hill.

[135] Sato, M. (1958) "On the generalization of the concept of a function." *Proc. Japan Acad.* **34**, pp. 126–130.

[136] Schmidt, R. (1981) *A Signal Subspace Approach to Multiple Emitter Location and Spectral Estimation*. PhD thesis, Stanford University.

[137] Schultz, L., Blanpied, G., Borozdin, K., *et al.* (2007) "Statistical reconstruction for cosmic ray muon tomography." *IEEE Transactions on Image Processing* **16(8)**, pp. 1985–1993.

[138] Schuster, A. (1898) "On the investigation of hidden periodicities with application to a supposed 26 day period of meteorological phenomena." *Terrestrial Magnetism* **3**, pp. 13–41.

[139] Shang, E. (1985) "Source depth estimation in waveguides." *Journal of the Acoustical Society of America* **77**, pp. 1413–1418.

[140] Shang, E. (1985) "Passive harmonic source ranging in waveguides by using mode filter." *Journal of the Acoustical Society of America* **78**, pp. 172–175.

[141] Shang, E., Wang, H., and Huang, Z. (1988) "Waveguide characterization and source localization in shallow water waveguides using Prony's method." *Journal of the Acoustical Society of America* **83**, pp. 103–106.

[142] Shieh, M., Byrne, C., and Fiddy, M. (2006) "Image reconstruction: a unifying model for resolution enhancement and data extrapolation: Tutorial." *Journal of the Optical Society of America A* **23(2)**, pp. 258–266.

[143] Shieh, M., Byrne, C., Testorf, M., and Fiddy, M. (2006) "Iterative image reconstruction using prior knowledge." *Journal of the Optical Society of America: A* **23(6)**, pp. 1292–1300.

[144] Shieh, M. and Byrne, C. (2006) "Image reconstruction from limited Fourier data." *Journal of the Optical Society of America: A* **23(11)**, pp. 2732–2736.

[145] Smith, C. Ray and Grandy, W.T., eds. (1985) *Maximum-Entropy and Bayesian Methods in Inverse Problems.* Dordrecht: Reidel Publ.

[146] Smith, C. Ray and Erickson, G., eds. (1987) *Maximum-Entropy and Bayesian Spectral Analysis and Estimation Problems.* Dordrecht: Reidel Publ.

[147] Sondhi, M. (2006) "The History of Echo Cancellation." *IEEE Signal Processing Magazine*, September 2006, pp. 95–102.

[148] Sondhi, M., Morgan, D., and Hall, J. (1995) "Stereophonic acoustic echo cancellation- an overview of the fundamental problem." *IEEE Signal Processing Letters* **2(8)**, pp. 148–151.

[149] Stark, H. and Woods, J. (2002) *Probability and Random Processes, with Applications to Signal Processing.* Upper Saddle River, NJ: Prentice-Hall.

[150] Stark, H. and Yang, Y. (1998) *Vector Space Projections: A Numerical Approach to Signal and Image Processing, Neural Nets and Optics.* New York: John Wiley and Sons, Inc.

[151] Strang, G. (1980) *Linear Algebra and Its Applications*. New York: Academic Press.

[152] Strang, G. and Nguyen, T. (1997) *Wavelets and Filter Banks*. Wellesley, MA: Wellesley-Cambridge Press.

[153] Therrien, C. (1992) *Discrete Random Signals and Statistical Signal Processing*. Englewood Cliffs, NJ: Prentice-Hall.

[154] Tindle, C., Guthrie, K., Bold, G., Johns, M., Jones, D., Dixon, K., and Birdsall, T. (1978) "Measurements of the frequency dependence of normal modes." *Journal of the Acoustical Society of America* **64**, pp. 1178–1185.

[155] Tolstoy, A. (1993) *Matched Field Processing for Underwater Acoustics*. Singapore: World Scientific.

[156] Twomey, S. (1996) *Introduction to the Mathematics of Inversion in Remote Sensing and Indirect Measurement*. New York: Dover Publ.

[157] Unser, M. (1999) "Splines: A perfect fit for signal and image processing." *IEEE Signal Processing Magazine* **16**, pp. 22–38.

[158] Van Trees, H. (1968) *Detection, Estimation and Modulation Theory*. New York: John Wiley and Sons, Inc.

[159] Walnut, D. (2002) *An Introduction to Wavelets*. Boston: Birkhäuser.

[160] Watson, J. (1968) *The Double Helix*. New York: Atheneum.

[161] Widrow, B. and Stearns, S. (1985) *Adaptive Signal Processing*. Englewood Cliffs, NJ: Prentice-Hall.

[162] Wiener, N. (1949) *Time Series*. Cambridge, MA: MIT Press.

[163] Wiener, N. (1964) *Generalized Harmonic Analysis and Tauberian Theorems*. Cambridge, MA: MIT Press.

[164] Wright, W., Pridham, R., and Kay, S. (1981) "Digital signal processing for sonar." *Proc. IEEE* **69**, pp. 1451–1506.

[165] Yang, T.C. (1987) "A method of range and depth estimation by modal decomposition." *Journal of the Acoustical Society of America* **82**, pp. 1736–1745.

[166] Yin, W. and Zhang, Y. (2008) "Extracting salient features from less data via $l_1$-minimization." *SIAG/OPT Views-and-News* **19(1)**, pp. 11–19.

[167] Young, R. (1980) *An Introduction to Nonharmonic Fourier Analysis*. Boston: Academic Press.

# Index

Printed in the United States
by Baker & Taylor Publisher Services